T0342506

Industry 4.0 Vision for the Supply of Energy and Materials

Industry 4.0 Vision for the Supply of Energy and Materials

Enabling Technologies and Emerging Applications

Edited by

Mahdi Sharifzadeh
Sharif University of Technology
Tehran, Iran

Registered Office
John Wiley & Sons, Inc., 111 River Street, Hoboken, NJ 07030, USA

Editorial Office
111 River Street, Hoboken, NJ 07030, USA

For details of our global editorial offices, customer services, and more information about Wiley products visit us at www.wiley.com.

Library of Congress Cataloging-in-Publication Data
Names: Sharifzadeh, Mahdi, editor.
Title: Industry 4.0 vision for the supply of energy and materials: enabling technologies and
    emerging applications / edited by Mahdi Sharifzadeh.
Other titles: Industry four point zero vision for energy and materials
Description: First edition. | Hoboken, NJ, USA : John Wiley & Sons, Inc., 2022. |
    Includes bibliographical references and index.
Identifiers: LCCN 2021061622 (print) | LCCN 2021061623 (ebook) | ISBN 9781119695936
    (hardback) | ISBN 9781119695967 (pdf) | ISBN 9781119695950 (epub) | ISBN 9781119695868 (ebook)
Subjects: LCSH: Industry 4.0. | Automation--Case studies. | Telecommunication--Case studies.
Classification: LCC T59.6 .I48 2022 (print) | LCC T59.6 (ebook) |
    DDC 658.4/038028563--dc23/eng/20220224
LC record available at https://lccn.loc.gov/2021061622
LC ebook record available at https://lccn.loc.gov/2021061623

Cover image: © Blue Planet Studio/Shutterstock, Alexander Tolstykh/Shutterstock, Phonlamai Photo/Shutterstock, Melena-Nsk/Shutterstock, MR.Cole_Photographer/Getty Images, Busakorn Pongparnit/Getty Images, chinaface/E+/Getty Images
Cover design by Wiley

Set in 9.5/12.5pt STIXTwoText by Integra Software Services Pvt. Ltd, Pondicherry, India

10 9 8 7 6 5 4 3 2 1

Contents

Preface

The Fourth Industrial Revolution, also known as Industry 4.0 or I4.0, refers to the recent accelerated uptake of a portfolio of technologies that enable a high degree of automation, integration, transparency, decentralization, and at the same time interconnectedness in the industrial sector, as well as other sectors such as health, education, and agriculture, enabling optimal and evidence-based decision-making. The emergence of Industry 4.0 is the result of dramatic advances and convergence in multiple disciplines. These include a sharp increase in the capabilities for artificial intelligence, wireless communication (e.g., Internet of things), cloud-based computations, smart transactions (blockchains), and robotics. Such capabilities have manifested themselves in the form of novel paradigms such as smart manufacturing, the Internet of things, predictive maintenance, and additive manufacturing.

The present contribution reports the collective endeavors of a multidisciplinary group of researchers to explore the emerging trends inspired by the aforementioned evolutions, especially with the focus on the flow of energy and materials in supply chains. The book has two main parts. In the first part, the key drivers of Industry 4.0, namely, artificial intelligence, wireless communication, blockchains and smart contracts, cloud computing, and robotics, are discussed. The second part explores the application of such advancements in the fields of energy networks, additive manufacturing, pharmaceutical industry, water distribution, renewable power generation, petroleum and gas industries, as discussed in the following.

Chapter 1 explores recent advancements in sensor and communication technologies and their contribution to the realization and commercial viability of industrial/ Internet of things (I/IoTs), smart manufacturing, and other Industry 4.0 paradigms. It also discusses the applications of communication technologies in Industry 4.0 and the required criteria. Other features of interest include the relevant standards and protocols, cellular and mobile technologies, the design of wireless systems for IIoT applications and relevant protocols, and smart sensors and their enabling role in Industry 4.0. The chapter concludes with predicting future trends in wireless communication for Industry 4.0.

Chapter 2 explores the concepts and methods that have enabled distributed and smart transaction systems such as blockchains. The features of interest include the blockchain taxonomy, desirable attributes, architecture, and most importantly the emerging applications in different sectors.

Chapter 3 is concerned with the application of robotics in Industry 4.0. First, a comprehensive survey of various classes of robots is presented. The classification is presented with respect to robots' geometry, actuator and control strategies, and kinematics, as well as more advanced features such as the number of agents and fabrication materials. Then the applications of various robotic systems are presented in different areas.

Chapter 4 focuses on the utilization of cloud computing in Industry 4.0. The main characteristics of cloud computing, as well as developed architectures and types of services, the model of cloud deployment, and the corresponding pros and cons are extensively discussed in this chapter. The last part of this chapter explores emerging paradigms such as edge computing and fog computing and elucidate their differences compared with traditional cloud and grid computing.

Chapter 5 provides a comprehensive review of artificial intelligence (AI) methods. The reviewed algorithms are categorized into supervised and unsupervised methods. In the supervised methods, two major groups of classification and regression methods are discussed. In the classification methods, the key features of decision trees (DT), the (naive) Bayesian classifier, K-nearest neighbors (KNNs), linear discriminant analysis (LDA), support vector machines (SVM) and kernel methods, relevance vector machines (RVMs), ensemble methods, and logistic regression are reviewed. In the regression methods, ordinary least squares (OLS) regression, ridge and lasso regression, support vector regression (SVR), Gaussian process regression (GPR), and thin plate spline (TPS) are discussed. A separate section is devoted to neural networks and deep learning, especially to recent developments in convolutional neural networks (CNNs), recurrent neural networks (RNNs), long short-term memory (LSTM), transformers, graph neural networks (GNNs), generative adversarial networks (GAN), autoencoders, and self-organizing maps (SOMs). Among unsupervised learning algorithms, special attention is paid to clustering methods such as k-means clustering and hierarchical clustering, DBSCAN, as well as linear and nonlinear dimensionality reduction methods (e.g., principal component analysis [PCA], linear discriminant analysis, manifold learning, Laplacian eigenmaps method). This chapter also provides a brief overview of semi-supervised and active Learning, bio-inspired methods, as well as reinforcement learning (RL).

The second part of the book begins with Chapter 6, in which the concept of the Internet of energy (IoE) is introduced. The discussion starts with the description of conventional energy grids, exposure of their limitations, and elucidation of the potentials that have become available through the incorporation of the Industry 4.0 technologies. It continues by reviewing the structure of IoE, with particular emphasis on energy routers, energy hubs, and software-defined networks (SDNs). The utilization of other I4.0 technologies such as big-data analytics and blockchains are also discussed with a special focus on energy trading and transactive IoE.

In Chapter 7, a more specific problem relevant to the Internet of energy is evaluated, which is concerned with the evolving nature of energy infrastructures. Considering the depleting fossil sources and their negative environmental impact, significant efforts are devoted to commercial utilization of renewable energies such as wind and solar power generation. However, the high penetration of renewable energies in the electricity grid also poses a significant challenge due to the intermittent and stochastic nature of wind and solar energies. The conventional approach to tackle such complexity is either through costly energy storage or provision of standby power plant capacity that is usually driven by fossil fuels. However, a paradigm shift has emerged using advanced control and communication systems that would deploy predictive analytics for real-time optimization, eliminating the need for expensive solutions such as energy storage and power plant extra capacity. The chapter explores the economic benefits of utilizing such intelligent systems, through the integration of artificial intelligence in the form time-series prediction, and optimization programming for electricity expansion planning and real-time dispatch.

Chapter 8 explores another specific utilization of Industry 4.0 technologies in the water distribution networks. Water pipelines are prone to operational issues such as aging, leakage, water theft, and sabotage attacks. The utilization of smart sensors quipped with wireless and cellular communication technologies opens up new avenues for monitoring and fault detection. Nonetheless, there is always a trade-off between the number of utilized sensors and selected technology (as well as associated costs) and the observability and possibility of fault detection. In this chapter, a systematic method based on multi-objective optimization is presented that, while minimizing the costs, ensures a certain degree of observability for the network, even in the case of multiple sensor failures.

Chapter 9 is focused on the evolution of oil and gas industries utilizing Industry 4.0 technologies. The discussions include the introduction of recent trends such as data acquisition and processing systems, smart and soft sensors, and digital twins, as well as challenges that need to be addressed for their commercial implementation. More attention is paid to the architecture that allows the fusion of Industry 4.0 technologies such as cloud computing, sensor-to-cloud connectivity, 5G, industrial Internet of things (IIoTs), and AI, with emphasis on standards that enable such integration. The chapter concludes by exploring potential future developments.

In recent years, the transformation of fossil-driven vehicles using electric engines has revolutionized the transportation industry. Chapter 10 studies the impact of Industry 4.0 on transportation electrification. The features of interest include the environmental, economic, and societal benefits that are achievable from such transformation, as well as corresponding barriers and challenges. A deep discussion of the electrification technologies is provided with special attention to the degree of electrification, types of electric motors, required battery and charging technologies, as well as connectivity of vehicles to grid (V2G), other vehicles (V2V), infrastructure (V2I), buildings (V2B), and clouds (V2C) technologies, which can promote energy efficiency as well as traffic safety. Other integrating Industry 4.0 technologies include blockchains, artificial intelligence, cyber-security, and robotics are also discussed in this chapter.

Dramatic advances in computational capabilities have also revolutionized the way that products and services are developed and dramatically have shortened their time to market. This is the focus of Chapter 11, with emphasis on the role that computer-aided molecular design (CAMD) plays in reducing the costs of designing new materials and products in the form of predicting their properties and reducing requirements for physical experimentation. Different CAMD methods are discussed, and implementation procedures are presented. This chapter also reviews the emerging and novel applications of CAMD in the industry.

Chapter 12 evaluates the impact of Industry 4.0 on the pharmaceutical industry. The discussions include the regulatory considerations in this sector, and the recent trends in smart manufacturing of pharmaceuticals in the form of continuous processing, analytical technologies, and digitalization.

The final chapter of the book explores additive manufacturing as a paradigm-shifting technology. Various types of 3D and 4D printing are reviewed and their advantages and disadvantages are presented. In addition, current challenges and future potential developments are discussed.

We sincerely hope that this contribution will open up new discussions and motivate novel research into the adaptation of the Industry 4.0 technologies in energy and material supply chains.

Dr. Mahdi Sharifzadeh,
On behalf of coauthors

Part I

Industry 4.0 Drivers

1

Connectivity through Wireless Communications and Sensors

Marzieh Jalal Abadi* and Babak Hossein Khalaj

School of Electrical Engineering, Sharif University of Technology, Tehran, Iran
* Corresponding author. School of Electrical Engineering, Sharif University of Technology, Tehran, Iran

1.1 Introduction: Key Technologies Enablers for Industry 4.0 and Supply Chain 4.0

The advent of digitalization in Internet era and mobile technologies yields changes in established business models and the global industry landscape [1]. Currently, manufacturing processes and operation environments are empowered by the Internet of things (IoT), autonomous robotics, and advanced data analytics. Moreover, the implementation of integrated automation and ubiquitous computing systems enables interconnection of human and machines in the context of a cyber-physical system (CPS). These successive technologies significantly improve the performance efficiency and customer satisfaction in the industrial sectors and lead to the fourth industrial revolution, termed as Industry 4.0 [2, 3].

Industry 4.0 is mainly identified by visibility, interconnectivity, autonomous performance, and predictive analysis. It improves agility in supply chain management, service efficiency, and cost reduction. For instance, introduction of Industry 4.0 into manufacturing induces optimized planning and decision-making across end-to-end (E2E) supply chains and delivers customized products to the end-users at competitive cost. In fact, Industry 4.0 brings the concept of Supply Chain 4.0 that changes the way supply chain operations are structured.

In this section, we briefly overview the background of Industry 4.0 followed by its vision on future manufacturing. We then review the key features and enablers of Industry 4.0. Since the focus of this chapter is on connectivity through wireless communication and sensors in the context of Industry 4.0, in the next sections we concentrate only on wireless connectivity in this era.

1.1.1 Background

The main idea of Industry 4.0 was first adopted by Germany in 2011 as a strategic initiative that determined future advanced manufacturing for 2020 [4, 5]. Thereafter, different countries have introduced similar programs for manufacturing research and innovation; for instance, Horizon2020 proposed factories of the future (FoF) in Europe [6, 7], Industrial IoT (IIoT) was introduced in the United States by General Electric (GE) [8], and the industrial value chain initiative (IVI) was founded in Japan [9].

Industry 4.0 Vision for the Supply of Energy and Materials: Enabling Technologies and Emerging Applications, First Edition. Edited by Mahdi Sharifzadeh.
© 2022 John Wiley & Sons, Inc. Published 2022 by John Wiley & Sons, Inc.

Industry 4.0 is defined as the integration of digitalization, intelligence, and communication technologies to industrial practices [10]. This new industrial paradigm automates work processes, optimizes products, and allows more granular customer services. Subsequently, a high level of productivity and efficiency is available to enterprises and organizations that embrace Industry 4.0 [11]. It employs innovations and disruptive developments such as IoT, CPS, big data, Fog computing, and virtual reality (VR) to accomplish this transformation. The main idea of Industry 4.0 has drawn on earlier perspectives and concepts and expanded over the years, although its landscape has been significantly altered lately [12].

1.1.2 Future Manufacturing Vision

Industry 4.0 has evolved conventional manufacturing systems a leap forward to smart factories. These advanced manufacturing systems, also termed interconnected factories and digital manufacturing, create smart products and processes. The term "smart" in Industry 4.0 framework is defined as an intelligent environment that offers real-time communication and cooperation of various devices to make decisions and act according to the obtained information [13]. In this manufacturing approach, all the components communicate autonomously, work without human intervention, and trigger operations from customers to suppliers [14]. This allows the creation of a smart environment capable of flexible and adaptive processes, enhances efficiency, and ensures integrated operations meet the requirements of sophisticated markets [5, 13].

In smart factories, the communication between all manufacturing resources is performed through standardized interfaces [15]. Additionally, new business models are introduced to create collaborative environments and better satisfy customers' changing requirements [16]. Thus, smart products are monitored as an active component of systems and at all stages of life cycle to deliver a high degree of customization. In Figure 1.1, we indicate the architecture of smart manufacturing in Industry 4.0.

1.1.3 Key Features of Industry 4.0

The innovative aspects of Industry 4.0 are expected to affect various industries such as manufacturing, operation companies, and service providers, yet there is no concise knowledge about challenges in its implementation and its implications and consequences. To better achieve this concept, its main features first must be understood. Based on the literature review, the main features of Industry 4.0 could be identified as:

- *Smart Environment*: The central aspect of Industry 4.0 adopts the expansion of intelligent ecosystems that lead to smart products, smart machines, and augmented operators [14]. In this context, smart machines promote independent and self-optimized processes that imply self-organizing production systems. Typically, smart products are self-aware and communicate individually with systems entities [11]. In response to the increasing variegated systems and their distinct requirements, the concept of augmented operators is introduced in Industry 4.0 as Operator 4.0 [17]. This advanced technology evolves the industrial workforce and creates new forms of cooperation among machines, agents, humans, and robots. These radical changes enable smart environments to automatically identify risks or exceptions and continuously require targets to adjust supply chain parameters.

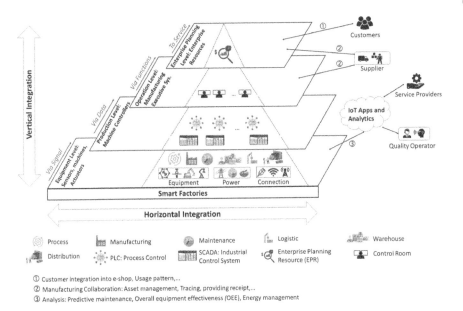

① Customer integration into e-shop, Usage pattern,...
② Manufacturing Collaboration: Asset management, Tracing, providing receipt,...
③ Analysis: Predictive maintenance, Overall equipment effectiveness (OEE), Energy management

Figure 1.1 Smart Factories in Industry 4.0.

- *Integration in Supply Chain*: A major characteristic of Industry 4.0 is integration that allows accelerated and flexible responses to systems changes. The integration is achieved at different levels of a system and through CPSs to provide system transparency in the value chain. Basically, there are three dimensions of integration, namely, horizontal, vertical, and end-to-end (E2E). Horizontal integration involves the value chain, such as resources, processes, and information flows within a company and among other enterprises [18]; in contrast, vertical integration concerns incorporating various technologies at different hierarchical levels of a company (e.g., from the equipment to the enterprise planning level; Figure 1.1). In fact, vertical integration allows autonomous CPSs to exchange information and trigger actions to build flexible and reconfigurable factories [19]. E2E integration refers to a holistic life cycle management that aims to facilitate true mass customization with lower operational costs and close the gap across the entire value chain (e.g., between product design, development, and customers).
- *Efficient Supply Chain*: Cooperation of diverse technologies with smart equipment and autonomous robots enhance efficiency in Industry 4.0. Given that an individual technology would have a limited impact on the supply chain, the collaboration of these digital technologies in industrial applications brings new possibilities and adds more values. To achieve this vision, visual computing technologies play an important role as facilitator and provide cohesion between technologies to further enhance efficiency [20]. The network configurations are also continuously monitored to ensure optimal fits to business requirements. In short, Industry 4.0 improves the entire value chain and quality of products, strengthens the cooperation of stakeholders, and offers advanced operations that subsequently provide a high level of efficiency in the supply chain.

1.1.4 Key Technologies Enablers for Industry 4.0

Industry 4.0 can be understood as a smart environment that is built by communication networks, automation technology, and production digitalization [21].

This vision succeeds through CPSs, IoT, and Internet of service (IoS) [22]. CPS controls and monitors processes in smart industries, while IoT enables real-time cooperation between every CPS. On the other hand, IoS offers services over the entire value chain in an organization or across companies. In Figure 1.2, we show the technology drivers of Industry 4.0 and briefly explain the key enablers, that is, CPS, IoT, and IoS.

1.1.4.1 Cyber-Physical Systems

A cyber-physical system is defined as an embedded system that could autonomously exchange information in an intelligent network [23]. These systems link physical elements (machines) and virtual entities (computing) to enable management of interconnected systems. CPS plays an important role in connecting autonomous components and subsystems across manufacturing environments, and it offers vertical and horizontal integration in the systems [24]. The connection of CPSs to the Internet is achieved through an IoT platform [23].

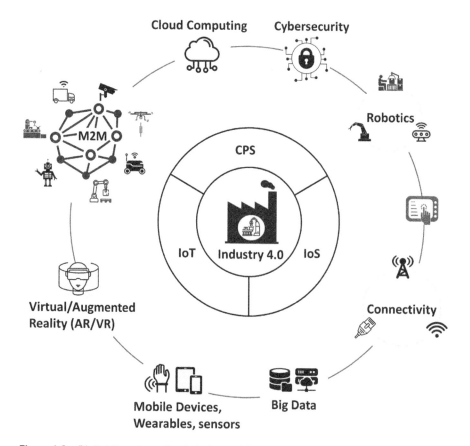

Figure 1.2 Digital Transformation in Industry 4.0.

1.1.4.2   Internet of Things

The concept of connected devices is a precursor to IoT and has appeared in wireless sensor networks (WSNs) with applications in agriculture, medical sciences, traffic management, and others [25]. After the birth of the Internet, the connection of physical things (e.g., computers) has been fulfilled. Nowadays, the Internet is expanded into the next level and interconnects everyday intelligent objects with each other and environments to exchange data and trigger actions. IoT systems have their foundation in a seamless connection of smart objects and active incorporation within business processes [26].

IoT is a key enabler of Industry 4.0 and realized through the holistic integration of smart equipment, smart systems, and intelligent decision-making in modern industrial systems [27]. This implies the provision of improved products and services using smarter, more reliable, and more autonomous things. The rapid expansion of IoT in industrial applications is called industrial IoT (IIoT) and opens many applications and new frontiers in Industry 4.0 [28].

1.1.4.3   Internet of Service

The term "service" is an ambiguous notion that may have very different meanings depending on the context. In business and economics, a service is an intangible activity offered by a provider to its consumers and creates value for them [29]. A service may just be a software with a defined interface available over the Internet.

Typically, enterprise systems utilize computing technologies and intelligence of smart objects over the Internet to propagate information and services [26]. Similar to IoT, the concept of Internet of Service is applied to services rather than physical (virtual) elements and creates information networks among users and service providers. IoS considers all basic business components (e.g., business objectives, processes, services) and technical basics such as capabilities of smart items in processing, sensing, and communication to support a service-oriented ecosystem and business applications modeling. Given that changes in the industrial environment such as connectivity disruption and reconfiguration of processes need reorganization of employed entities, new and optimized business processes in IoS help to flexibly model a self-organized service environment that adopts the model during deployment and according to specified policies. IoS is the innovative service model of information and is considered an important pillar of Industry 4.0. For instance, in the context of smart factories and future manufacturing, IoS uses the Internet and its software to produce services as customized products for individual users and based on the system's ability.

1.2   Drivers, Motivations, and Applications for Communication

Generally, the major driver in introducing any new technology is the potential increased revenue and its benefit to the principal sectors such as health, safety, and industry. Communication is a key technology that integrates the digital and physical world and has redefined many consumer-oriented businesses in health care, finance, and industry. The new wave in communication technology provides real-time and seamless connections between industrial assets (the things, machines, sites, and environments) and enables intelligent industrial operations for extensive and heterogeneous production line, instrumentation, and process monitoring. Thus, different industrial sectors are motivated to deploy suitable communication technologies and infrastructure with flexible installation and high availability. Typically, wired communication in industrial systems is built on an IEEE 802.3 standard (Ethernet) that offers sophisticated solutions within strict requirements. Furthermore, nonhierarchical networks, increases in number of sensors, and extended connection of operating equipment change the network topology of wired systems and increase data traffic. This highlights the needs of future systems for broadband and deterministic communication that assist functions and provides synchronicity between the production processes. Hence, the communication systems used today will either broaden or be replaced by new developments.

With the growing demand and advance in communication technologies, wireless network becomes a concise, optimized, and widely popular solution for connectivity. Another promising candidate technology for communication is cellular networks (e.g., 4G/LTE and 5G) that enable ubiquitous access to information in disparate vertical applications. Wireless technology reinforces Industry 4.0 movement, and its successful deployment enables the possibility for fully autonomous systems and platforms. Notable advantages of wireless communication are system simplicity, reduced system size, and mass as well as improved system resilience to hazards through communication diversity. The principal benefits of wireless technologies in industrial applications could be associated to the following distinct areas.

1.2.1 Wireless Instrumentation

Contrary to wired communication, wireless technologies for instrumentation are often at lower expenses, with reduced installation time, and minimum disruption. They also contribute to extend coverage into remote or hostile areas. Therefore, wireless technologies offer better insight to prospective safety issues and operational requirements in industrial plants and facilities. Some technical requirements such as long battery life and optimized data rates should be adopted to specific applications for these networks. Since wireless apparatus are integrated with available operational systems via standard industrial interfaces, they should preserve the security of system for various cyberattacks. Ultimately, major beneficial factors of wireless instrumentation are their reliability, flexibility, and cost efficiency in resources and time management.

1.2.2 Mobile Technologies

Wireless access in buildings or public spaces mainly utilizes wireless local area network (WLAN) that enables Internet access for mobile devices in the vicinity, through access points deployed within the limited area. For instance, in some industrial applications, the control room is brought to the field and local onsite WLAN provides real-time line access, fault diagnostics, and maintenance for remote centers.

Despite the proliferation of systems and standards adhering to WLANs, it might not be efficient for all industrial mobile applications. In some industries, implementing local instruments and infrastructure for expansion of wireless coverage in the processing zones is substantial. This is strongly applicable to oil, gas, and mining industries, which enact strict regulations to all electrical apparatus. For instance, there are governing rules on classification of areas and workplaces safety for hazardous environments [30, 31]. Given that network equipment and their installation in these environments must be certified to comply with standard rules, the equipment costs will drastically increase in such industries. This motivates industrial applications to embrace mobile technologies such as cellular communication, public networks, and private and dedicated mobile networks as alternative communication solutions.

The oil and gas industries have employed GPRS[1] and UMTS[2] as public networks for wireless connectivity. To deliver an acceptable level of service experience in IIoT, a number of performance requirements such as latency, bitrate, density, mobility, availability, and permitted level of packet loss should be set. Besides, the high diversity of devices in oil and gas enterprises generates various requirements for communication solutions. For instance, underground/open-cast mining, and offshore oil rigs require unmanned platforms; the communication networks in these complex Industry 4.0 use cases should offer denser connectivity and handle huge data transfer with low latency. Real-time data streaming also underpins effective monitoring and is a critical success factor for these use cases. In its support, modern long-term evolution (LTE)-based cellular systems could deploy optimizations and reduce latency according to the service requirements. Private LTE network is another solution that exploits dedicated radio equipment and serves the premises of enterprises' exclusive network with customized configurations to meet their exact performance requirements. For instance, licensed-assisted access (LAA) is a standard variant of LTE-unlicensed (LTE-U) [32], which leverages the combination of free 5 GHz unlicensed spectrum and licensed spectrum in new radio (NR) operations to deliver a performance boost for mobile users [33]. However, some stringent requirements fall under the category of ultra-reliable and low latency communication (URLLC) that are not achievable with 4G/LTE networks, at least not with high speed and scale. Thereby, 5G technologies and its proclaimed benefits can meet these critical requirements through 5G, its NR access technology and its expansion to the unlicensed spectrum [34].

5G new radio unlicensed (NR-U) is a potential candidate for next generation communication for Industry 4.0 scenarios, whereas an enterprise could install NR-U access points in place of Wi-Fi gateways and provide LTE coverage for the smart industry [35, 36]. 5G NR-U is a transformation in LTE-U/LAA from 4G/LTE to 5G NR; it opens up the 6 GHz bands for unlicensed access to alleviate the spectrum scarcity [37]. The 6 GHz bands in 5G NR-U offer the additional spectrum of 1.2 GHz and are deemed critical in supporting emerging bandwidth-intensive and latency-sensitive applications such as wireless augmented/virtual reality (AR/VR).

NR access technology in 5G includes the development of a new flexible air interface, denoted as 5G NX-radio, to attain the extreme communication requirements in terms

---

1 General packet radio service (GPRS) is a packet switching technology that enhances GSM cellular communication network to transmit mobile data [224]. GPRS is a 2.5G mobile cellular technology and offers data rates up to 172 Kbps.

2 Universal mobile telecommunications system (UMTS) is a 3G mobile cellular system developed within 3GPP and is exploited as a GSM network to offer packet-based data transmission at data rates up to 2 Mbps [225].

of latency and reliability. 5G NX-radio is not compatible with previous 5G air inter-faces and because of the availability of larger bandwidth, it will be initially imple-mented at new spectrum (primarily above 6 GHz) [38]. Sophisticated signaling methods are involved in 5G NX-radio, where new coding techniques along with the short and flexible radio frames structure are employed in the design of 5G-air interface [39]. 5G NX-radio encompasses two key technologies: (1) massive multiple-input, multiple-output (MIMO) and beamforming; and (2) ultra-lean design. Beamforming combats the challenging radio propagation and enables reduced interference along with very high throughput at high frequencies [40]. Low latency is achieved by ultra-lean design of 5G NX-radio and through minimizing the transmission time of the con-trol command over the radio interface [38]. Consequently, the transmission time of a single packet over the air is short and expected to be a fraction of LTE's [39]. In addition, the ultra-lean design in 5G NX-radio facilitates retransmission of packets within the constrained latency.

Ultimately, mobile technologies comprise various applications with different requirements and purposes. General requirements of mobile communication technol-ogies are seamless integration, reliability, and scalability. Some requirements in industrial settings such as suitable bandwidth, and security mechanism are highly application dependent.

1.2.3 Asset and Personnel Management

Nowadays, industrial operations have become more complex, and managing resources and information flow is becoming increasingly important. One class of successful industrial applications is connected logistic processes that precisely organize flow of materials, related information, and activities between the points of origin and con-sumption to meet the requirements of customers or corporations. Communication technology plays an important role in this area and connects different sectors and levels of a smart process.

Assets and personnel management provides real-time visibility of fleet, processes, and value networks. This reduces labor, transportation, and installation costs while highly improving operation in remote sites. Currently, there are various services and techniques for asset management, personnel tracking, and identification of materials, inventory, and tools. GPS[3] and public navigation services cannot meet the require-ments applicable to most industrial applications and are not viable solutions for asset management in industrial applications. Some applications use wireless technologies such as Bluetooth, Wi-Fi, and ultra-wideband (UWB) for positioning and should be chosen according to the specific usage scenario [41, 42, 43]. In such techniques, preci-sion, infrastructure demands, maintenance, and real-time requirements with respect to position update should be optimized. Since asset management is performed in smart environments where devices are often connected, four main features should be con-sidered: ubiquitous connectivity, power efficiency, security, and reliability [44]. In this context, the usage of wireless communications is highly beneficial, especially in remote areas.

---

3 Global positioning system [226].

1.2.4 Safety Management and Systems

Safety-related applications are always of critical nature. Such systems are of two main types: safety monitoring and safety integration systems [45]. The former relates to quick safety reaction times to prevent physical damages to humans or materials, and the latter requires real-time communication and immediate reaction. Safety monitoring applications (e.g., gas leakage monitoring) may tolerate certain levels of delay and loss; however, safety integration systems (e.g., steam pressure control) require a communication network with very low latency, ultra-high reliability, and resiliency. Due to the impact of interference on wireless communication, safety integration systems rarely use wireless networks as their primary communication channel.

In addition, the transmission channel for safety systems must use a so-called safety protocol as the supervision mechanism [46, 47]. The protocol constantly verifies performance metrics of the transmission channel (e.g., latency, synchronicity, reliability). If the safety protocol detects any violation in its key metrics, the safety application is switched into an unsafe or fail-safe state. Considering that functional safety is a part of safety system that acts in response to risky circumstances, it exploits industrial automation and smart techniques to actively minimize system risk and failure.

1.2.5 Security and Surveillance

This class of applications often relies on commercial communication technologies such as Wi-Fi when wired solutions are cost prohibited to transmit voice, video, and identification information related to the security of industry space. In wireless video surveillance systems, cameras are mounted on drones, land vehicles, and remote fixed locations. The transmitted video delivers footages to enhance critical awareness and to assist in decision-making in a wide range of applications such as seismic changes and natural disasters, harbor inspection, object detection on assembly lines, and rescue operations. Naturally, wireless video surveillance systems require high bandwidth, scalable and robust networks, and video analysis algorithms to properly address security requirements.

1.3 Design Criteria and Communication Requirements in the Industry 4.0 Era

Wireless technologies are an essential aspect of Industry 4.0 implementation, where all instances involved in value creation are properly interconnected. To provide better communication between service users, marketplace, and service providers, cellular networks supplement communication systems to promote potential transformation to Industry 4.0. Obviously, provisioning of wireless communication throughout a large industrial site with wide range of heterogeneous applications creates trade-offs between performance parameters and complicates the design of the overall wireless solution. Therefore, communication technologies should take into account certain design criteria specifically adapted for the industrial space. In this section, we present the main selection criteria and requirements for industrial wireless technologies.

1.3.1 Reliability

In the context of wireless communication, reliability could be identified as the capability of a wireless technology to seamlessly communicate in the presence of wide range of obstructions in space. The reliability of a wireless system is quantified by a number of key performance indexes such as its radio frequency (RF) spectrum usage, RF agility, and link budget [48]. Because of the physical nature of RF waves, the utilization of RF spectrum is fully regulated by government bodies to minimize interference. RF agility improves reliability through interference reduction or avoidance techniques in the RF spectrum. The link budget metric estimates the power of the received signal, accounting for the transmit power and gain wireless medium gain and loss. A higher link budget value corresponds to reduced interference and offers increased level of reliability.

Reliability of a wireless network also depends on network topology, media access control (MAC) design, and the adopted modulation, and coding scheme [49]. Other factors may also affect the reliability of a given wireless system. For instance, if reliability is characterized in terms of power consumption, the longevity of wireless network should be adjusted above a certain threshold level to ensure no significant degradation in system performance and reliability.

1.3.2 Latency

Industry 4.0 applications often require low latency networks to collect, store, or analyze data and to make decisions based on updated information. In the wireless medium, the variation of link quality deteriorates uplink/downlink transmissions over time, leading to higher latency and subsequent system failure [50]. In time-critical applications, if the communication does not meet the low latency requirement, data may lose its original value [51]. MAC design has a significant impact on latency of the network [49].

1.3.3 Coverage

Transmission range is defined as the maximum distance a signal is sent by a transceiver and can be reached and interpreted at the receiver. Given that industrial environment is RF hostile and constantly changing, the coverage range is affected by transmission power, complexity, and propagation properties [49]. Generally, higher data rates degrade penetration capabilities in an industrial site full of obstacles but at the cost of reduced coverage for network. Coverage of wireless solutions can be extended by deployment of routing, peer-to-peer (P2P) communication, and increasing transmission power using repeaters and power amplifiers. Nevertheless, these techniques increase system complexity and result in less reliability. To summarize, it is necessary to identify the requirements of an individual application and identify the proper trade-off between key performance metrics correspondingly.

1.3.4 Power Efficiency

Power efficiency is a qualitative metric that represents energy consumption of system components. In wireless networks, energy consumption depends on data rate, network topology, MAC protocol, and hardware design [49]. One possible solution to optimize system power consumption is using dynamic power management to boost system efficiency. Power efficiency and reliability are also closely interconnected. In fact, while better power efficiency will lead to higher reliability, system demands on ultra-reliable low latency communication may cause significant energy consumption [52]. It is thus necessary to consider well-thought trade-offs under the acceptable level of system complexity.

1.3.5 Simplicity

Although adopting Industry 4.0 unveils great potentials in industrial landscape, interconnected supply chain will be highly affected by additional complexity in mobility, sensory systems, machines, and information [53]. Recently, a number of technologies have been standardized, and many proprietary alternatives are constantly being offered for wireless communication [54]. Wireless solutions and their heterogeneous connectivity foreseen in the context of Industry 4.0 result in increased amount of generated data, parameters, and variables in networks. Integrating novel automation systems and advanced monitoring techniques increases the overall degree of complexity in products, enterprises, and the underlying network infrastructure. For example, an industrial site is now remotely accessible through the Internet, and its generated information should be integrated for adaptive and effective decision-making. Decentralized decision-making is one possible solution to reduce complexity. Such approach is based on independent decision-making at lower layers and provides more flexibility for customized products and flexible environments. In conflicts between decision-makers at such levels, decision-making tasks are delegated and handled to other levels.

1.3.6 Scalability

Wireless networks in industrial environments often rely on solid infrastructures in which large number of devices are randomly spread over a wide deployment area. In addition, communication networks and protocols should be dynamic and support different types of network changes such as device replacement, service modifications, and system upgrades without degrading the desired efficiency and operation. To establish connections and maintain autonomous network connectivity in such environments, wireless systems should be scalable and comply with real-time changes. Scalability is primarily determined by MAC design, and developing scalable MAC protocols is highly challenging [49]. Various techniques are employed in MAC layer that utilize diversity in frequency band, time, and space to deal with scalability. For instance, contention-free conventions such as time-division multiple access (TDMA) protocols are considered to impose strict constraint on network density and fulfill desired quality of service (QoS).

1.3.7 Security

Industrial systems are often equipped with both advanced and aging machines. The equipment might be located in remote areas, and it is often hard to be replaced or moved. In such systems, connectivity of old machines is not fully supported by the modern and state-of-the-art device monitoring capabilities. Therefore, interconnecting and integrating equipment can provide a target entry point for cyberattacks. Security should also be considered as an essential feature in high-level design of industrial wireless networks to ensure secure data aggregation and authentication. Therefore, security overheads should always be considered delicately in connection with other design criteria, and a proper balance should be pursued with respect to other QoS performance requirements in heterogeneous industrial networks.

1.4 Wireless Technology and International Standards

Industrial use of wireless communication is evolving and requires development of innovative technologies that enable industrial services. Different wireless technologies and standards across multiple industries are often adapted (modified or even redesigned) by an industry to fulfill its diverse requirements in terms of respective key performance metrics. Consequently, the standardization efforts aim at addressing such requirements according to the levels of QoS and working channels [55, 56]. Although wireless networks offer diverse and dynamic approaches for connectivity, it should be noted that none of these technologies have become prominent in the market. This is mainly due to the limitations of such technologies and lack of reliable business models [57].

Wireless communication technologies in industrial IoT span over a wide range of carrier frequencies and could support connectivity in the order of microsecond latency. For instance, the latency of Wireless Industrial Automation and Process Automation (WIA-PA) network is at least 10 ms with the range of up to 100 m [58]. Even though the communication link of this technology is reliable and fast, its range is limited. The existing trade-offs in wireless technologies are not just between latency and range, and other parameters such as channel bandwidth, data rate, number of nodes per network, and energy consumption are not the same across different wireless technologies. This wide variety of options for broad and diversified use cases makes selection of potential technologies even more challenging. As shown in Figure 1.3, industrial wireless technologies can be categorized into short-range and long-range technologies. Many technologies have arisen in both markets, as shown in Figure 1.3.

1.4.1 Short-Range Wireless Communication

Short-range communication technologies use radio waves to transmit data to a receiving node within a very short distance. Their offered range is generally suited for wireless personal area network (WPAN) and wireless local area network (WLAN). The distinct feature of short-range wireless communication is the low power required to communicate, making them suitable for battery-equipped mobile devices. A number of such protocols discussed here are Zigbee, WirelessHART, ISAl00.11a, Bluetooth and WIA-PA, and all are governed by IEEE 802.15 standards. In addition, we review Wi-Fi networks regulated by the IEEE 802.11 standard.

1.4.1.1 IEEE 802.15.4 Standard
The IEEE 802.15.4 standard was initially developed for WPAN and utilizes transceivers with low power consumption to offer low to moderate data rates (Kbps to several Mbps). Over time, it has undergone a number of improvements to address the

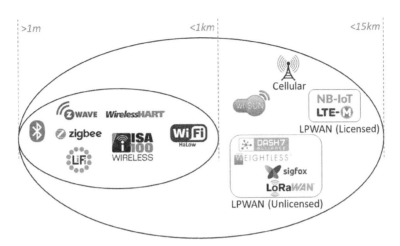

Figure 1.3 Wireless Technologies Range Coverage.

requirements of various use cases [59]. Many industrial IoT applications employ this standard for slow-paced continuous processes (up to 250 Kbps) and long durability transmissions to carry process automation data.

For transmission of mission-critical data, this standard defines a physical layer (PHY) and a medium access control (MAC) sublayer for low rate WPAN. Two node classes are introduced to operate on the MAC layer: (1) full function device (FFD); and (2) reduced function device (RFD). The FFD node is more effective and implements full protocol in the network. It can play three roles in the network, namely, the personal area network (PAN), coordinator, coordinator of the network, or act as a device (functional mode). On the other hand, the RFD node has more limited resources and performs extremely simple tasks such as sensing the environment. In this standard, the power consumption of bit transmission is governed by its deployed hardware [57]. Since forwarding nodes (e.g., routers) should be always available and set to "on," irrespective of their actual load of traffic, energy consumption of IEEE 802.15.4 is high [60]. In addition, the single channel property of the IEEE 802.15.4 MAC layer is not reliable and suffers from a high level of signal fading and destructive interference, particularly in multi-hop scenarios.

To overcome such limitations, IEEE 802.15.4e has been proposed as an extension of this standard [59]. It modifies the existing MAC sublayer to better comply with WSN and emerging IoT applications, with the same PHY layer. Therefore, it is compatible with IEEE 802.15.4 hardware and can be deployed without any adjustment. Various MAC channel access protocols are established by IEEE 802.15.4e standard and present different MAC layer modes. Among them, time-slotted channel hopping (TSCH) becomes highly appropriate for industrial process automation and control applications [61]. Dust Network initially adopted TSCH for its proprietary MAC [62]. Because of TSCH time and frequency diversity characteristics, its core idea was later adopted in several industrial standards: WirelessHART [63], ISA100.11a [64], and WIA-PA [58]. An open standard protocol stack called 6TiSCH was later designed that combined lower layers of IEEE 802.15.4e with higher layers of the Internet Engineering Task Force (IETF) to fulfill stringent requirements of industrial IoT networks with respect to ultra-low jitter, low latency, and high reliability [65, 66]. To summarize, the IEEE 802.15.4e standard enhances its performance in terms of reliability, deterministic latency, network capacity, and power consumption to support seamless communication of embedded devices in design of IoT systems [67].

In the following, we elaborate some well-known industrial variants of the IEEE 802.15.4 standard.

Zigbee. Developed by Zigbee Alliance and primarily targeted residential and commercial applications to provide wireless connectivity [68], this wireless technology is based on PHY and MAC specifications of IEEE 802.15.4. However, it has its own specifications for all upper layers of the protocol reference model [69]. Zigbee defines an application framework (AF) for application layer that demultiplexes incoming data between registered applications. Network nodes can operate as either FFDs or RFDs and exclusively transmit data toward the router or coordinator. Since it offers various features for protocols of all the layers above MAC and PHY, it can be properly adapted to large deployments.

Zigbee operates on the frequency band of 2.4 GHz (i.e., the Industrial Scientific and Medical [ISM] band) and shares the spectrum with IEEE 802.15.1 and IEEE 802.11. This makes it susceptible to high mutual interference and noise and leads to frequent backoff for Zigbee MAC protocol [70]. Zigbee is not highly suited to provisioning easy channel access and delivery of data in delay-sensitive applications. Consequently, it is not suitable for industrial applications that require deterministic delay and high reliability [71].

To combat limitations of its MAC layer, Zigbee Alliance developed a variant called Zigbee PRO, which specifically supports industrial process and control applications [72]. Zigbee PRO can change network operating channels and enhances security features if it faces significant levels of interference or noise [73]. Zigbee Smart Energy (Zigbee SE) is another related protocol that relies on Zigbee IP and effectively manages the power consumption of nodes. Considering the cost efficiency, low-powered structure, and redundancy capability of Zigbee SE, it is suitable for demand–response and load control systems such as smart grids [74].

WirelessHART. This open standard specifically targets wireless instrumentation for factory automation [75]. The main motive behind WirelessHART was to deliver feasible solutions that address stringent timing requirements and severe interference conditions of the industrial ecosystem [76]. The PHY layer of WirelessHART is based on IEEE 802.15.4; however, it applies TDMA channel access for MAC protocol to guarantee collision-free channel access [62]. Therefore, WirelessHART communication is scheduled and offers strict time slots as well as network-wide time synchronization due to the adoption of the TDMA-based MAC layer. WirelessHART MAC protocol features some prominent properties such as channel blacklisting and channel hopping schemes, which enhance data bandwidth and system robustness. More importantly, the WirelessHART network layer supports self-organizing mesh networks to assist in automatic configuration, optimization, diagnostician, and healing networks.

Different from generic WSNs, the WirelessHART network design and employment includes eight different types of devices. The sensors and actuators are denoted as field devices, which are usually attached to the plant equipment or processes to collect data such as pressure, humidity, and fluid flow from physical environment. The remaining seven are deployed to assist in network management and functionalities such as network interoperation, security, and optimization. WirelessHART is mainly considered as a centralized wireless network and utilizes a central network management to keep communication and route scheduling up to date. Consequently, it is more suited to industrial applications where network graphs should be continuously adapted to network changes and demands. WirelessHART can serve a large number of devices and high network data rates by using multiple gateways connected to a HART over IP backbone or multiple access points [77].

ISA100.11a. The International Society of Automation (ISA) developed industry standards that offer reliable and secure systems for automation and control applications. ISA100.11a is an industrial wireless communication standard ratified in 2009 and operates in a 2.4 GHz band (ISM band) [64]. A PHY layer of ISA100.11a is based on the IEEE 802.15.4 standard; however, its MAC layer is built on a modified, noncompliant MAC protocol of IEEE 802.15.4 and utilizes a combination of contention- and scheduled-based MAC scheme. To achieve real-time networking, a MAC protocol of ISA100.11.a exploits TDMA and carrier-sense multiple access (CSMA) along with additional spatial, frequency, and temporal diversity. Channel blacklisting and frequency hopping are also leveraged to address mutual interference from coexisting wireless systems and enhance network robustness [70].

The network of ISA100.11a consists of field and infrastructure devices. Both classes of devices are further divided into multiple types to assist in achieving a proper network architecture. Similar to WirelessHART, it exploits self-healing networks. The configuration of monitoring (e.g., slot allocation and scheduling), network runtime configurations, and execution of security standards policies are performed by the system manager as an infrastructure device. ISA100.11a connects field and plant networks via gateway devices. It embraces either distributed or centralized management; however, the proposed distributed management does not specify how network resources should be coordinated [78]. Different from WirelessHART, ISA100.11a provides graph and source routing while also offering options for configuration-based time-slot sizes, explicit congestion notification, and dual acknowledgment [77].

Wireless networks for industrial automation-process automation (WIA-PA). The Chinese Industrial Wireless Alliance introduced WIA-PA in 2008 as the national standard that proposes architecture and communication specifications for industrial automation and process use cases. Later, this standard was approved by the IEC (International Electrotechnical Commission) [58]. The WIA-PA PHY layer complies with the IEEE 802.15.4 standard, though its MAC layer applies a hybrid scheme (a combination of scheduled- and contention- based mechanisms) on IEEE 802.15.4 MAC protocol. It also exploits TDMA, CSMA, and frequency-division multiple access (FDMA) approaches for channel access. Given that WIA-PA MAC layer leverages adaptive frequency hopping, time slot hopping, and adaptive frequency switching, it can cope with varying network conditions and is considered a self-healing network [78]. Similar to WirelessHART and ISA100.11a, this standard employs a reactive approach that exploits redundant routing and gateway devices to prevent failures in networks, further enhancing its reliability and self-organizing characteristics [78].

Physical devices in WIA-PA networks are categorized into five classes [76]: (1) handheld devices to monitor and control the production plants and configure network devices; (2) field devices (sensors and actuators) located in the field to control or monitor industrial processes; (3) routing devices; (4) gateway devices that connect WIA-PA networks to various plant networks; and (5) a host computer as the user interface for management and maintenance. Both centralized and distributed mechanisms are deployed in WIA-PA networks to perform network and security management. Typically, network manager configures the network, schedules communication, handles routing tables, and protects the overall network, whereas the security manager is responsible for security and authentication management in the network. To conserve energy in WIA-PA networks, two-level aggregation mechanism is exploited: packet aggregation at network and application layers. This is different from WirelessHART and ISA100.11a, which utilizes only a one-level packet aggregation scheme. Additionally, WIA-PA adopts two-stage communication resource allocation.

1.4.1.2 IEEE 802.15.1 Standard

It was approved in 2002 and designed for short-range, low-power, and low-cost connectivity in WPAN applications. IEEE 802.15.1 is established on Bluetooth v1.1 foundation specifications that govern Bluetooth technology [79]. There are two main variations of this technology: classic Bluetooth and Bluetooth low energy (BLE). Classic Bluetooth supports devices with high demand of small transmission and low energy consumption, whereas BLE is ideal for applications that require communication of small quantities of data on an occasional or periodic basis.

Bluetooth classic. The Bluetooth PHY layer is adapted from the IEEE 802.15.1 standard and operates in the frequency band of 2.4 GHz. Its transmission technique exploits frequency hopping spread spectrum (FHSS). This results in the reduction of interference from nearby systems sharing the same frequency band (ISM band) and increases system robustness. Bluetooth defines two types of network topologies, namely, piconet and scatternet. Piconet is a single-hop topology that enables communication between one master node and multiple slave nodes. A scatternet topology is a cluster of Bluetooth piconets overlapping in space and time; there is only one master node, while a slave node could operate as slave in different piconets. Bluetooth MAC layer mainly focuses on establishing physical connections between the master and slaves, synchronizing network nodes with the master node's clock, packet transmission on physical channels, and device management for energy-saving modes [80].

Bluetooth low energy (BLE). A smart variation of the IEEE 802.15.1 standard, also known as Smart Bluetooth, BLE supports industrial wireless communication [81]. It is designed for short-range communication and, compared with classic Bluetooth, has much smaller latency, lower energy consumption, and increased range with higher data rates [49]. BLE adopts master–slave architecture, and role of a device (as master or slave) is defined in its MAC layer. To decrease channel sensitivity to interference and multipath fading, BLE employs TDMA-based MAC protocol and frequency hopping [82]. Additionally, the center frequency of channels has been assigned in a way to minimize interference with IEEE 802.11 channels [57]. Even though there are options to increase BLE's data rate at the expense of range, the limited range of BLE makes it inept for extensive IIoT deployments. To address this problem, a Bluetooth mesh network is introduced to increase the number of relaying neighbors, making it applicable to industrial use cases such as alerting and logging systems [83]. Recently most wearables are equipped with BLE interfaces. Overall, considering its low power consumption, BLE could be utilized as a practical and effective wireless technology for IoT applications that require short-range communication [68, 81].

1.4.1.3 IEEE 802.11 Standard (Wi-Fi)

Wi-Fi is a family of wireless network protocols regulated by the IEEE 802.11 standard [84], originally designed to offer high data rate connectivity and Internet access to a limited number of devices for WLAN. The early versions of the standard (IEEE 802.11b/g/n/ac) are limited by high power consumption and frame overhead. In addition, they do not fully support small and deterministic payloads required for mission-critical IoT applications. Thus, a set of modifications is proposed by IEEE 802.11 working group to amend WLAN to IoT scenarios. The first of such efforts was IEEE 802.11ah, known as Wi-Fi HaLow, which fulfills IoT requirements and could achieve denser deployment, lower overhead, and less energy consumption compared with legacy Wi-Fi networks [85]. The main features of the IEEE 802.11ah PHY layer are inherited from IEEE 802.11ac, which accommodates relatively narrow channel bandwidths. Its MAC layer has adopted some enhancements to improve power-saving features, to assist large number of devices, and to increase data rate. IEEE802.11ah supports machine-to-machine (M2M) communications for IoT services such as smart metering and industrial automation [86].

The IEEE 802.11ax standard, also called Wi-Fi6, is another effort that supports mission-critical data transmissions [87]. It underpins complex applications such as VR and robotic motion control [88]. To ensure distinct transmission scheduling for time-critical and delay-sensitive use cases, the MAC layer of Wi-Fi6 adopts orthogonal frequency division multiple access (OFDMA). Moreover, adaptive modulation and coding schemes are used in MAC protocol to tackle high levels of variations in wireless links [89].

1.4.2 Long-Range Wireless Communication

Long-range wireless networks are adapted for scenarios that require long-distance data transmission (up to 100 km). These wireless communications include very small aperture terminal (VSAT) technology, cellular networks (2G/3G/4G, LTE, and 5G), and low power wide area networks (LPWANs). Since VSAT

technology is based on satellite communication and employed in hard-to-reach places or remote sites, it is beyond the scope of this chapter. Since we will review cellular networks in Section 1.5, in this section we will focus on a number of main LPWAN technologies: long range (LoRa) [90], LTE-M [91], and narrowband Internet of things (NB-IoT) [92].

1.4.2.1 Low Power Wide Area Network (LPWAN)

The term LPWAN, also known as low power wide area (LPWA) network or low power network (LPN), was primarily designed for M2M networking. LPWAN denotes energy-efficient, low-cost, and especially wide area coverage communication [93]. "Wide area" in LPWAN relates to radio links of over 1 kilometer range. The applications of LPWAN are limited to low bandwidth communication with low data rate and infrequent transmissions [94]. Therefore, it is well-suited for IoT services that require small data transmission over a wide area. For industrial IoT systems, LRWAN is suitable for retrieving data from field devices that transmit low traffic (a few bytes in the payload) over long distances for a short period of air time for each node per day.

Wireless LPWANs have emerged in licensed and unlicensed frequencies. They include open standards and proprietary options such as Huawei's cellular IoT (CIoT) [95] and NWave [96] and differ in coverage area and other technical characteristics. A set of LPWAN standards has been specified by GSMA wireless industry association in 2015 to assist network operators to meet the requirements of IoT use cases in terms of coverage, energy consumption, and cost [97]. Consequently, LPWAN became a preferable choice for IoT use cases and gained significant attention as a complementary technology to the existing cellular networks. LPWAN comes in various ranges, sizes, and operational properties. In this section, we overview three LPWAN technologies with different features.

Long range (LORA). As a promising wireless standard for IoT use cases, LoRa is a physical layer technology for LPWAN developed by Semtech Corporation [98]. It operates in an unlicensed band and offers seamless connection with wide range deployment (several kilometers) and minimum investment and maintenance costs. LoRa has adopted enhanced modulation and optimum network protocols for finite energy sensors.

LoRa modulation achieves bidirectional communication through a proprietary spread spectrum technique in the sub-GHz frequency band: the chirp spread spectrum (CSS) scheme [99, 100]. This technique in LoRa generates a narrowband (NB) signal and spreads it over a wider channel bandwidth to provide a signal with wide bandwidth.[4] This results in larger signal-to-noise ratio where the received signal is difficult to be jammed, making it resilient to communication channel degradation arising from Doppler effect, multipath fading, and a high level of noise [101, 102]. Importantly, LoRa modulation contributes to greater maximum coupling loss (MCL) compared with existing sub-GHz communications and enlarges the capacity of the network as well as extending the coverage distance [103]. LoRa modulation utilizes variable spreading factors (SF) to balance between the data rate and range: a lower SF offers higher data rate at the cost of shorter range; by contrast, a high SF provides a low data rate, implying a larger range. Depending on the SF and channel bandwidth, a LoRa network achieves the data rate of 22 bps–50 kbps. In addition to adaptive data rates, spreading enables the simultaneous transmission of multiple spread signals over identical frequency channel. Therefore, LoRa base stations could simultaneously receive the transmitted messages with different SFs [104]. The largest length of payload for every message is 243 bytes [93].

LoRa's performance has been verified in multiple countries on smart meters, traffic monitoring, and smart health care [105, 106]. It also optimizes the protocol for power-restricted sensors and introduces three modes of operation for LoRa-based terminals in IoT environments. According to various application scenarios, these modes of operation identify data-receiving windows for each class of end devices and determine how these terminals communicate with the network. LoRa technology utilizes AES-128 for data encryption to ensure channel security.

LoRaWAN. As LoRa is mainly focuses on physical layer and its specification for connection of devices to infrastructures, LoRaWAN focuses on MAC protocol [107]. It is maintained by the LoRa Alliance and acts as the protocol that manages LPWAN

---

4 According to the Shannon theorem, a possible solution for low signal-to-noise ratio scenarios is to increase transmission channel bandwidth.

communication from end devices to gateways. LoRaWAN is based on the pure Additive Links Online Hawaii Area (ALOHA) protocol [49] to increases the success rate of the reception of the messages. This is because in ALOHA all base stations within the range could receive each message sent by an end device.

NB-IoT. Standardized by the Third Generation Partnership Project (3GPP) as a narrowband IoT communication technology [92], the NB-IoT is built on the prevailing LTE functionalities and works on the licensed frequency bands. Since NB-IoT could coexist with GSM and LTE, its deployment is rather simple, particularly in the existing LTE networks. The protocol of NB-IoT is derived from the LTE protocol; however, many LTE functionalities are reduced to make it simple and more suited to IoT applications. Thus, from the perspective of a protocol stack, NB-IoT could be seen as a novel air interface built on LTE infrastructure. NB-IoT could deploy LTE backend systems and broadcast signals for all end apparatuses within a cell. To minimize battery (and resource) consumption of the end devices, the cell is designed for short and sporadic data messages. Additionally, properties such as monitoring the quality of channel, dual connectivity, and carrier aggregation requiring a higher amount of battery are not permitted. The NB-IoT PHY layer is designed to conform to a subset of LTE standards; however, it exploits bandwidth of 180 KHz for narrowband transmission over uplink and downlink. FDMA and OFDMA are utilized for channel access in uplink and downlink, respectively [108]. An extensive review of NB-IoT PHY and MAC layers is discussed in [109].

The design objectives of NB-IoT encompass extended coverage, the multitude of devices with low data rate, and long battery life for delay-tolerant applications [110]. This technology is promising for indoor coverage and provides long-range and high sensitivity at the cost of adaptive throughput [111]. Three operation modes are provided for deployment of NB-IoT: stand-alone operation, guard-band operation, and in-band operation [103].

Connectivity of NB-IoT performs better than most of the competing technologies in terms of range, availability, and robustness. However, the latency of NB-IoT is unpredictable, and the procedure of random resource reservation increases the connection latency in dense networks [49]. Therefore, NB-IoT is not applicable to time-critical use cases such as safety systems. The main employments of NB-IoT for industrial applications are smart fleet management, smart logistic, and smart manufacturing [112, 113]. Recently, NB-IoT is exploited by telecom industry for smart lighting in the major cities [114].

The efficient utilization of existing cellular networks motivates different telecom manufacturers and vendors to promote NB-IoT standardization and commercialization. NB-IoT was initially proposed in 3GPP Release 13; further features and improvements such as localization methods, mobility, multicast services, and more technical details were specified in 3GPP Release 14 and beyond to satisfy the requirements of NB-IoT applications.

LTE-M. LTE machine type communication (MTC), termed as LTE-M, was proposed by 3GPP as a LPWAN standard to enable services and devices for M2M communication in IoT systems [91]. This standard adopts licensed frequency bands and relies on LTE-based protocols with the specifications of M2M communication in LTE advanced (LTE-A). Later, 3GPP defined a new profile for implementing MTC resources in LPWAN called Category 0 or CAT-0 [115]. In addition, two special categories were defined in the context of LTE that underpin IoT technology and the features of M2M communication: CAT-M for MTC and CAT-N for NB-IoT. CAT-M counts on mobile cellular network infrastructure to keep the coverage while reducing complexity. Different notions are used for each category; for instance, CAT-N standard is mainly referred to as NB-IoT, whereas CAT-M is known as CAT-M1, LTE eMTC, LTE-M2M, and LTE-M [116]. In this chapter, we will use LTE-M collectively for CAT-M.

The LTE-M standard supports both CAT-0 and CAT-M modes and takes advantage of existing LTE networks. It allows the LTE-installed infrastructure to be reused via a software upgrade to the existing LTE base stations [117]. LTE-M technology is highlighted by its efficient energy consumption that prolongs the battery lifetime of networks (more than 10 years). It also provides speeds of 300 Kbps and 375 Kbps for downlink and uplink, respectively [118]. LTE-M offers wireless network with low complexity and extended coverage for both indoors and underground. The performance of LTE-M for M2M communication is evaluated in [119] and discussed based on the network metrics. LTE-M supports a diverse range of vertical industries, applications, and deployment scenarios. For instance, Telstra and AT&T have used LTE-M in asset tracking and maintenance to provide wireless connection in IoT environments [120].

LTE-M has some advantages over NB-IoT, such as enhanced mobility, higher bandwidth and data rate, lower latency, and supporting voice over Internet (VoLTE) for simple use cases. In [121], LTE-M and NB-IoT are analyzed and compared for rural applications in terms of coverage and capacity. New network technologies such as NB-IoT and LTE-M, which are established by mobile technology and support LPWAN, are referred to as mobile IoT.

1.4.3 Comparative Study of Wireless Standards for Industrial IoT

As noted already, various wireless technologies and standards provide connectivity in industrial systems. To choose the appropriate wireless technology for an industrial IoT application, different factors should be considered. Tables 1.1 and 1.2 present the main technical differences among the aforementioned technologies. The comparison considers the PHY and MAC layer features along with various performance measures that each technology aims to fulfill. Such comparative study would assist in specifying potential wireless technologies for an industrial application.

1.5 Cellular and Mobile Technologies

Section 1.4 elaborated on different wireless technologies and standards that serve a variety of industrial applications. However, these wireless technologies are not sufficient for industrial applications that utilize data-intensive machines. In this context, cellular and mobile networks open up new opportunities in industrial applications. Cellular networks are empowered with ubiquitous presence, reliable communication links, widespread coverage, and mobility. These characteristics enhance operations of local networks and tailor them precisely to industrial applications for better leveraging the potentials of Industry 4.0 [123].

Historically, the primary focus of cellular communication was human-centric communication. With the rapid development of embedded devices and smart equipment,

Table 1.1 Comparison of Wireless Technologies: Short-Range Technologies.

	Zigbee	Wireless HART	ISA100.11a	WIA-PA	BLE	Wi-Fi HaLow
Standard	IEEE 802.15.4	PHY: IEEE 802.15.4 MAC: HART	IEEE 802.15.4	IEEE 802.15.4	IEEE 802.15.1	IEEE 802.11.ah
Frequency band	2.4 GHz	2.4 GHz	2.4 GHz	2.4 GHz	2.4 GHz	Sub-1GHz
Number of Channels	16	16	16	16	40[5]	7[6]
Topology	Star, Tree, Mesh	Star, Mesh	Star, Mesh, Star-Mesh	Hybrid Star-Mesh	P2P, Star, Mesh[7]	Star, Tree
Spreading	DSSS	DSSS, FHSS	DSSS, FHSS	DSSS	FHSS	MIMO-OFDM
MAC channel access	GTS, CSMA; Time slot is flexible	TSMP (TDMA, CSMA); Time slot of 10 ms	TDMA, CSMA; Time slot of 10–12 ms	TDMA, CSMA and FDMA; Time slot is configurable	TDMA	Hybrid EDCA/DCF
Channel bandwidth	2 MHz	2 MHz	2 MHz	20 MHz	2 MHz[8]	1/2/4/8/16 MHz
Range	10–100 m	<600 m	<600 m (100 m[9])	1–100 m	<100 m (<300 m[7])	90 m-1 km
Data rate	<250 Kbps	<250 Kbps	<250 Kbps	<250 Kbps	< 1 Mbps (0.125/1/2 Mbps[7])	0.15-78Mbps[10]
Nodes per network	64,000	Hundreds per AP	Thousands per gateway	100	Piconet: 7	8192
Power profile	~ 3 years	4–10 years	6 years	1 year	1 year	~1.5–13 years
Latency	Enumeration 30 ms	>10 ms	> 100 ms	>10 ms	>6 ms	>5 ms
Encryption	128-bit AES	128-bit AES	128-bit AES	128-bit AES	128-bit AES	WPA

5 37 Data Channels and 3 Advertising Channels.

6 The channels depend on the bandwidth spectrum that is available in a given country.

7 For BLE5.

8 For GFSK.

9 For indoor.

10 Depending on channel bandwidth.

Table 1.2 Comparison of Wireless Technologies: Long-Range Technologies.

	NB-IoT	LTE-M	LoRa/LoRaWAN
Standard	3GPP Rel.13 (planned)	3GPP Rel.13 (planned)	LoRa-Alliance (De-facto Standard)
Frequency band	Licensed LTE band	Licensed LTE band	Unlicensed sub-GHz[11]
Modulation	LTE-based OFDMA(DL) & SC-FDMA(UL)	LTE-based OFDMA(DL) & SC-FDMA(UL)	Proprietary CSS
Spreading	FDD/TDD	FDD	FHSS(ALOHA)
bidirectional	Yes/Half-duplex	Yes/Half-duplex	Yes/Half-duplex
Maximum payload length	256 bytes	1600 bytes	243 bytes
Maximum coupling loss (MCL)	155.7 dB	165 dB	169 dB
Channel bandwidth	1.4–20 MHz	180/200 KHz	125/250/500 KHz
Data rate	DL: 300 Kbps; UL: 375 Kbps[12]	DL: 200–300 Kbps; UL: 144 Kbps	22 bps–50 Kbps[13]
Range	11 Km	1 Km(urban), 11 Km(rural)	2 Km(urban),15 Km(rural)
Latency	10 ms–4 s	1.4–10 s (UL: < 10 s)	Not Guaranteed
Nodes per network	300–1500 per cell	~ 52000 per cell	200 per gateway[14]
Mobility	Connected mobility with some limitation (inter frequency handover)	No connected mobility (only idle mode reselection)	Better than NB-IoT
Energy efficiency	>10 years battery life of devices	>10–20 years battery life of devices	>10 years battery life of devices
Interference Immunity	Low	Low	Very high
Encryption	LTE encryption, 128/256-bit AES	LTE encryption, 128/256-bit AES	128-bit AES
Attack	Active and passive Eavesdropping attacks, sniffing attacks, and DoS	Active and passive Eavesdropping attacks, sniffing attacks, and DoS	Replay attack, DoS, Eavesdropping, Bit-Flipping attack, and LoRa class B attacks [122]

---

11 868*MHz* in Europe, 915*MHz* in North America, and 433MHz in Asia.

12 Peak data rates is 1*Mbps* for both UL & DL.

13 Depending on channel bandwidth and spreading factor.

14 5000 motes per gateway.

new communication standards were introduced to focus not only on the connectivity of people but also on communications between devices and machines in IoT. In this context, MTC has been proposed as a compelling solution that offers connectivity for diverse growing smart services such as smart meters, remote patient monitoring, smart manufacturing, boat tracking, and other similar cases [124, 125]. Within the cellular context, MTC is usually known as machine-to-machine (M2M) communication [57]. We will use MTC and M2M interchangeably in this chapter.

The MTC landscape uses both wireless and fixed networks to provide Internet access for a number of diverse applications [126]. This leads to diverse network protocols and data formats that exhibits different behavior in MTC systems [127]. MTC suffers from some fundamental limitations such as low coverage, and limited scalability. Cellular systems such as 4G, LTE, and 5G could be recognized as alternative technologies that extensively support MTC networks.

In this section, we first focus on a review of the current status of MTC in 3GPP cellular standards. We shall subsequently review LTE, 4G, and 5G and their enhanced features for communication in industrial environments.

1.5.1   3GPP Cellular: MTC

3GPP characterized MTC as a form of data communication between machines in an autonomous manner that does not necessarily require human intervention [128]. MTC denotes two communications scenarios: (1) communication between MTC devices and MTC servers (e.g., in utility smart metering); and (2) direct communication between MTC devices without intermediate server (e.g., IoT) [128]. Although MTC could utilize different types of radio access technologies [129], MTC solutions based on mobile access technologies are of vital importance because cellular MTC offers viable benefits such as mobility, roaming support, robustness against single point of failures, and immediate and reliable data delivery [126, 128]. Moreover, scalability and ease of deployment in cellular MTC can be accomplished via an untethered method. Cellular MTC also excels the ability of connecting devices to the core enterprise systems through a standardized application programming interface (API), in a scalable, real-time, and secure way.

Although MTC offers compelling advantages, it exhibits shortcomings that impact the level of networking and viability of business models. The main challenges are relevant to the diversity of M2M applications and their requirements, energy consumption, and radio resources cost [130]. It also suffers from limited resources in MTC devices (e.g., computation, power resources) and traffic characteristics of MTC applications that are dependent on specific use cases [131]. Such characteristics bring up new technical issues that must be effectively addressed to fully support MTC in cellular systems [126, 130].

1.5.1.1   3GPP MTC Standardization

As mentioned earlier, a number of standard bodies have collaborated on MTC architectures to provide connectivity between shared MTC devices. 3GPP has already specified standardization to promote adaptation and requirements of MTC. There are multiple groups in 3GPP for MTC functions, requirements, and interfaces. The continuous enhancements of 3GPP have appeared in several releases that present ongoing amendments and progresses in standardization works, facilitate introduction of new features, and establish a uniform platform for technologies deployment.

The standardization efforts also focus on optimizing the core network infrastructure to provide efficient delivery of M2M services and minimize operational costs. The first study of 3GPP system architecture (SA1) on MTC was first released in 2007 [132]. In Release 10, 3GPP SA2 put its efforts into identifying MTC communications requirements and system optimizations to address two important challenges in mass-market MTC services: MTC signaling congestion and network overload [133]. Release 10 supports MTC in the Universal Mobile Telecommunications System (UMTS) and LTE core networks. Logical analysis, requirement refinement, and protocol implementation were later introduced in Release 11 [134]. In addition, Release 11 studied network improvements for M2M gateways, P2P communications, co-located M2M devices, and M2M group development. In Release 12, the focus was on identifying key enablers for RF and PHY layer to facilitate LTE deployment in IoT environments [91]. Normative works were pursued in Release 13 to extend MTC coverage and reduce its cost (e.g., bandwidth, transmit power) for cellular IoT deployment. It also focused on identifying multiple categories for new user equipment (UE) [135]. These techniques, namely, NB-IoT and LTE-M, further strengthened in Release 14 and provided novel features such as mobility for service continuity, reduced overhead in network, and support of IoT data in mission-critical use cases [136]. Release 15 and beyond studied additional MTC enhancements for LTE. In 2018, 3GPP foresaw the standardization initiatives and subset of 5G requirements for MTC applications and services.

1.5.1.2 MTC Technical Requirements

MTC is a promising technology for connection of intelligent devices and appliances to the Internet and other networks. Given that 3GPP cellular systems were not primarily designed for machine-type communications, all MTC technical requirements in mobile and cellular technologies should be identified in advance. Some key requirements are as follows:

- Low complexity: MTC networks consist of heterogeneous connected devices from multiple vendor equipment and protocols [127]. Hence, a scalable MTC network architecture in a standard format is required to manage system heterogeneity and associated complexity [137]. 3GPP reduces MTC devices complexity by removing the unnecessary features of these devices. For instance, in 3GPP Release 12 and 13, a number of complexity reductions were identified for LTE. Such changes do not impact interoperability with normal 3GPP devices while maintaining IoT requirements.
- Increased energy efficiency: A majority of MTC devices are in small size, battery-powered, and located in remote areas. These features imply that recharging and replacement of batteries are infeasible. To prolong the MTC systems' life cycle, optimization techniques are used to achieve power efficiency in MTC nodes' sensing and data transmission [138]. These energy-efficient techniques could be applied in the application, network, and link layers.
- High coverage: Most industrial applications, such as smart metering and factory automation, require high levels of coverage, and their connectivity model succeeds where nearly the entire network elements are reachable. On the other hand, the large number of network nodes within a cell impacts the achieved QoS. In addition, the extended coverage of the wireless networks in indoor and industrial spaces is

challenging and requires large number of base stations that would be very costly. 3GPP proposed a viable approach in Release 12 that improves MTC devices coverage, facilitates a scalable IoT system, and stipulates low complexity without significant increase in overall cost.

- Reliability: MTC wireless networks might be unreliable because of interference and noise from adjacent equipment, RF channel fluctuations, and machine interconnections [127]. Given that delivery of sensory data to applications should be reliable in terms of E2E delay [139], some possible solutions such as software reconfiguration of cognitive radios and spatial-temporal redundancy techniques are utilized to improve network reliability. For instance, cognitive radio software reconfiguration enables equipment and terminals to dynamically switch between various wireless modes and adapt to environment changes to decrease RF self-interference and external noise [127]. This technique is exploited in software-defined radio (SDR) systems.

- MTC user identification and control: Almost all MTC devices in the market are equipped with subscriber identification module (SIM) that contains crucial information about device profile, identity, and subscribed services. Since network operators support customized MTC services according to the subscription profile, it is essential to regulate access of MTC devices individually and based on the prior defined SIM profiles. MTC user identification impacts decisions of network operators for serving and access of MTC devices. 3GPP defines multiple UE categories in LTE to identify, isolate, and restrict access of MTC devices.

- Service enablement and exposure: To ensure connectivity and scalability of MTC systems in emerging industrial IoT, 3GPP requires third-party support to enable required services for MTC. In this context, ETSI and the oneM2M Global Initiative are standardization organizations that collaborate on E2E service enablement. Whereas ETSI focuses on enabling services across servers, gateways, devices, and standard service interfaces, oneM2M develops practical details to tackle requirements of the service layer for M2M communication [140, 141]. Since MTC systems should be able to deal with heterogeneous sensing, it is necessary that application platforms be connected to 3GPP core network via secure interfaces. Additionally, a privacy-preserving approach should be adopted to manage availability of personal information for IoT customers and applications [142].

1.5.1.3 MTC Trade-Off for Different Cellular Generations

Various smart devices are constantly released into IIoT market, and it is difficult to integrate all these parts through communication systems. An important aspect of MTC is to connect devices regardless of type of their cellular and mobile networks, business industries, or machine types. This section briefly reviews different cellular technologies and their trade-offs for MTC deployments.

- The 2G family (GSM, GPRS, and EGPRS) is ideal for M2M communication as its power consumption and cost is low. In addition, M2M transmission requires very few bytes that could be effectively handled by 2G. Compared with newer generations, 2G is not spectrum efficient; therefore, its data rate and device management communication are less efficient for the same wireless bandwidth. Furthermore, spectrum sharing is not a viable option for 2G.

- The 3G family (UMTS and HSPA) supports the data rate of 1–3 Mbps that exceeds the requirements of most M2M applications. For automotive M2M applications that need a broad range of data rates, 3G is an appropriate wireless network. Compared with 2G, devices, network equipment, and connectivity are more expensive and less power efficient.

- 4G technologies (LTE and LTE-A) have an "all IP" technology that makes network infrastructure deployments simpler and less expensive than older cellular networks. It offers improved spectral efficiency, greater longevity, bandwidth flexibility, and scalability, all of which meet the requirements for MTC applications. To cope with the increased complexity of the protocols in 4G, high-performance processors in the radios are required. This leads to higher cost in 4G and makes large-scale M2M deployments difficult.

- 5G technology supports the requirements of MTC design as the forefront of IoT by offering lower cost, better power efficiency, and increased data rate for both terminals and systems. It also offers minimum latency for delay sensitive applications, massive MTC access, and seamless integration of IoT devices, all without QoS deterioration [143]. A number of 5G features also fit well with the M2M path, namely, service creation, service provisioning, and dense deployments.

1.5.2 LTE Features Enhancement

LTE is introduced in 3GPP Release 8 [144]. The foundation of the LTE network is an IP network architecture used for mobile, fixed, and portable broadband access. The all-IP architecture of LTE enables new converging services based on the IP multimedia core subsystem. To achieve high data rates, LTE utilizes a combination of orthogonal frequency division multiple access (OFDMA), multiple-input, multiple-output (MIMO) antenna systems, scalable bandwidth, higher order modulation schemes, and spatial multiplexing in the downlink. LTE works in two modes: time division duplex (TDD) and frequency division duplex (FDD). In 3GPP Release 10, LTE Advanced was introduced as a more advanced version of LTE [145].

LTE enhancements have been included in subsequent releases of LTE standard for different LTE-based devices (i.e., Cat-M, Cat-0) to meet the requirements of M2M/IoT devices [146]. The new versions are called LTE-eMTC, or LTE-M by 3GPP, and promise new generation of devices with lower cost, ubiquitous coverage, and ultra-low battery life.

1.5.3 4G Features Enhancement

The emerging IoT environment is composed of various types of wireless access nodes that require seamless connectivity [147]. The 4G network exploits interoperability and integration between multiple radio access networks (RAN) and radio access technologies (RAT) to create a more solid heterogeneous networking paradigm. This section briefly summarizes the evolution of 4G and its features that accommodate connectivity in IoT.

- Coverage extension: Relaying is a key radio access technology in 4G that extends base station range beyond its coverage area [148]. A relaying service is composed of chains of relaying nodes that adopt a suitable transmission scheme (and spectrum) based on the required latency and reliability [149]. With reference to IoT systems, the adoption of relaying services decreases network overload, leading to improved network scalability. It also alleviates the single point of failure issue and provides some mean of fault-tolerant communications in IoT systems.

- Enhanced data rate and throughput: 4G networks utilize the concept of licensed and unlicensed carriers' aggregation, where control-related traffic and non-critical transmissions are sent via licensed and unlicensed bands, respectively. Such techniques can be beneficial for scalable IoT systems that require high throughput.
- Power saving: As elaborated in 3GPP Release 12, frequent link quality measurements at the device side are a main source of energy consumption in networks. To address this issue, the power-saving mode was introduced as a viable solution to manage data transmission. In power-saving mode, a device could transmit uplink data at any time. However, for downlink communication, the device is reachable only either when it is active in uplink or is at configurable time instances.
- RAN as a service: RAN utilizes radio resource virtualization to create various virtual functions and to expose them via cloud platforms to distribute networks functionalities and management [150]. Also known as RAN as a service, it improves the flexibility of the communication infrastructure and allows IoT systems to self-heal and self-configure.
- Device-to-device (D2D) communication: It entails the possibility of data exchange between two devices in the unlicensed band without involving base stations or with just its partial aid [151]. In this technology, devices serve as mobile relays to communicate in IoT environment.

1.5.4 5G Features Enhancement

The 5G cellular network provides advanced wireless connectivity for various use cases and vertical industries and paves the way for Industry 4.0. It exploits different technologies such as network slicing, network function virtualization (NFV), software defined network (SDN), and mobile edge computing (MEC) to pre-allocate resources for both communications and computing [152]. The broadband capability of 5G mobile network facilitates direct and seamless wireless communication from the field level to the cloud and enables new operating models without redesigning the production line for smart manufacturing [153]. In addition, private cellular connections could help in assigning spectrum bands and base stations such that data transmission is performed via secure and reserved channels [154, 155]. A main differentiator between 5G and previous generations of cellular networks lies in its significant emphasis on MTC and IoT. For instance, 5G air interface has enacted novel techniques in PHY and MAC layers that accommodate MTC [156]. It seems that 5G leads to convergence of the many different communication technologies as it becomes the standard wireless technology.

5G specifically focuses on supporting communication at very low latencies, unparalleled reliability, and massive IoT connectivity. Based on these distinct features in 5G networks, the emerging diversified telecommunication services are categorized into three main classes [157]: (1) ultra-reliable and low latency communications (URLLC); (2) enhanced mobile broadband (eMBB); and (3) massive machine-type communications (mMTC).

1.5.4.1 Ultra-Reliable and Low Latency Communications

Compared with 4G network, E2E radio network latency in 5G is reduced to 1 ms with peak data transfer rate of 20 Gbps; this offers an ultra-responsive connection with ultra-low latency [158].[15] URLLC is expected to play a key role in a wide range of mission-critical and industrial automation applications such as remote medical assistant, autonomous vehicle control, and robot and drone control that rely on high data rates and low latency [159, 160]. One important aspect of URLLC scenarios is to communicate in real time or within a very short time. Consequently, applications that require highly reliable communication can be implemented in 5G, such as remote-controlled plants or smart factories. At the same time, 5G exploits lower frequency bands that propagate farther in the environment, providing a more robust means of communication between IoT devices in buildings. Such a feature also leads to prolonged devices battery life (about years).

In addition to highly reliable communications [161], 5G new radio introduces the concept of extra transmission redundancy to better support industrial applications in URLLC scenarios [162]. Design of URLLC services heavily relies on a number of factors including more reliable channel coding techniques, effective resource sharing (control-flow and data), and grant-free transmission for uplink data, all supported by 5G new radio standard [163].

1.5.4.2 Enhanced Mobile Broadband

5G is expected to support around 29 billion devices connected to IoT by 2022 [164, 165]. eMBB delivers high data rates across 5G coverage area, where downlink data rates of at least 100 Mbps per device is supported over a typical dense urban environment. With the increased bandwidth, the high-performance connectivity is also sustained for indoor spaces such as industrial campuses [166]. Therefore, bandwidth hungry applications such as virtual reality for remote maintenance worker assistance and production line imaging for quality inspection significantly benefit from the higher bandwidth available through 5G eMBB. In addition to fast data rate, 5G eMBB provides low latency of 4 ms over the air, enabling real-time data transmission, processing, and decision-making for autonomous warehousing robots and industrial machine control [167]. One approach for providing eMBB is to implement millimeter wave (mm-wave) technology and install high-frequency mm-wave antennas [168].

1.5.4.3 Massive Machine Type Communication

The 5G solution that serves a large number of MTC applications is mMTC, which deals with scalable connectivity for massive number of devices. 5G offers wireless communication to over 1 million IoT terminals (with diverse QoS requirements) per square kilometer of area [169]. mMTC data communication is typically infrequent, making it ideal for alerting systems or periodic sampling (e.g., indicating manufacturing equipment failure and low frequency environmental sensing). This feature is suitable for a wide range of use cases across utilities, industrial campuses, and logistics. mMTC should handle different exceptional challenges, such as varied and intermittent traffic, QoS provisioning, large signaling overhead, and congestion in RAN [137].

---

15 On 4G networks, the typical round-trip latency for radio transmission is about 15 ms with a typical E2E system latency of some 30–100 ms [227, 228].

1.6 Wireless System Design Enablers and Metrics for Emerging IIoT Applications

IIoT is one of the key components of the Fourth Industrial Revolution. It provides customized architectures and standardized interfaces for data acquisition, transmission, and analytics in industrial applications [25]. Diverse industrial applications differ in terms of operational settings, technical requirements, and service environments. Therefore, it is not possible to provide one multi-purpose wireless solution for all IIoT use cases. Each wireless system design requires theoretical and experimental measures based on its expected performance. In this section, we first review conventional technical enablers in design of wireless networks for IIoT and then discuss the metrics on the desired performance.

1.6.1 General Technical Enablers in Design of Wireless Network for IIoT

The National Institute of Standards and Technology (NIST) proposed a reference framework as a guideline that helps users select and design a given wireless system, customize its configuration based on the specific application requirements, successfully deploy it, and finally ensure its performance for industrial environments [55]. Based on this framework, the design of industrial wireless communication should be evaluated from three major aspects: system modeling and verification, radio resource management (RRM) schemes, and protocol interfaces design.

1.6.1.1 System Modeling and Verification

The connectivity in IIoT systems could exploit well-established communication protocols to reduce network configurations and customizations. However, the increasing integration of communication into automation aspects makes IIoT systems more complex and prone to errors (e.g., device failures, mistakes in configuration). Given that failure in communication may be catastrophic in industrial applications using IIoT, it is essential to use proper system models and verification schemes to increase the level of certainty in IIoT systems. There are different approaches to create system models, such as theoretical inference and simulation tests.

An important reference in IIoT wireless system design is modeling data traffic patterns to capture and predict the dynamic behavior of systems and handle system complexity. However, we initially need to understand the theoretical basis of inference models and the conditions underlying their effectiveness before choosing the method apt to IIoT environments and service characteristics. This justifies a rising need for network simulation platforms and testbeds that emulate real-world industrial systems and perform system-level verification [170]. The simulation frameworks could also assist in performance evaluation of wireless networks for the next-generation factories and process automation systems [171].

1.6.1.2 Radio Resource Management

The rise of ultra-dense and dynamic wireless networks in the Industry 4.0 paradigm implies a further number of simultaneous transmissions. Therefore, it is necessary to efficiently utilize limited wireless resources such as RF spectrum resources and radio network infrastructure to fulfill strict QoS requirements and achieve a reasonable level of performance across IIoT systems. Radio resource management (RRM) involves strategies and algorithms that manage radio transmission for reliable service delivery in dynamic and diverse wireless networks.

The fundamental challenge in IIoT wireless communication is cross-technology interference combined with harsh signal propagation conditions in industrial systems. This results in deficient networks performance and service failures regardless of prudent initiatives [172]. One possible solution for interference mitigation is coexistence mechanisms. There are two principal concerns in the design of coexistence mechanisms: interference management and load balancing [173]. Since RRM includes transmission power management, radio resources scheduling, user allocation, and preventive–reactive congestion control, effective RRM procedures could be exploited in coexistence mechanisms to mitigate the interference level. Another possible approach to avoid wireless networks interference is the employment of cognitive radio channel sensing of the ambient

radio environments to detect availability of channels. Subsequently, RRM is performed on the clear channels through optimizing proper channel features for data transmission [174, 175]. In addition to reliable link, RRM involves strategies for controlling power transmission, which is particularly important for IIoT devices working on batteries.

1.6.1.3 Protocol Interface Design

IIoT involves diverse industrial assets such as manufacturing control systems, industrial process controllers, devices, and components. These controllers and equipment are distributed throughout an industrial site and use both proprietary and open protocols for communication. Basically, proprietary protocols belong to a particular product line within the systems [176, 177, 178], while open protocols are governed by standard organizations and utilized across products. Considering the advent of communication technologies and various protocols for both proprietary and open standards, there are technical challenges in the interconnection of different industrial protocols and assets. This calls for the creation of systems and techniques that integrate different protocol interfaces in manufacturing and industrial processes and that allow the interoperation of devices, equipment, and services from various vendors and operators. In this context, Open Platform Communication-Unified Architecture (OPC-UA[16]) is a possible solution that offers interconnection and easy integration of newer standards (wired and wireless) and middleware technologies in Industry 4.0 [179]. Moreover, a communication interface for IIoT is proposed in [180] that provides communication among programmable controllers and devices for operating and monitoring processes and equipment.

Design of protocol interface could be discussed in horizontal and vertical directions [55]. Figure 1.1 indicates the concept of integration in Industry 4.0. Typically, horizontal protocol interfaces provide interconnectivity among nodes to meet certain network functionalities such as clock synchronization among nodes. Time-sensitive network (TSN)

---

16 OPC-UA is an open standardized software interface developed by the OPC Foundation. It enables M2M communication protocols to interconnect production control systems and industrial automation for the purposes of Industry 4.0 [229]. OPC-UA satisfies the desired requirements of industrial automation and establishes an interface standard that bridges the heterogeneity of industrial communication systems on both communication and information levels.

protocols are examples of such interfaces that provide real-time connectivity for mission-critical and time-sensitive use cases [162, 181]. On the other hand, vertical protocol interfaces assist in secure data and service integration and offer consistent information flow via protocol stacks [55]. IETF has released encapsulation and compression techniques that unify such operations for devices [182]. 6LoWPAN is an example of networking technology from IETF that determines fragmentation mechanism for IPv6 headers [183].

1.6.2 Metrics for Wireless System Design in IIoT

The autonomous communication in the future IIoT environment significantly relies on wireless networks. There are different wireless technologies competing with each other for various industrial services and use cases. Since network connectivity should be robust and efficient, crucial challenges are raised in network design and integration of automation and control systems. Furthermore, real-time analysis of IIoT systems is a major issue for smart environments and services. In designing a system, we need to actively evaluate system performance and its efficiency. Multiple factors such as outage probability, power consumption, and spectral efficiency could verify the performance and efficiency of IIoT wireless design. These metrics can be categorized into two broad classes, namely, resources and services indicators.

1.6.2.1 Resource Indicators

Wireless solutions in IIoT should specify the required resources to create connectivity within the systems. To identify the level of network resource utilization in wireless systems, a set of metrics are introduced as resource indicators: spectrum, time, power, and network computing [55].

1) Spectrum: In the frequency domain, the wireless channels are identified by the channel bandwidth. Given that the data rate of each wireless channel is related to its bandwidth [184], different services in IIoT environment are highly dependent on spectrum resources to accommodate the desired data rate. For instance, in mission-critical communications such as safety systems, the radio spectrum resource should be assigned properly for effective deployment (i.e., high data rate and low latency) [185]. Given that spectrum resources are shared in wireless domain for a large variety of applications, wireless networks suffer from spectrum scarcity. This highlights the need for communication protocol interoperation within different frequency bands and channels [186].

2) Time: Wireless spectrum resource usage could be defined in terms of the transmission time and length. Time manages the temporal operation of wireless resources based on the desired performance such as QoS [187]. In this context, the access time of channel spectrum could be considered a resource indicator to be optimized for wireless networks.

3) Energy consumption: Power consumption of wireless nodes at transmission side is closely dependent on transmission range, quality of transmission channel, and transmission power at RF front ends. The aforementioned resources are essential in the design of a wireless system. Various energy-efficient resource allocation techniques (with given QoS constraints) are proposed to assist in design of wireless systems for IIoT.

4) Computing resources: Network resources are in both hardware and software types (physical and virtual). Basically, computing resources include all network resources tied up to the aforementioned resources (i.e., spectrum, time, and energy consumption). For instance, in TDMA-based transmission, different communication links are assigned to users at different time slots and through computing resources.

To achieve the diversity gain for different applications in IIoT, the resource space is used in multiple dimensions. For example, a wireless solution could exploit a TDMA scheme in the multi-antenna wireless systems, while both power and spectrum are considered in the system design.

1.6.2.2 Service Indicators

In addition to wireless network resources that specify the cost to build connections, service performance highly affects the design of IIoT networks. Several service measures are identified to indicate the level of service satisfaction (while ensuring QoS), and they are extensively utilized in formulating wireless design. Service performance factors that form a set of basic metrics are data throughput, delay, coverage, and scalability.

1) Data throughput: Service requirements are specified by several data rate metrics such as real-time throughput, aggregated throughput, raw data rate, good throughput (goodput), peak and minimum throughput, average throughput, and instantaneous data rate. Throughput is the key indicator for high-performance services such as autonomous driving and robot-assisted surgery.
2) Delay: It specifies the latency over the communication link and is composed of the transmission delay, queuing delay, and jitter. As a service indicator, delay defines the speed of data delivery across networks and verifies the timeline for data packets transmission. This is an essentially important metric for mission-critical industrial applications such as real-time control and safety that are featured by stringent latency requirements.
3) Coverage: It is defined as the wireless network range that connectivity is provided for services. Generally, the worst-case channel for all terminals in the coverage area is considered as the performance metric of wireless network.
4) Scalability: It stands for the number of nodes that could be connected to the wireless network. Given that there are various IIoT services with different levels of scalability, this performance metric is highly associated with the traffic load of network.

Generally, the combination of multiple indicators forms a composite service metric in wireless systems, and the selected indicators are optimized for the given service requirements.

1.7 MAC Protocols in IIoT

Industrial environments have some special properties such as equipment noise, electromagnetic interference, and complex coordination among devices that pose unique challenges for wireless connectivity. IIoT systems often require support of seamless communication and diverse objectives of functions for a wide range of applications. Therefore, wireless communication protocol design that meets these performance criteria is essential to fulfill service QoS and ensure prolonged lifetime.

MAC layer is a key sublayer of the data link layer that exploits use of specific protocols to administer nodes privilege to access shared wireless medium according to application requirements. MAC protocol controls the radio and channel sensing scheme and defines nodes duty cycle, communication mode between devices, data rates, transmission power, and range. Given that radio is the foremost source of power consumption in networks, MAC protocols could significantly impact nodes' overall power consumption and their lifetime [188]. In addition to the aforementioned key features, the MAC layer also controls other wireless settings such as frames synchronization, source–destination address management, error detection for physical layer transmission, collision reduction, and mitigation of idle listening.

With the goal of meeting the requirements of IIoT applications, the existing protocols could be adapted, evolved, or developed for different performance characteristics. MAC protocol schemes can be categorized into two broad classes: scheduled based and contention based [189]. Hybrid schemes are also proposed as a combination of these schemes.

1.7.1 Scheduled-Based Schemes

In scheduled-based protocols, also known as fixed reservation–based schemes, a fixed duration of time, frequency, or other domain is scheduled and assigned to nodes for network resources access. The scheduling assignment algorithm is conducted by a centralized base station and aims at avoiding channel collisions. In addition to a collision-free schedule, a device is simply set to sleep when it is not using its time slot to prevent idle listening and message overhearing [190]. This scheme is more suitable for networks that deploy low-mobility nodes and require infrequent topology changes and scheduling adjustment. It also tends to be more predictable and offers deterministic E2E delay. However, in dense networks, nodes should wait to gain access to the wireless medium, and additional queuing delay shall be incurred. Synchronization is an important issue in this approach leading to higher complexity and additional traffic due to additional control packets. The following multiple access schemes are utilized in typical multi-user wireless communication systems.

- TDMA: Time is divided among nodes for a given and identical frequency channel. Therefore, a fixed portion of time is assigned to every node to transmit data. For successful TDMA slot assignment and collision-free communication, tight clock synchronization should be established between nodes. GinMAC [191] and wireless arbitration (WirArb) [192] are some collision-free TDMA-based MAC protocols in time-sensitive IIoT. For instance, GinMAC provides reliable data delivery as well as deterministic time delay for industrial process automation such as closed-loop control systems.[17] Given that only one node is allowed to transmit data during the scheduled time slots, TDMA suffers from relatively high delay.

---

17 GinMAC protocol has restricted scalability due to the exclusive TDMA slot usage, and it supports up to 25 nodes for in-field control stations.

- FDMA: As the name infers, accessible frequency bandwidth is partitioned into non-overlapped sub-channels, where each individual sub-channel is adequate to accommodate transmission of a signal spectrum. Ideally, through proper frequency assignment algorithms, a unique physical frequency is dedicated to every node to offer a collision-free protocol. The FDMA-based protocols support multiple frequencies and require more costly hardware. They generally are not useful for IoT systems because of a high level of power consumption and more complicated design [193].
- CDMA: A MAC channel access method that enables transmission of multiple signals in a single transmission channel. A combination of special encoding scheme and spreading spectrum technology is exploited to send multiple signals through a single channel. The basic principle is that users have access to the whole bandwidth for the entire duration, but they utilize different CDMA codes; this assists the receiver to distinguish among different users. Given that the entire bandwidth is allocated to a CDMA channel, this scheme suffers from limited flexibility in adapting bandwidth, particularly for M2M communication in IIoT systems.
- OFDMA: A multiple access scheme that divides the entire channel resources into small time-frequency resource units. Since the available bandwidth is divided into multiple mutually orthogonal narrowband sub-carriers, several users could share these sub-carriers and simultaneously transmit data. In other words, the signal is first split into multiple smaller sub-signals, and resource units are allocated to them. Then, each data stream is modulated and transmitted through the assigned resource units. OFDMA allows several users with various bandwidth requirements simultaneously to transmit data at different (orthogonal) frequencies. Therefore, channel resources can be assigned with much more flexibility for different types of traffic. In addition to high spectral efficiency, OFDMA can effectively overcome interference and frequency- selective fading caused by multipath. OFDMA is a promising multiple access scheme adopted for wide range of mobile broadband wireless networks such as LTE, Wi-Fi6, and 5G [194–196].

1.7.2 Contention-Based Schemes

In contention-based protocols, nodes perform random access competition with each other to access medium on demand. Before transmission, nodes perform channel sensing to verify whether the medium is clear and wait for a specified backoff period to transmit data. Each node performs the channel sensing independently, and the allocated channel will be accessible for the required duration. Once data communication is complete, the occupied channel is released. Compared with scheduled-based protocols, this class of protocols does not require centralized control and precise time synchronization [197]. Moreover, these protocols are adequately simple, adaptive toward change in network topology, and robust to variation in nodes traffic load and density [198]. They also do not require extra message exchange overhead, thanks to the independent decision-making process for channel access. There are several protocols under this class, such as ALOHA [199] and variants of CSMA [200]. BREATH is a self-adapting CSMA MAC protocol that provides reliable, energy efficient, and timely data transmission for industrial control applications [52].

A consequential drawback of contention-based schemes is that the probability of collisions and idle listening grows with increased node density, leading to unacceptable performance in terms of latency. To alleviate the effect of collisions, additional control packets may be added to the MAC protocol, resulting in noticeable control overhead in IIoT systems. Contention-based MAC protocols could be performed through synchronous and asynchronous protocols.

- Synchronous protocols: This class of schemes employs local time synchronization between nodes to alternately switch their operation mode between active and sleep modes. In these protocols, a node operates in active mode for packets listening or sleeping mode to decrease overhearing and idle listening. To prevent overload from frequent synchronization messages, the protocol could use infrequent synchronization, although it may decrease network adaptability to nodes mobility [189].
- Asynchronous protocols: Unlike synchronous protocols, this method does not require explicit scheduling between nodes. Instead, a low power listening (LPL) concept could be employed, where each node transmits data with a long enough preamble so that receiver is guaranteed to wake up during preamble transmission [201]. Basically, the receiver is often in sleep mode and wakes up shortly to sense the channel for every preamble. If a sender has data, it will send preamble to the receiver until it is awake and properly acquires the preamble. Then, the receiver remains in active mode to receive incoming data. After the transmission or reception period, all nodes check their data queue before going to sleep mode. Duty cycle and idle listening of asynchronous protocols could be decreased through dynamic preamble sampling [202]. The advantages of asynchronous protocols are flexibility to topology changes, less synchronization overhead, and a reduction in a receiver's idle listening. Nevertheless, asynchronous protocols suffer from transmitters' overemission before sending data, extra power consumption in unintentional receivers, and increased latency [189]. It also does not fully resolve the issue of channel collisions.

1.7.3 Hybrid Schemes

The combination of scheduled- and contention-based techniques is adopted in the so-called hybrid MAC schemes. In this class of schemes, time is divided into two periods, and, based on the requirements, it cooperatively switches between contention- or scheduled-based techniques to serve both modes. During contention period, nodes access to the medium and utilize contention approaches to broadcast in a common broadcast frequency. Then, during schedule-based period, each node performs reliable unicast transmission. Random–distributed scheduling algorithms are exploited in base stations to allocate the schedule of each node. For IoT use cases, some hybrid schemes jointly integrate FDMA with contention-based protocols and TDMA. Several MAC protocols are proposed in the literature based on such hybrid schemes [203, 204].

1.8 Smart Sensors

Similar to industrial development, sensors and instrumentation development could be classified into four categories: mechanical indicators, electrical sensors, electronic sensors, and smart sensors [205]. Recent evolutions and advancements of sensing technology in the field of Industry 4.0 are exclusively entitled as Sensor 4.0 [205].

Smart sensors are advanced platforms often associated with intelligent sensing and adaptive communication with physical and computational environments. They connect many different physical and informational subsystems that create the necessity for algorithms to quickly assess streamline analysis. Smart sensors could be linked together over wireless and cellular networks and carry larger volumes of data at reduced latency. Therefore, they are considered a key component for developing IIoT applications and providing efficient, reliable, and robust functionalities for a given system.

Along with the increased capabilities of smart sensors, they have also become more flexible, power efficient, and miniaturized. Fusing sensing and local computing capabilities provides a solid framework for intelligent machines used in smart environments.

1.8.1 Benefits of Smart Sensors in the Supply Chain

Smart sensors change the way systems collect data. Aside from the aforementioned features, they offer three key benefits to Supply Chain 4.0:

- Operational efficiency: Smart sensors offer valuable added value to the system in real time, which enables the company to analyze and respond without human intervention. This results in operational efficiency through automation, improved demand planning, inventory control, asset management, and product life cycle management.
- Management and visibility: Rapid deployment of various smart sensors in IIoT enhances visibility across systems and assists in E2E supply chain management. This leads to cost reduction and generates incremental revenue. In addition, smart sensors connect end users more closely to the businesses and provide critical insight into customers' behavior to enhance services. Another primary benefit offered by smart sensors within the Industry 4.0 framework is associated with the increased visibility of workflows and processes. The sensor measurements help real-time monitoring of equipment and assist in proactively receiving advance notices from potential problems or anomalies.
- Self-care and predictive maintenance (PM): Smart sensors could take advantage of artificial intelligence in Industry 4.0 and create self-identification, -diagnosis and -configuration sensors, often collectively identified under the term self-X [206]. In addition to improving the overall performance of the systems, smart sensors allow for quicker response to modifications, repairs, and failures. They are also considered as an essential component in the prospect of predictive maintenance and could identify the service time of machines before they break down. This alleviates the problems of unplanned production outages and reduces maintenance downtime through better monitoring.

Altogether, the connection of smart devices and systems in Industry 4.0 environment organically pertains to every stage of the supply chain. This leads to decreased cost and more efficiencies in the value chain.

1.8.2 Criteria for Adoption of Smart Sensors for Industry 4.0

Currently, various advanced sensors are utilized in different industries such as logistics, agriculture, rail and traffic control, and shipping. The principal objective for adoption of smart sensors is to enhance systems quality, reliability, and precision [207].

Despite advancement of smart sensors, their implementation in Industry 4.0 is generally limited by noise and signal attenuation. There are two principal measures for adopting smart sensors in Industry 4.0:

- Interoperation and interconnection: Smart sensors are of multi-vendor nature, and their interoperability is necessary, particularly for crucial sensors metadata such as timestamp, validity of data, sensor's geo-location, and device status. Thus, it is essential to ensure their integrity and compatibility with current and emerging IIoT systems. New standards for smart sensors provide effective configurations, integrations, and improved calibration [208, 209]. Similarly, interconnections between multiple smart sensors and communication technologies hamper interoperation and lead to system complexity and deficiency. Ultimately, successful deployment should be contingent on legacy ecosystems, and strong implementation plans are required based on the business, industry, and circumstances. To achieve this, some advanced technologies provide solutions for nonsafety applications [210].
- Security and trust: An important criterion for smarts sensors adoption in industrial applications is trust and security. The notion of trusting a sensor and its performance is important, particularly for control and safety applications. Therefore, both sensors and communication protocols that collect sensory data should be secure, trusted, accurate, calibrated, reliable, and timely.

There are trade-offs with respect to smart sensors selection such as complexity, ease of deployment, cost, and maintenance. To facilitate deploying smart sensors, the concept of sensing as a service is a possible solution where equipment, data capture, and management are leased or offered to assist in using smart sensors; however, full control over the sensor features will be compromised as a drawback [211].

1.8.3 Key Leverage for Smart Sensors in Supply Chain 4.0

Based on the supply chain operations reference (SCOR) [212], Supply Chain 4.0 has four stages: plan, source, make, and deliver. Implementing smart sensors is beneficial to all stages and can provide E2E insights for the company. Smart sensors enable real-time inventory management and identify performance metrics that improve inventory, supply planning, product design, and development. They also provide an efficient, transparent, and traceable flow of raw material from sources to customers and ensure accurate and consistent supply. Monitoring and predictive maintenance of machinery leverage smart sensors data and improve performance of fully connected production facility.

1.9 Future Trends in Wireless Communication for Industry 4.0

The digital transformation to Industry 4.0 and FoF is not easily implemented, as there are diverse use cases, connectivity requirements, and multiple levels of QoS that should be considered in designing and deploying each application. Therefore, Industry 4.0 faces critical challenges that suggest future research trends for 2022 and beyond. Some major topics in this domain are:

- Development of new AI-enabled solutions: The integration of deep learning, data analysis, and artificial intelligence technologies, along with the industrial Internet, assists in offering smart monitoring and intelligent production and services.
- Implementing Edge and Fog computing: Since cloud computing systems may suffer from capacity scarcity and experience high latency, Edge- and Fog-based computing are promising solutions for low-latency and time-critical applications. In this context, a novel mobile network architecture is presented in [213], where the radio access network (RAN) relies on Fog computing to address latency issue and leads to a more reliable system.
- Cybersecurity and privacy: It is an important concern in IIoT scenarios as a heterogeneous connected environment and becomes more critical in post-Covid working trend due to worker expansion and remote working. Recent advancements in blockchain technology, wireless communication, and Edge computing offer trusted, distributed, and P2P network for failure prediction in IIoT that could improve security and intelligence of such systems [204].

Despite the fact that wireless communication is well established for some industrial use cases, heterogeneity of the connectivity landscape and its integration in the system pose critical issues in the industrial Internet. In this section, we identified some principal challenges of future wireless communication in Industry 4.0 and discuss them in the following sections.

1.9.1 Diverse Communication Requirements for Different Industrial Use Cases

As we discussed in Section 1.3, the requirements of wireless industrial Internet for various types of applications widely vary in terms of energy efficiency, deployment complexity, latency, and MAC protocols. Many wireless technologies, standards, and protocols are used in IIoT and smart manufacturing systems, and managing their coexistence in a system is still an open question. For instance, deterministic transmissions are highly important in control, process, and operation systems and should be guaranteed in coexistence with various wireless networking technologies. Additionally, issues arise due to the nature of wireless medium such as limited spectrum, shared bandwidth, reliable durability, and availability.

Altogether, E2E communication is highly challenging in industrial environment, and it highlights the necessity for management and optimization of wireless networks in Industry 4.0. The main objectives of network management for wireless industrial Internet are (1) real-time and dynamic wireless network optimization and management to offer flexible communication; (2) network resource allocation at different levels of system (equipment, production, operation and enterprise planning levels) to ensure required QoS; and (3) monitoring workflows to improve a network's visibility and performance.

1.9.1.1 Possible Solutions
One approach to guaranteeing the required QoSs in compliance with service-level agreements is to deploy a unified middle-ware that integrates various wireless technologies tailored for individual applications. Emerging technologies such as NFV, SDN, and distributed Edge computing could be leveraged for a smart and uniform platform

to enhance network management and visibility [214]. The middle-ware could also provide standard interface to integrate workflows, cellular technologies, and private networks [215]. A 5G cellular network is an example of standardized technology that copes with diverse requirements and QoS levels.

1.9.2 Challenges in Cellular and Mobile Technologies for Industrial Networking

A major wireless communication technology in the industrial Internet is cellular and mobile technologies such as 5G and B5G (Beyond fifth-generation), which are better suited to high-performance and fast motion applications in harsh environments. 5G offers E2E communication through public and private/dedicated connectivity in a highly flexible, reconfigurable, reliable, and power efficient manner. This makes it suitable for a wide range of distributed industrial use cases. Even though 5G significantly transforms wireless communication in terms of the technical requirements such as low latency, high bandwidth and data rate, and dynamically adapts with a proliferation of equipment, operations, and processes in IIoT environments, additional efforts are needed to encounter synergy between communication networks, operations, and system maintenance. In Section 1.5, we discussed cellular and mobile technologies in detail, and next we focus on the future vision to address its challenges.

1.9.2.1 Possible Solutions

Industrial 5G is not still fully widespread and available; however, various operators and companies propose solutions for future smart manufacturing based on 5G. For instance, an industrial 5G router is proposed in [216] that could provide private stand-alone 5G network in an industrial environment. A 5G starter kit is also developed in [217] as a future-proof wireless and cellular networking solutions for industrial communication and IIoT that could be deployed within sites and buildings and between factories. Various frameworks offer 5G connectivity fully aligned with the vision of Industry 4.0 and offer advanced discrete automation, flexible control over smart robot motion, and AR lenses for remote monitoring in future mining and ports [218, 219].

5 Gang is another novel networking architecture for future industrial communication that leverages SDN, Edge, and slicing technologies to combine 5G, wireless communication standards, and wired technologies in production facilities [220]. It retrofits conventional machines and advanced equipment to the network via minimal human intervention and efficiently adapts their configuration to the system requirements. Based on the capacity and needs of a smart factory, 5 Gang could work on both private and public cellular networks, and its architecture could be deployed on existing 5G architectures.

Apart from the distinguishing features of 5G and the rapid deployment of Industry 4.0, the connection density and throughput of 5G is expected to fall short of the stringent requirements of the upcoming Industry X.0. Furthermore, an increasing number of smart devices and applications in industrial environments require better power efficiency in the next generation of cellular networks. To address these challenges and to fill capability gaps of 5G, a 6G mobile network is proposed to support future cellular networks by the year 2030. To highlight the vision of connectivity with 6G, Table 1.3 compares the main parameters of 5G and 6G mobile networks. Since the standard performance metrics of 6G have not yet been identified by standardization bodies, we have shown only some provisional values [221, 222]. It should be noted that 5G assists in the deployment of Industry 4.0, but 6G will foster the potential use cases of smart industry and will exploit advanced technologies to resolve 5G limitations in the future.

1.9.3 Interoperability of Wireless Communication in Industry 4.0

Given that different communication technologies offer specific equipment and often use proprietary protocols and RFs, the interoperability of various networks is not yet guaranteed in industrial deployment. This results in complexity for both systems and end users. In addition, some current industrial networks are not practical in the near future. For instance, the standard bandwidth of an industrial control network is 100 Mb, which will not be able to handle the required data rates just a few years from now [223]. On the other hand, most of existing wireless communications target control systems and non-critical monitoring, and it is necessary to propose a viable wireless network for real-time services and process systems.

1.9.3.1 Possible Solutions

The future trend to use wireless communication in Industry 4.0 use cases is to standardize network protocols and technologies and to provide feasible deployment across equipment vendors, applications, and geo-locations. In the coming years, some wireless standards will be reviewed, standardized, and enhanced for some application domains. Currently, several wireless communication technologies offer partially modified and updated versions to cope with the industry and technology demands [214]. For instance, Wi-Fi6 is an initiative of the IEEE 802.11 family that will offer extended range and higher bandwidth in the coming years [87].

Table 1.3 Comparison of 5G and 6G Cellular Networks [221, 222].

Key Performance Indicators (KPI)	5G	6G
Peak data rate	20 Gbps	≥ 1 Tbps
Peak spectral efficiency	30 b/s/Hz	60 b/s/Hz
Area traffic capacity	10 Mb/s/m^2	1 Gb/s/m^2
Connection density	10^6 devices/km^2	10^7 devices/km^2
Network energy efficiency	not specified	1 Tb/J
Latency	1 ms	10–100 μs
Jitter	not specified	1μs
Mobility	500 km/h	≥ 1000 km/h

1.10 Conclusion

Industry 4.0 is a blueprint that digitizes the value chain and is highly instrumental in bringing physical production processes and their inherent real-time control functionalities to life. An essential aspect of implementing Industry 4.0 is the seamless connectivity of all value creation factors: service user, marketplace, and service provider. This concept envisages smart environments that offer a certain degree of automated control and processing, with minimum human interventions. At the core of this networking and integrated data concept is seamless communication that connects industrial environments and production areas. Diverse wireless standards and technologies are available to accelerate the deployment of smart technologies in process control and automation applications.

In this chapter, we first reviewed the concept of Industry 4.0 along with its technological requirements and applications. Subsequently, we detailed wireless technologies and international standards in this context. Since each wireless solution has its own pros and cons, we compared them based on a set of technical performance metrics to assist in choosing the appropriate wireless communication for a given application. Considering that existing wireless systems will be complemented or replaced by new developments such as cellular networks and MTC communications, such technologies were also fully covered in this chapter. A list of objectives in terms of resource and service indicators was also provided for an efficient design of wireless networks in IIoT environment. Finally, MAC protocols and smart sensors were discussed to address the key issues in the design process of wireless connectivity in the industrial Internet. We also reviewed a number of future research trends and directions in wireless communication for Industry 4.0.

References

1 Roblek, V., Bach, M.P., Mesko, M., and Berton- Celj, A. (May 2013). The impact of social media to value added in knowledge-based industries. *Kybernetes* 42 (4): 554–568.

2 Schlechtendahl, J., Keinert, M., Kretschmer, F., Lechler, A., and Verl, A. (Feb 2015). Making existing production systems industry 4.0- ready. *Prod. Eng.* 9: 143–148.

3 Almada-Lobo, F. (2016). The Industry 4.0 revolution and the future of manufacturing execution systems (MES). *J. Innov. Manag.* 3: 16–21.

4 Kagermann, H., Lukas, W.-D., and Wolfgang, W. (2011). Industrie 4.0 – mitdem Internet er dinge auf dem wegzur 4. Industriellen revolution. *VDI Nachrichten* 13 (1): 2–3.

5 Kagermann, H., Wahlster, W., and Helbig, J. (Apr 2013). Recommendations for Implementing the Strategic Initiative Industrie 4.0 – Securing the Future of German Manufacturing Industry. Final report of the Industrie 4.0 working group, ACATECH – National Academy of Science and Engineering, Miinchen. https://www.din.de/blob/76902/e8cac883f42bf28536e7e8165993f1fd/recommendations-for-implementing-industry-4-0-data.pdf.

6 Horizon. (2020). European commission. Call for factories of the future. https://ec.europa.eu/programmes/horizon2020/en/news/call-factories-future-1.

7 European Factories of the Future Research Association (EFFRA). Factories of the Future. https://www.effra.eu/factories-future.

8 Evans, P. and Annunziata, M. (Jan 2012). Industrial Internet: Pushing the boundaries of minds and machines. General Electric. https://www.ge.com/news/sites/default/files/5901.pdf.

9 Industrial Valuechain Initiative. What is IVI (Industrial Value Chain Initiative)? https://iv-i.org/wp/en/about-us/whatsivi.

10 Pereira, A.C. and Romero, F. (2017). A review of the meanings and the implications of the industry 4.0 concept. *Procedia Manuf.* 13: 1206–1214.

11 Schmidt, R., Mohring, M., Harting, R.-C., Reichstein, C., Neumaier, P., and Jozinovic, P. (Jun 2015). Industry 4.0 – potentials for creating smart products: Empirical research results. In: *Business Information Systems* (ed. W. Abramowicz), 16–27. Springer.

12 Kagermann, H. (2015). Change through digitization—value creation in the age of Industry 4.0. In: *Management of Permanent Change* (ed. H. Albach, H. Meffert, A. Pinkwart, and R. Reichwald), 23–45. Springer Fachmedien Wiesbaden.

13 Radziwon, A., Bilberg, A., Bogers, M., and Madsen, E.S. (2014). The smart factory: Exploring adaptive and flexible manufacturing solutions. *Procedia Eng.* 69: 1184–1190.

14 Weyer, S., Schmitt, M., Ohmer, M., and Gorecky, D. (2015). Towards industry 4.0 – standardization as the crucial challenge for highly modular, multi-vendor production systems. *IFAC- PapersOnLine* 48 (3): 579–584.

15 Qin, J., Liu, Y., and Grosvenor, R. (2016). A categorical framework of manufacturing for industry 4.0 and beyond. *Procedia CIRP* 52: 173–178.

16 Glova, J., Sabol, T., and Vajda, V. (2014). Business models for the Internet of things environment. *Procedia Econ. Fin.* 15: 1122–1129.

17 Romero, D., Bernus, P., Noran, O., Stahre, J., and Fast-Berglund, A. (Sep 2016). The operator 4.0: human cyber-physical systems & adaptive automation towards human-automation symbiosis work systems. In: *Advances in Production Management Systems. Initiatives for a Sustainable World* (ed. I. Naas, O. Vendrametto, J.M. Reis, R.F. Goncalves, M.T. Silva, V.C. Gregor, and D. Kiritsis), 677–686. Springer International Publishing.

18 Hahn, T. (Aug 2014). Future of manufacturing: view on enabling technologies. Siemens Corporate Technology. https://opcfoundation.org/wp-content/uploads2014/09/3_140805_OPC_Foundation_Redmond_v7a_incl_Siemens_Slides_20140731.pdf.

19 Taiwan, D. (Sep 2015). Challenges and Solutions for the Digital Transformation and Use of Exponential. https://www2.deloitte.com/tw/en/pages/manufacturing/articles/industry4-0.html.

20 Posada, J., Toro, C., Barandiaran, I., Oyarzun, D., Stricker, D., De Amicis, R., Pinto, E. B., Eisert, P., Dollner, J., and Vallarino, I. (2015). Visual computing as a key enabling technology for Industrie 4.0 and industrial Internet. *IEEE Comput. Graph. Appl.* 35 (2): 26–40.

21 Zhou, K., Liu, T., and Lifeng, Z. (2015). Industry 4.0: towards Future Industrial Opportunities and Challenges. *2015 12th International Conference on Fuzzy Systems and Knowledge Discovery (FSKD)*, 2147–2152.

22 Hermann, M., Pentek, T., and Otto, B. (2016). Design Principles for Industrie 4.0 Scenarios. *2016 49th Hawaii International Conference on System Sciences (HICSS)*, 3928–3937.

23 Lee, J., Bagheri, B., and Kao, H.-A. (2015). A cyber-physical systems architecture for industry 4.0-based manufacturing systems. *Manuf. Lett.* 3: 18–23.

24 Francalanza, E., Borg, J., Constantinescu, C. (2017). A knowledge-based tool for designing cyber physical production systems. *Comput. Ind.* 84: 39–58.

25 Xu, L.D., He, W., and Li, S. (2014). Internet of Things in Industries: a Survey. *IEEE Trans. Industr. Inform.* 10 (4): 2233–2243.

26 Haller, S., Karnouskos, S., and Schroth, C. (2009). The Internet of things in an enterprise context. In: *Future Internet – FIS 2008* (ed. J. Domingue, D. Fensel, and P. Traverso), 14–28. Springer Berlin Heidelberg.

27 Mourtzis, D., Vlachou, E., and Milas, N. (2016). Industrial big data as a result of IoT adoption in manufacturing. *Procedia CIRP* 55: 290–295.

28 Sisinni, E., Saifullah, A., Han, S., Jennehag, U., and Gidlund, M. (2018). Industrial Internet of things: Challenges, opportunities, and directions. *IEEE Trans. Industr. Inform.* 14 (11): 4724–4734.

29 Baida, Z., Gordijn, J., and Omelayenko, B. (2004). A shared service terminology for online service provisioning. *Proceedings of the 6th International Conference on Electronic Commerce*, ICEC'04, 1–10, New York, NY. Association for Computing Machinery.

30 Equipment for Potentially Explosive Atmospheres (ATEX). https://ec.europa.eu/growth/sectors/mechanical-engineering/atex_en.

31 National Electrical Code@. National Fire Protection Association- NFPA 70@. https://www.nfpa.org/codes-and-standards/all-codes-and-standards/list-of-codes-and-standards/detail?code=70.

32 Bajracharya, R., Shrestha, R., Zikria, Y.B., and Kim, S.W. (2018). LTE in the unlicensed spectrum: A survey. *IETE Tech. Rev.* 35 (1): 78–90.

33 3GPP. (2015). Feasibility study on licensed-assisted access to unlicensed spectrum (Release 13). TR 36.889, 3rd Generation Partnership Project (3GPP). https://portal.3gpp.org/desktopmodules/Specifications/SpecificationDetails.aspx?specificationId=2579.

34 Nadas, J.P.B., Zhao, G., Souza, R.D., and Muhammad, A.I. (2020). Ultra reliable low latency communications as an enabler for industry automation. In: *Wireless Automation as an Enabler for the Next Industrial Revolution* (ed. S. Hussain, M.A. Imran, and Q.H. Abbasi), 89–107. John Wiley & Sons, Ltd.

35 Karaki, R., Cheng, J., Obregon, E., Mukherjee, A., Kang, D.H., Falahati, S., Koorapaty, H., and Drugge, O. (2017). Uplink Performance of Enhanced Licensed Assisted Access (eLAA) in Unlicensed Spectrum. In *2017 IEEE Wireless Communications and Networking Conference (WCNC)*, 1–6.

36 Bajracharya, R., Shrestha, R., and Jung, H. (May 2020). Future is unlicensed: Private 5G unlicensed network for connecting industries of future. *Sensors* 20 (10): 2774.

37 Lu, X., Petrov, V., Moltchanov, D., Andreev, S., Mahmoodi, T., and Dohler, M. (2019). 5G-U: Conceptualizing integrated utilization of licensed and unlicensed spectrum for future IoT. *IEEE Commun. Mag.* 57 (7): 92–98.

38 Tombaz, S., Frenger, P., Athley, F., Semaan, E., Tidestav, C., and Fu- Ruskar, A. (2015). Energy Performance of 5G-NX Wireless Access Utilizing Massive

Beamforming and an Ultra-Lean System Design. In *2015 IEEE Global Communications Conference (GLOBECOM)*, 1–7.

39 Torsner, J., Dovstam, K., Miklos, G., Skubic, B., Mildh, G., Mecklin, T., Sandberg, J., Nyqvist, J., Neander, J., Martinez, C., Zhang, B., and Wan, J. (Nov 2015). Industrial Remote Operation: 5G Rises to the Challenge. Ericsson Technology Review. https://www.ericsson.com/en/reports-and-papers/ericsson-technology-review/articles/industrial-remote-operation-5g-rises-to-the-challenge.

40 Sahlli, E., Ismail, M., Nordin, R., and Abdulah, N. (Jun 2017). Beamforming techniques for massive MIMO systems in 5G: Overview, classification, and trends for future research. *Front. Inf. Technol. Electron. Eng.* 18: 753–772.

41 Anandhi, S., Anitha, R., and Sureshkumar, V. (2019). IoT enabled RFID Authentication and secure object tracking system for smart logistics. *Wirel. Pers. Commun.* 104 (2): 543–560.

42 Growindhager, B., Stocker, M., Rath, M., Boano, C.A., and Romer, K. (2019). SnapLoc: an Ultra-Fast UWB-Based Indoor Localization System for an Unlimited Number of Tags. In: *2019 18th ACM/IEEE International Conference on Information Processing in Sensor Networks (IPSN)*, 61–72.

43 Lee, C.K.M., Ip, C.M., Park, T., and Chung, S.Y. (2019). A Bluetooth Location- Based Indoor Positioning System for Asset Tracking in Warehouse. In: *2019 IEEE International Conference on Industrial Engineering and Engineering Management (IEEM)*, 1408–1412.

44 Thales. Asset Tracking. https://www.thalesgroup.com/en/markets/digital-identity-and-security/iot/industries/asset-tracking.

45 Frotzscher, A., Wetzker, U., Bauer, M., Rentschler, M., Beyer, M., Elspass, S., and Klessig, H. (2014). Requirements and Current Solutions of Wireless Communication in Industrial Automation. In: *2014 IEEE International Conference on Communications Workshops (ICC)*, 67–72.

46 PROFIBUS & PROFINET International (PI). (Apr 2016). PROFISafe. https://www.profibus-profinet.cz/images/Dokumenty/PROFINET/2812_PROFIsafe_SystemDescription_ENG__2016_web.pdf.

47 ODVA. CIP Safety™—common Industrial Protocol. ODVA, Inc. (Open DeviceNet Vendors Association). https://www.odvaorg/technology-standards/distinct-cip-services/cip-safety/.

48 Davis, J. (Dec 2008). Top Five Selection Criteria for Industrial Wireless Technologies. Cypress Semiconductor Corp. https://www.eetimes.com/top-five-selection-criteria-for-industrial-wireless-technologies/#.

49 Seferagic, A., Famaey, J., Eli, D.P., and Hoebeke, J. (Jan 2020). Survey on wireless technology trade-offs for the industrial Internet of things. *Sensors* 20 (2): 488.

50 Zhu, J., Zou, Y., and Zheng, B. (2017). Physical-layer security and reliability challenges for industrial wireless sensor networks. *IEEE Access* 5: 5313–5320.

51 Rao, S.K. and Prasad, R. (May 2018). Impact of 5G technologies on industry 4.0. *Wirel. Pers. Commun.* 100 (1): 145–159.

52 Park, P., Fischione, C., Bonivento, A., Johansson, K.H., and Sangiovanni-Vincent, A. (2011). Breath: An adaptive protocol for industrial control applications using wireless sensor networks. *IEEE Trans. Mob. Comput.* 10 (6): 821–838.

53 Porter, M.E. and Heppelmann, J.E. (Nov 2014). How smart, connected products are transforming competition. *Harvard Business Review*. https://hbr.org/2014/11/how-smart-connected-products-are-transforming-competition.

54 Huang, V.K.L., Pang, Z., Chen, C.A., and Tsang, K.F. (2018). New trends in the practical deployment of industrial wireless: from noncritical to critical use cases. *IEEE Ind. Electron. Mag.* 12 (2): 50–58.

55 Liu, Y., Kashef, M., Lee, K.B., Benmohamed, L., and Candell, R. (2019). Wireless network design for emerging IIoT applications: Reference framework and use cases. *Proc. IEEE* 107 (6): 1166–1192.

56 Al-Fuqaha, A., Guizani, M., Mohammadi, M., Aledhari, M., and Ayyash, M. (2015). Internet of things: A survey on enabling technologies, protocols, and applications. *IEEE Commun. Surv. Tutor.* 17 (4): 2347–2376.

57 Palattella, M.R., Dohler, M., Grieco, A., Rizzo, G., Torsner, J., Engel, T., and Ladid, L. (2016). Internet of things in the 5G Era: Enablers, architecture, and business models. *IEEE J. Sel. Areas Commun.* 34 (3): 510–527.

58 IEC. (Dec 2015). Industrial Networks-Wireless Communication Network and Communication Profiles-WIA-PA. Standard IEC 62601, International Electrotechnical Commission (IEC). https://webstore.iec.ch/publication/23902.

59 IEEE Standard for Low-Rate Wireless Networks. (2016). *IEEE Std 802.15.4-2015 (Revision of IEEE Std 802.15.4-2011)*, 1–709.

60 Palattella, M.R., Accettura, N., Vilajosana, X., Watteyne, T., Grieco, L.A., Boggia, G., and Dohler, M. (2013). Standardized protocol stack for the Internet of (important) things. *IEEE Commun. Surv. Tutor.* 15 (3): 1389–1406.

61 Shi, K., Zhang, L., Zhiying, Q., Tong, K., Chen, H., and Berder, O. (Jan 2019). Transmission scheduling of periodic real-time traffic in IEEE 802.15.4e TSCH-based industrial mesh networks. *Wireless Communications and Mobile Computing*.

62 Pister, K. and Doherty, L. (2008). TSMP: Time Synchronized Mesh Protocol. *Proceeding of the IASTED International Symposium- Distributed Sensor Networks*, 391, 61. https://people.eecs.berkeley.edu/~pister/publications/2008/TSMP%20 DSN08.pdf.

63 IEC. (Mar 2016). Industrial Communication Networks – Wireless Communication Network and Communication Profiles – WirelessHART™. Standard IEC 62591, International Electrotechnical Commission (IEC). https://webstore.iec.ch/ publication/24433.

64 IEC. (Oct 2014). Industrial networks – Wireless communication network and communication profiles – ISA 100.11a. Standard IEC 62734, International Electrotechnical Commission (IEC). https://webstore.iec.ch/publication/7409.

65 Dujovne, D., Watteyne, T., Vilajosana, X., and Thubert, P. (2014). 6TiSCH: deterministic IP-enabled industrial Internet (of things). *IEEE Commun. Mag.* 52 (12): 36–41.

66 Vilajosana, X., Pister, K., and Watteyne, T. (May 2017). RFC 8180 – Minimal IPv6 over the TSCH Mode of IEEE 802.15.4e (6TiSCH) Configuration. Internet Engineering Task Force (IETF). https://datatracker.ietf.org/doc/pdf/rfc8180.pdf.

67 Bartolomeu, P., Alam, M., Ferreira, J., and Fonseca, J.A. (2018). Supporting deterministic wireless communications in industrial IoT. *IEEE Trans. Industr. Inform.* 14 (9): 4045–4054.

68 Siekkinen, M., Hiienkari, M., Nurminen, J.K., and Nieminen, J. (2012). How Low Energy Is Bluetooth Low Energy? Comparative Measurements with ZigBee/802.15.4. In: *2012 IEEE Wireless Communications and Networking Conference Workshops (WCNCW)*, 232–237.

69 IEEE. (2007). Approved IEEE Draft Amendment to IEEE Standard for Information Technology-Telecommunications and Information Exchange Between Systems-Part 15.4: wireless Medium Access Control (MAC) and Physical Layer (PHY) Specifications for Low-Rate Wireless Personal Area Networks (LR-WPANS): amendment to Add Alternate Phy (Amendment of IEEE Std 802.15.4). *IEEE Approved Std P802.15.4a/D7*, January.

70 Lo Bello, L. and Toscano, E. (2009). Coexistence issues of multiple co-located IEEE 802.15.4/ZigBee networks running on adjacent radio channels in industrial environments. *IEEE Trans. Industr. Inform.* 5 (2): 157–167.

71 Lennvall, T., Svensson, S., and Hekland, F. (2008). A Comparison of WirelessHART and ZigBee for Industrial Applications. In: *2008 IEEE International Workshop on Factory Communication Systems*, 85–88.

72 Zigbee Alliance, Zigbee PRO with Green Power. https://zigbeealliance.org/wp-content/uploads/2019/11/docs-09-5499-26-batt-zigbee-green-power-specification.pdf.

73 Radmand, P., Domingo, M., Singh, J., Arnedo, J, Talevski, A., Petersen, S., and Carlsen, S. (2010). ZigBee/ZigBee PRO Security Assessment Based on Compromised Cryptographic Keys. *2010 International Conference on P2P, Parallel, Grid, Cloud and Internet Computing*, 465–470.

74 Souza, G.B.D.C., Vieira, F.H.T., Lima, C.R., Deus, G.A.D.J., De Castro, M.S., De Araujo, S.G., and Vasques, T.L. (2016). Developing smart grids based on GPRS and ZigBee technologies using queueing modeling-based optimization algorithm. *ETRI J.* 38 (1): 41–51.

75 Hassan, S.M., Ibrahim, R., Bingi, K., Chung, T.D., and Saad, N. (2017). Application of wireless technology for control: A wirelesshart perspective. *Procedia Comput. Sci.* 105: 240–247.

76 Raza, S., Faheem, M., and Guenes, M. (2019). Industrial wireless sensor and actuator networks in industry 4.0: Exploring requirements, protocols, and challenges—A MAC survey. *Int. J. Commun. Syst.* 32 (15): e4074.

77 Nixon, M. and Round Rock, T. (Sep 2012). A Comparison of WirelessHART and ISA100. 11a. Emerson Process Management. https://www.emerson.com/documents/automation/white-paper-a-comparison-of-wirelesshart-isa100-11a-en-42598.pdf.

78 Liang, W., Zhang, X., Xiao, Y., Wang, F., Zeng, P., and Haibin, Y. (2011). Survey and Experiments of WIA-PA Specification of industrial wireless network. *Wirel. Commun. Mob. Comput.* 11 (8): 1197–1212.

79 IEEE 802.15 WPAN Task Group 1 (TG1). https://www.ieee802.org/15/pub/TG1.html.

80 Bruno, R., Conti, M., and Gregori, E. (2002). Bluetooth: Architecture, protocols and scheduling algorithms. *Cluster Comput.* 5 (2): 117–131.

81 Patti, G., Leonardi, L., and Lo Bello, L. (2016). A Bluetooth Low Energy RealTime Protocol for Industrial Wireless Mesh Networks. In: *IECON 2016 – 42nd Annual Conference of the IEEE Industrial Electronics Society*, 4627–4632.

82 Gomez, C., Oller, J., and Paradells, J. (Aug 2012). Overview and evaluation of Bluetooth low energy: An emerging low-power wireless technology. *Sensors* 12 (9): 11734–11753.

83 Baert, M., Rossey, J., Shahid, A., and Hoebeke, J. (Jul 2018). The Bluetooth mesh standard: An overview and experimental evaluation. *Sensors* 18 (8): 2409.

84 IEEE 802.11 Wireless Local Area Networks. https://www.ieee802.org/11.

85 Banos-Gonzalez, V., Afaqui, M., Lopez-Aguilera, E., and Garcia-Villegas, E. (Nov 2016). IEEE 802.11ah: A technology to face the IoT challenge. *Sensors* 16 (11): 1960.

86 Hazmi, A., Rinne, J., and Valkama, M. (2012). Feasibility Study of IEEE 802.11 ah Radio Technology for IoT and M2M Use Cases. In: *2012 IEEE Globecom Workshops*, 1687–1692.

87 Siemens. (Oct 2020). Boost in Efficiency with WiFi6 – New WLAN Standard Makes It Easier to Handle Large Numbers of Participants. https://press.siemens.com/global/en/news/boost-efficiency-wi-fi-6.

88 Cavalcanti, D., Perez-Ramirez, J., Rashid, M.M., Fang, J., Galeev, M., and Stanton, K.B. (2019). Extending accurate time distribution and timeliness capabilities over the air to enable future wireless industrial automation systems. *Proc. IEEE* 107 (6): 1132–1152.

89 Ali, R., Kim, S.W., Kim, B.-S., and Park, Y. (2018). Design of MAC layer resource allocation schemes for IEEE 802.11ax: Future directions. *IETE Tech. Rev.* 35 (1): 28–52.

90 LoRa Alliance@, https://portal.3gpp.org/desktopmodules/Specifications/SpecificationDetails.aspx?specificationId=2578.

91 3GPP. (Jun 2013). Study on Provision of Low-Cost Machine-Type Communications (MTC) User Equipments (UEs) based on LTE (Release 12). TR 36.888, 3rd Generation Partnership Project (3GPP). https://portal.3gpp.org/desktopmodules/Specifications/SpecificationDetails.aspx?specificationId=2578.

92 3GPP. (Nov 2015). New WI proposal: NB-IoT (Release 13). RP 151619, 3rd Generation Partnership Project (3GPP). https://portal.3gpp.org/ngppapp/TdocList.aspx?meetingId=31198.

93 Mekki, K., Bajic, E., Chaxel, F., and Meyer, F. (2019). A comparative study of LPWAN technologies for large-scale IoT deployment. *ICT Express* 5 (1): 1–7.

94 Rebbeck, T., Mackenzie, M., and Afonso, N. (2014). *Low-Powered Wireless Solutions Have the Potential to Increase the M2M Market by over 3 Billion Connections.* Analysys Mason.

95 Huawei. CIoT: Cellular Internet of Things. https://carrier.huawei.com/en/products/wireless-network/lte/c-iot.

96 NWAVE. Nwave Smart Parking Company. https://www.nwave.io.

97 GSMA. (Dec 2015). GSMA Welcomes Mobile Industry Agreement on Technology Standards for Global Low Power Wide Area Market. https://www.gsma.com/newsroom/press-release/gsma-welcomes-mobile-industry-agreement-on-technology-standards.

98 Vangelista, L., Zanella, A., and Zorzi, M. (2015). Long-range IoT technologies: The dawn of LoRa. In: *Future Access Enablers for Ubiquitous and Intelligent Infrastructures* (ed. V. Atanasovski and A. Leon-Garcia), 51–58. Springer International Publishing.

99 Sforza, F. (2013). Communications System. US Patent US8406275B2, application filed 09 March 2010 and granted 26 March.

100 Reynders, B., and Pollin, S. (2016). Chirp Spread Spectrum as a Modulation Technique for Long Range Communication. In: *2016 Symposium on Communications and Vehicular Technologies (SCVT)*, 1–5.

101 Reynders, B., Meert, W., and Pollin, S. (2016). Range and Coexistence Analysis of Long Range Unlicensed Communication. In: *2016 23rd International Conference on Telecommunications (ICT)*, 1–6.

102 Petajajarvi, J., Mikhaylov, K., Pettissalo, M., Jan- Hunen, J., and Iinatti, J. (2017). Performance of a low-power wide-area network based on LoRa technology: Doppler robustness, scalability, and coverage. *Int. J. Distrib. Sens. Netw.* 13 (3).

103 Song, Y., Lin, J., Tang, M., and Dong, S. (2017). An Internet of energy things based on wireless LPWAN. *Engineering* 3 (4): 460–466.

104 Mikhaylov, K., Petaejaejaervi, J., and Haenninen, T. (2016). Analysis of Capacity and Scalability of the LoRa Low Power Wide Area Network Technology. In: *European Wireless 2016; 22nd European Wireless Conference*, 1–6.

105 Petric, T., Goessens, M., Nuaymi, L., Toutain, L., and Pelov, A. (2016). Measurements, Performance and Analysis of LoRa FABIAN, a Real-World Implementation of LPWAN. In: *2016 IEEE 27th Annual International Symposium on Personal, Indoor, and Mobile Radio Communications (PIMRC)*, 1–7.

106 Petajajarvi, J., Mikhaylov, K., Hamalainen, M., and Iinatti, J. (2016). Evaluation of LoRa LPWAN Technology for Remote Health and Wellbeing Monitoring. In: *2016 10th International Symposium on Medical Information and Communication Technology (ISMICT)*, 1–5.

107 LoRa Alliance. About LoRaWAN@. https://lora-alliance.org/about-lorawan.

108 Wang, Y.E., Lin, X., Adhikary, A., Grovlen, A., Sui, Y., Blankenship, Y., Bergman, J., and Razaghi, H.S. (2017). A primer on 3GPP narrowband Internet of things. *IEEE Commun. Mag.* 55 (3): 117–123.

109 Mwakwata, C.B., Malik, H., Alam, M.M., Yannick, L.M., Parand, S., and Mumtaz, S. (Jun 2019). Narrowband Internet of things (NB-IoT): From physical (PHY) and media access control (MAC) layers perspectives. *Sensors* 19 (11).

110 Adhikary, A., Lin, X., and Wang, Y.E. (2016). Performance Evaluation of NB-IoT Coverage. In: *2016 IEEE 84th Vehicular Technology Conference (VTC- Fall)*, 1–5.

111 Gozalvez, J. (2016). New 3GPP standard for IoT [Mobile Radio]. *IEEE Veh. Technol. Mag.* 11 (1): 14–20.

112 Singtel. Smart Logistics, Get Better Visibility into Your Operations. https://www.singtel.com/business/solutions/iot-solutions/use-cases/smart-logistics.

113 GSMA. (Jun 2019). NB-IoT Commercialisation Case Study. https://www.gsma.com/iot/wp-content/uploads/2019/08/201902_GSMA_NB-IoT_Commercialisation_CaseStudy.pdf.

114 Huawei. Huawei LiteOS-assisted Smart Lighting Solution. https://www.huawei.com/minisite/liteos/en/lighting.html.

115 Rico-Alvarino, A., Vajapeyam, M., Xu, H., Wang, X., Blankenship, Y., Bergman, J., Tirronen, T., and Yavuz, E. (2016). An overview of 3GPP enhancements on machine to machine communications. *IEEE Commun. Mag.* 54 (6): 14–21.

116 Oliveira, L., Rodrigues, J., Kozlov, S., Rabelo, R., and Albuquerque, V. (Jan 2019). MAC layer protocols for Internet of things: A survey. *Future Internet* 11 (1): 16.

117 Ali, A., Hamouda, W., and Uysal, M. (2015). Next generation M2M cellular networks: Challenges and practical considerations. *IEEE Commun. Mag.* 53 (9): 18–24.

118 Thales, L.T.E.-M. Connectivity Optimized for IoT. https://www.thalesgroup.com/en/markets/digital-identity-and-security/iot/resources/innovation-technology/lte-m.

119 Dawaliby, S., Bradai, A., and Pousset, Y. (2016). In Depth Performance Evaluation of LTE-M for M2M Communications. In: *2016 IEEE 12th International Conference on Wireless and Mobile Computing, Networking and Communications (WiMob)*, 1–8.

120 GSMA. (Jan 2019). LTE-M Commercialisation Case Study. https://www.gsma.com/iot/wp-content/uploads/2019/11/201901_GSMA_LTE-M_Commercial_Case_Study-ATT_Telstra.pdf.

121 Lauridsen, M., Kovacs, I.Z., Mogensen, P., Sorensen, M., and Holst, S. (2016). Coverage and Capacity Analysis of LTE-M and NB-IoT in a Rural Area. In: *2016 IEEE 84th Vehicular Technology Conference (VTC-Fall)*, 1–5.

122 Anani, W., Ouda, A., and Hamou, A. (2019). A survey of wireless communications for IoT echo-systems. *2019 IEEE Canadian Conference of Electrical and Computer Engineering (CCECE)*, 1–6.

123 Christmann, D. (Jul 2020). Hannover Messe: Bosch Solutions for Manufacturing. Bosch Media Service. https://www.bosch-presse.de/pressportal/de/en/hannover-messe-bosch-solutions-for-manufacturing-215168.html.

124 Potter, C.H., Hancke, G.P., and Silva, B.J. (2013). Machine-to-Machine: possible Applications in Industrial Networks. *2013 IEEE International Conference on Industrial Technology (ICIT)*, 1321–1326.

125 Tan, S., Sooriyabandara, M., and Fan, Z. (Aug 2011). M2M communications in the smart grid: Applications, standards, enabling technologies, and research challenges. *Int. J. Digit. Multimed. Broadcast.*

126 Taleb, T. and Kunz, A. (2012). Machine type communications in 3GPP networks: Potential, challenges, and solutions. *IEEE Commun. Mag.* 50 (3): 178–184.

127 Zhang, Y., Yu, R., Nekovee, M., Liu, Y., Xie, S., and Gjessing, S. (2012). Cognitive machine-to-machine communications: Visions and potentials for the smart grid. *IEEE Netw.* 26 (3): 6–13.

128 ETSI. (Mar 2016). Digital cellular telecommunications system (Phase 2+) (GSM); Universal Mobile Telecommunications System (UMTS); LTE; Service requirements for Machine-Type Communications (MTC); Stage 1 (V13.1.0). TS 122 368, European Telecommunications Standards Institute (ETSI). https://www.etsi.org/deliver/etsi_ts/122300_122399/122368/13.01.00_60/ts_122368v130100p.pdf.

129 Andreev, S., Galinina, O., Pyattaev, A., Gerasimenko, M., Tirronen, T., Torsner, J., Sachs, J., Dohler, M., and Koucheryavy, Y. (2015). Understanding the IoT connectivity landscape: A contemporary M2M radio technology roadmap. *IEEE Commun. Mag.* 53 (9): 32–40.

130 Chen, K.-C. and Lien, S.-Y. (2014). Machine-to-machine communications: Technologies and challenges. *Ad Hoc Netw.* 18: 3–23.

131 Shariatmadari, H., Ratasuk, R., Iraji, S., Laya, A., Taleb, T., Jantti, R., and Ghosh, A. (2015). Machine-type communications: Current status and future perspectives toward 5G systems. *IEEE Commun. Mag.* 53 (9): 10–17.

132 3GPP. (Mar 2007). Study on Facilitating Machine to Machine Communication in 3GPP Systems (Release 8). TR 22.868, 3rd Generation Partnership Project (3GPP). https://portal.3gpp.org/desktopmodules/Specifications/SpecificationDetails.aspx?specificationId=671.

133 3GPP. (Jun 2011). Service Requirements for MTC (Release 10). TR 22.368, 3rd Generation Partnership Project (3GPP). https://portal.3gpp.org/ChangeRequests.aspx?q=1&versionId=39548&release=184.

134 3GPP. (Sep 2012). System Improvements for MTC (Release 11). TR 23.888, 3rd Generation Partnership Project (3GPP). https://portal.3gpp.org/desktopmodules/Specifications/SpecificationDetails.aspx?specificationId=968.

135 3GPP. (Sep 2015). GERAN Study on Power Saving for MTC Devices (Release 13). TR 43.869, 3rd Generation Partnership Project (3GPP). https://portal.3gpp.org/desktopmodules/Specifications/SpecificationDetails.aspx?specificationId=501.

136 3GPP. (Jun 2018). LTE; 5G; Release Description (Release 14). TR 21.914, 3rd Generation Partnership Project (3GPP). https://www.etsi.org/deliver/etsi_tr/121900 _121999/121914/14.00.00_60/tr_121914v140000p.pdf.

137 Sharma, S.K. and Wang, X. (2020). Toward Massive Machine Type Communications in Ultra-Dense Cellular IoT Networks: Current Issues and Machine Learning-Assisted Solutions. *IEEE Commun. Surv. Tutor.* 22 (1): 426–471.

138 Fu, H., Chen, H.-C., Lin, P., and Fang, Y. (2012). Energy-Efficient Reporting Mechanisms for Multi-Type Real-Time Monitoring in Machine-to-Machine Communications Networks. In: *2012 Proceedings IEEE IN- FOCOM*, 136–144.

139 Accettura, N., Palattella, M.R., Dohler, M., Grieco, L.A., and Boggia, G. (2012). Standardized Power-Efficient Internet-Enabled Communication Stack for Capillary M2M Networks. In: *2012 IEEE Wireless Communications and Networking Conference Workshops (WCNCW)*, 226–231.

140 ETSI. (Jul 2020). oneM2M; Industrial Domain Enablement (V2.5.1). TR 118 518, European Telecommunications Standards Institute (ETSI). https://www.etsi.org/ deliver/etsi_tr/118500_118599/118518/02.05.01_60/tr_118518v020501p.pdf.

141 oneM2M. (May 2019). Functional Architecture. TS- 0001-V3.15.1, oneM2M. https:// onem2m.org/images/files/deliverables/Release3/TS-0001-Functional_ Architecture-V3_15_1.pdf.

142 Sadeghi, A., Wachsmann, C., and Waidner, M. (2015). Security and privacy challenges in industrial Internet of things. In: *2015 52nd ACM/EDAC/IEEE Design Automation Conference (DAC)*, 1–6.

143 Biral, A., Centenaro, M., Zanella, A., Vangelista, L., and Zorzi, M. (2015). The challenges of M2M massive access in wireless cellular networks. *Digit. Commun. Netw.* 1 (1): 1–19.

144 3GPP. (Mar 2009). Technical Specifications and Technical Reports for a UTRAN-based 3GPP System (Release 8). TR 21.101, 3rd Generation Partnership Project (3GPP). https://portal.3gpp.org/ChangeRequests.aspx?q=1&versionId=36836&relea se=182.

145 3GPP. (Oct 2010). Evolved universal terrestrial radio access (E-UTRA); Carrier Aggregation; Base Station (BS) Radio Transmission and Reception (Release 10). TR 36.808, 3rd Generation Partnership Project (3GPP). https://portal.3gpp.org/ desktopmodules/Specifications/SpecificationDetails.aspx?specificationId=2487.

146 3GPP. (Sep 2013). Study on Enhancements to Machine-Type Communications (MTC) and Other Mobile Data Applications; Radio Access Network (RAN) Aspects (Release 12). TR 37.869, 3rd Generation Partnership Project (3GPP). https:// portal.3gpp.org/desktopmodules/Specifications/SpecificationDetails. aspx?specificationId=2631.

147 Andrews, J.G. (2013). Seven ways that hetnets are a cellular paradigm shift. *IEEE Commun. Mag.* 51 (3): 136–144.

148 Zhang, X., Shen, X.S., and Xie, L. (2014). Joint subcarrier and power allocation for cooperative communications in LTE-advanced networks. *IEEE Trans. Wirel. Commun.* 13 (2): 658–668.

149 Yongsheng, H., Chen, Z., and Hao, Z. (2018). Relay node, distributed network of relay node and networking method thereof. European Patent EP09846416A, application filed 22 December 2009 and granted 28 March.

150 Sabella, D., Rost, P., Sheng, Y., Pateromichelakis, E., Salim, U., Guitton- Ouhamou, P., Di Girolamo, M., and Giuliani, G. (2013). RAN as a service: challenges of designing a flexible RAN architecture in a cloud-based heterogeneous mobile network. In: *2013 Future Network Mobile Summit*, 1–8.

151 Gandotra, P. and Jha, R.K. (2016). Device-to-device communication in cellular networks: A survey. *J. Netw. Comput. Appl.* 71: 99–117.

152 Bizanis, N., and Kuipers, F.A. (2016). SDN and virtualization solutions for the Internet of things: A survey. *IEEE Access* 4: 5591–5606.

153 Wang, X. and Gao, L. (2020). *When 5G Meets Industry 4-0*. Springer.

154 Qualcomm. (Oct 2017). Private LTE networks create new opportunities for industrial IoT. Qualcomm Technologies, Inc. https://www.qualcomm.com/media/documents/files/private-lte-network-presentation.pdf.

155 Ericsson. (Jul 2020). Private networks for industries. https://www.ericsson.com/en/networks/offerings/mission-critical-private-networks/private-networks.

156 Bockelmann, C., Pratas, N., Nikopour, H., Au, K., Svensson, T., Ste- Fanovic, C., Popovski, P., and Dekorsy, A. (2016). Massive machine-type communications in 5G: Physical and MAC-layer solutions. *IEEE Commun. Mag.* 54 (9): 59–65.

157 ITU-R. (Sep 2015). IMT Vision–Framework and overall objectives of the future development of IMT for 2020 and beyond. Recommendation ITU-R M.2083-0, International Telecommunication Union (ITU). https://www.itu.int/dms_pubrec/itu-r/rec/rn/R-REC-M.2083-0-201509-IllPDF-E.pdf.

158 3GPP. (Mar 2019). Study on physical layer enhancements for NR ultra reliable and low latency case (URLLC) (Release 16). TR 38-824, 3rd generation partnership project (3GPP). https://portal.3gpp.org/desktopmodules/Specifications/SpecificationDetails.aspx?specificationId=3498

159 Oleshchuk, V. and Fensli, R. (2011). Remote patient monitoring within a future 5G infrastructure. *Wirel. Pers. Commun.* 57 (3): 431–439.

160 Chen, H., Abbas, R., Cheng, P., Shirvanimoghaddam, M., Hardjawana, W., Bao, W., Li, Y., and Vucetic, B. (2018). Ultra-reliable low latency cellular networks: Use cases, challenges and approaches. *IEEE Commun. Mag.* 56 (12): 119–125.

161 Yilmaz, O.N.C., Wang, Y.E., Johansson, N.A., Brahmi, N., Ashraf, S.A., and Sachs, J. (2015). Analysis of ultra-reliable and low-latency 5g communication for a factory automation use case. In: *2015 IEEE International Conference on Communication Workshop (ICCW)*, 11901195.

162 Farkas, J., Varga, B., Miklos, G., and Sachs, J. (Aug 2019). 5G-TSN integration meets networking requirements for industrial automation. Ericsson. https://www.ericsson.com/4a4cb4/assets/local/reports-papers/ericsson-technology-review/docs/2019/5g-tsn-integration-for-industrial-automation.pdf.

163 O'Connell, E., Moore, D., and Newe, T. (Jun 2020). Challenges associated with implementing 5G in manufacturing. *Telecom* 1 (1): 48–67.

164 Thales. (Dec 2020). Introducing 5G technology and networks (speed, use cases and rollout). https://www.thalesgroup.com/en/markets/digital-identity-and-security/mobile/inspired/5G.

165 Ericsson.(Sep 2019). Ushering in a better connected future. https://www.ericsson.com/en/about-us/company-facts/ericsson-worldwide/india/authored-articles/ushering-in-a-better-connected-future.

166 GSMA. (Feb 2020) 5G implementation guidelines: NSA Option 3. https://www.gsma.com/futurenetworks/wp-content/uploads/2019/03/5G-Implementation-Guidelines-NSA-Option-3-v2.1.pdf.

167 GSMA. (Oct 2020). 5G IoT Private and dedicated networks for Industry 4.0. https://www.gsma.com/iot/wp-content/uploads/2020/10/2020-10-GSMA-5G-IoT-Private-and-Dedicated-Networks-for-Industry-4.0.pdf.

168 Qualcomm. Deploying 5G NR mmWave to unleash the full 5G's potential. https://www.qualcomm.com/media/documents/files/deploying-mmwave-to-unleash-5g-s-full-potential.pdf.

169 Bockelmann, C., Pratas, N.K., Wunder, G., Saur, S., Navarro, M., Gregoratti, D., Vivier, G., De Carvalho, E., Ji, Y., Stefanovic, C., et al. (2018). Towards massive connectivity support for scalable mMTC communications in 5G networks. *IEEE Access* 6: 28969–28992.

170 Glabowski, M., Hanczewski, S., Stasiak, M., Weis- Senberg, M., Zwierzykowski, P., and Bai, V. (2020). Traffic modeling for industrial Internet of things (IIoT) networks. In: *Image Processing and Communications* (ed. M. Choras and R.S. Choras), 264–271. Springer International Publishing.

171 Candell, R., Zimmerman, T., and Stouffer, K. (Dec 2015). An industrial control system cybersecurity performance testbed. NIST Interagency/Internal Report (NISTIR) 8089, US Department of Commerce, National Institute of Standards and Technology (NIST). https://www.nist.gov/publications/industrial-control-system-cybersecurity-performance-testbed.

172 Wetzker, U., Splitt, I., Zimmerling, M., Boano, C.A., and Romer, K. (Aug 2016). Troubleshooting wireless coexistence problems in the industrial Internet of things. In: *2016 IEEE Intl Conference on Computational Science and Engineering (CSE) and IEEE Intl Conference on Embedded and Ubiquitous Computing (EUC) and 15th Intl Symposium on Distributed Computing and Applications for Business Engineering (DCABES)*.

173 Liu, Y., and Moayeri, N. (Sep 2017). Wireless Activities in the 2 GHz Radio Bands in Industrial Plants. Technical Note (NIST TN) 1972, US Department of Commerce, National Institute of Standards and Technology (NIST). https://www.nist.gov/publications/wireless-activities-2-ghz-radio-bands-industrial-plants.

174 Lien, S., Tseng, C., Chen, K., and Su, C. (2010). Cognitive radio resource management for QoS guarantees in autonomous femtocell networks. In: *2010 IEEE International Conference on Communications*, 1–6.

175 Chiwewe, T.M., Mbuya, C.F., and Hancke, G.P. (2015). Using cognitive radio for interference-resistant industrial wireless sensor networks: An overview. *IEEE Trans. Industr. Inform.* 11 (6): 1466–1481.

176 ODVA. EtherNet/IP™. ODVA, Inc. (Open DeviceNet Vendors Association). https://www.odva.org/technology-standards/key-technologies/ethernet-ip.

177 PROFIBUS & PROFINET International (PI). PROFINET: the leading industrial ethernet standard. https://www.profibus.com/technology/profinet.

178 Modbus Organization. Modbus specifications and implementation guides. https://www.modbus.org/specs.php.

179 Burke, T.J. (2017). OPC unified architecture – interoperability for Industrie 4.0 and the Internet of things. OPC Foundation. https://opcfoundation.org/wp-content/uploads/2016/05/OPC-UA-Interoperability-For-Industrie4-and-IoT-EN-v5.pdf.

180 Prakash, S. and Kinage, A. IIoT (Industrial Internet of Things) Communication Interface. US Patent 15/995079, application filed 09 April 2019.

181 Mahmood, A., Exel, R., Trsek, H., and Sauter, T. (2017). Clock synchronization over IEEE 802.11—A survey of methodologies and protocols. *IEEE Trans. Industr. Inform.* 13 (2): 907–922.

182 Nieminen, J., Savolainen, T., Isomaki, M., Patil, B., Shelby, Z., and Gomez, C. (Oct 2015). RFC 7668 – IPv6 over BLUETOOTH(R) Low Energy. Internet Engineering Task Force (IETF). https://tools.ietf.org/html/rfc7668.

183 Thubert, P., Bormann, C., Toutain, L., and Cragie, R. (Apr 2017). IPv6 over low-power wireless personal area network (6LoWPAN) routing header. Internet Engineering Task Force (IETF). https://tools.ietf.org/html/rfc8138.

184 Tse, D., and Viswanath, P. (2005). *Fundamentals of Wireless Communication.* Cambridge University Press.

185 Suriyachai, P., Roedig, U., and Scott, A. (2012). A survey of MAC protocols for mission-critical applications in wireless sensor networks. *IEEE Commun. Surv. Tutor.* 14 (2): 240–264.

186 Beltran, F. (Sep 2017). Accelerating the introduction of spectrum sharing using market-based mechanisms. *IEEE Commun. Stand. Mag.* 1: 66–72.

187 Samanta, A., and Misra, S. (2018). Dynamic connectivity establishment and cooperative scheduling for QoS-aware wireless body area networks. *IEEE Trans. Mob. Comput.* 17 (12): 2775–2788.

188 Kiran, M.P.R.S., Subrahmanyam, V., and Rajalakshmi, P. (2018). Novel power management scheme and effects of constrained on-node storage on performance of MAC layer for industrial IoT networks. *IEEE Trans. Industr. Inform.* 14 (5): 2146–2158.

189 Kumar, A., Zhao, M., Wong, K., Guan, Y.L., and Chong, P.H.J. (2018). A comprehensive study of IoT and WSN MAC protocols: Research issues, challenges and opportunities. *IEEE Access* 6: 76228–76262.

190 Ullah, F., Abdullah, H., Kaiwartya, O., Kumar, S., and Arshad, M.M. (Dec 2017). Medium access control (MAC) for wireless body area network (WBAN): Superframe structure, multiple access technique, taxonomy, and challenges. *Hum.-Centric Comput. Inf.* 7. 10.1186/s13673-017-0115-4

191 Suriyachai, P., Brown, J., and Roedig, U. (2010). Time-critical data delivery in wireless sensor networks. In: *Distributed Computing in Sensor Systems* (ed. R. Rajaraman, T. Moscibroda, A. Dunkels, and A. Scaglione), 216–229. Springer Berlin Heidelberg.

192 Zheng, T., Gidlund, M., and Akerberg, J. (2016). WirArb: a new MAC protocol for time critical industrial wireless sensor network applications. *IEEE Sens. J.* 16 (7): 2127–2139.

193 Akyildiz, I. and Vuran, M.C. (2010). Medium access control. In *Wireless Sensor Networks*, chapter 5, 77–116. John Wiley & Sons, Ltd.

194 Xiao, X., Tao, X., and Lu, J. (2015). Energy-efficient resource allocation in LTE-based MIMO-OFDMA systems with user rate constraints. *IEEE Trans. Veh. Technol.* 64 (1): 185–197.

195 Bankov, D., Didenko, A., Khorov, E., and Lyakhov, A. (2018). OFDMA Uplink Scheduling in IEEE 802.11ax Networks. In: *2018 IEEE International Conference on Communications (ICC)*, 1–6.

196 Jacob, S., Menon, V.G., Joseph, S., Vinoj, P.G., Jolfaei, A., Lukose, J., and Raja, G. (2020). A novel spectrum sharing scheme using dynamic long shortterm memory with CP-OFDMA in 5G networks. *IEEE Trans. Cogn. Commun. Netw.* 6 (3): 926–934.

197 Karl, H. and Willig, A. (2005). MAC Protocols. In: *Protocols and Architectures for Wireless Sensor Networks* (ed. H. Karl and A. Willig), 111–148. John Wiley & Sons, Ltd.

198 Doudou, M., Djenouri, D., Badache, N., and Bouabdallah, A. (2014). Synchronous contention-based mac protocols for delay- sensitive wireless sensor networks: A review and taxonomy. *J. Netw. Comput. Appl.* 38: 172–184.

199 Abramson, N. (1985). Development of the ALOHANET. *IEEE Trans. Inf. Theory* 31 (2): 119–123.

200 Georgiadis, L. (2003). Carrier-sense multiple access (CSMA) protocols. In: *Wiley Encyclopedia of Telecommunications* (ed. J.G. Proakis). American Cancer Society.

201 Daabaj, K. and Ahmeda, S. (2011). Real-time cross-layer routing protocol for ad hoc wireless sensor networks. In: *Advances in Computer Science and Engineering* (ed. M. Schmidt). Intechopen.

202 El-Hoiydi, A. (2002).ALOHA with preamble sampling for sporadic traffic in ad-hoc wireless sensor networks. In: *2002 IEEE International Conference on Communications- Conference Proceedings- ICC 2002 (Cat- No-02CH37333)*, 5, 3418–3423.

203 Liu, Y., Yuen, C., Cao, X., Hassan, N.U., and Chen, J. (2014). Design of a scalable hybrid MAC protocol for heterogeneous M2M networks. *IEEE Internet Things J.* 1 (1): 99–111.

204 Nguyen, V., Oo, T.Z., Chuan, P., and Hong, C.S. (2016). An efficient time slot acquisition on the hybrid TDMA/CSMA multichannel MAC in VANETs. *IEEE Commun. Lett.* 20 (5): 970–973.

205 Schutze, A., Helwig, N., and Tizian, S. (May 2018). Sensors 4.0 – smart sensors and measurement technology enable industry 4.0. *J. Sens. Sens. Syst.* 7: 359–371.

206 Johar, M. and Koenig, A. (2011). Case study of an intelligent AMR sensor system with self-x properties. In: *Soft Computing in Industrial Applications* (ed. A. Gaspar-Cunha, R. Takahashi, G. Schaefer, and L. Costa), 337–346. Springer Berlin Heidelberg.

207 Kanoun, O. and Trankler, H. (2004). Sensor technology advances and future trends. *IEEE Trans. Instrum. Meas.* 53 (6): 1497–1501.

208 Goncalves, G., Reis, J., Pinto, R., Alves, M., and Correia, J. (2014). A step forward on intelligent factories: A smart sensor-oriented approach. *Proceedings of the 2014 IEEE Emerging Technology and Factory Automation (ETFA)*, 1–8.

209 Lee, K. (2000). IEEE 1451: A standard in support of smart transducer networking. *Proceedings of the 17th IEEE Instrumentation and Measurement Technology Conference [Cat. No. 00CH37066]*, 2, 525–528.

210 OPC Foundation. What is OPC? https://opcfoundation.org/about/what-is-opc.

211 Gaddis, B. (Aug 2015). How sensors will revolutionize service businesses. https://www.wired.com/insights/2013/05/how-sensors-will-revolutionize-service-businesses/.

212 Association for Supply Chain Management (ASCM). (2019). *Digitising Your Supply Chain Provides the Agility Your Company Needs to Survive and Thrive*. SCOR Digital Standard. https://scor.ascm.org/processes/introduction.

213 Shih, Y.-Y., Chung, W.-H., Pang, A.-C., Chiu, T.-C., and Wei, H.-Y. (2017). Enabling low-latency applications in fog-radio access networks. *IEEE Netw.* 31 (1): 52–58.

214 Graf, U., Heidel, R., Kadel, G., Karcher, B., Mildner, F., Schulz, D., and Tenhagen, D. (2016). Network-based communication for Industrie 4.0. Federal ministry for economic affairs and energy (BMWi). https://www.plattform-i40.de/IP/Redaktion/ EN/Downloads/Publikation/network-based-communication-for-i40. pdf?__blob=publicationFile&v=8.

215 5G-PPP. (Oct 2015). 5G and the factories of the future. 5G infrastructure public private partnership (5G-PPP). https://5g-ppp.eu/wp-content/uploads/2014/02/5G-PPP-White-Paper-on-Factories-of-the-Future-Vertical-Sector.pdf.

216 Siemens. (Nov 2020). Industrial 5G–The wireless network of the future. www. siemens.com/press/industrial-5g.

217 HMS Networks. HMS and Ericsson-enabling smart manufacturing and Industry 4.0. https://www.hms-networks.com/about/partner-program/hms-in-customer-partner-programs/ericsson/hms-and-ericsson__enabling-smart-manufacturing-and-industry-4.0.

218 Ericsson. (Dec 2020). The future of mining and hitting paydirt with private cellular. https://www.ericsson.com/en/blog/2020/12/the-future-of-mining-hitting-paydirt-with-private-cellular.

219 Ericsson. (Feb 2021). Why connected ports are smarter with private cellular networks. https://www.ericsson.com/en/blog/2021/2/connecting-future-ports-with-private-cellular-networks.

220 Karrenbauer, M., Ludwig, S., Buhr, H., Klessig, H., Bernardy, A., Huanzhuo, W., Pallasch, C., Fellan, A., Hoffmann, N., Seelmann, V., et al.et al. (2019). Future industrial networking: From use cases to wireless technologies to a flexible system architecture. *Automatisierungstechnik* 67 (7): 526–544.

221 Rajatheva, N., Atzeni, I., Bjornson, E., Bourdoux, A., Buzzi, S., Dore, J.-B., Erkucuk, S., Fuentes, M., Guan, K., Hu, Y., et al. et(Apr 2020). White paper on broadband connectivity in 6G. *arXiv Preprint arXiv:2004-14247*.

222 Akhtar, M.W., Hassan, S., Ghaffar, R., Jung, H., Garg, S., and Shamim Hossain, M. (Dec 2020). The shift to 6G communications: vision and requirements. *Hum.-Centric Comput. Inf.* 10 (1): 53.

223 Mitsubishi Electric. Industry 4.0-the road to digitalisation in future manufacturing. Mitsubishi Electric. https://gb3a.mitsubishielectric.com/fa/en/news/ content?id=2989

224 Romero, J., Martinez, J., Nikkarinen, S., and Moisio, M. (2003). GPRS and EGPRS performance. In: *GSM, GPRS and EDGE Performance: Evolution Towards 3G/UMTS*, 2e, chapter 7 (ed. T. Halonen, J. Melero, and J.R. Garcia), 235–305. John Wiley & Sons, Ltd.

225 Holma, H., Kristensson, M., Salonen, J., and Toskala, A. (2002). UMTS services and applications. In: *WCDMA for UMTS: Radio Access for Third Generation Mobile Communications*, 3e, chapter 2, (ed. H. Holma and A. Toskala), 11–45. John Wiley & Sons, Ltd.

226 GPS.GOV. The global positioning system. https://www.gps.gov.

227 Bennis, M., Debbah, M., and Poor, H.V. (2018). Ultrareliable and low-latency wireless communication: Tail, risk, and scale. *Proc. IEEE* 106 (10): 1834-1853.

228 Ji, H., Park, S., Yeo, J., Kim, Y., Lee, J., and Shim, B. (2018). Ultra-reliable and low-latency communications in 5G Downlink: Physical layer aspects. *IEEE Wirel. Commun.* 25 (3): 124–130.

229 VDMA. (Dec 2017). Industrie 4.0 communication guideline based on OPC UA. https://industrie40.vdma.org/documents/214230/20743172/Leitfaden_OPC_UA_ Englisch_1506415735965.pdf/a2181ec7-a325-44c0-99d2-7332480de281.

2

Blockchain and Smart Contracts

Yahya Kabiri and Mahdi Sharifzadeh*

Sharif Energy, Water and Environment Institute (SEWEI), Sharif University of Technology, Tehran, Iran
* Corresponding author. Sharif Energy, Water and Environment Institute (SEWEI), Sharif University of Technology, Tehran, Iran

2.1 Blockchain Overview

Blockchain is an enabling technology for implementing distributed ledgers. A distributed ledger is a common database that is distributed and stored over different nodes of a network of computers [1]. Desirable features such as immutability, transparency, anonymity, nonrepudiation characteristic, and decentralized architecture, pose block chains at the leading technological edge for secure peer-to-peer transactions, without the need for a costly third authorization party [2, 3]. The first commercial application of blockchains was in Bitcoin, for recording peer-to-peer transactions in its electronic cash system [4]. As shown in Figure 2.1, a block in Bitcoin as a simplified blockchain consists of a set of recorded data including a block header and transaction records. Every node in a blockchain network keeps a copy of the ledger and therefore prevents tampering of the ledger's data [2].

As shown in Figure 2.1, *block headers* are used to identify the blocks and include the previous *block hash, timestamp, Merkle root,* and *nonce*. Hash cryptography is a mathematical one-way function to transform block data into a specific length called hash output to enhance security in blockchain. In the hash cryptographic, it is almost impossible to regenerate the original input data from the hash output. Using the previous block hash in the block header of each block, the subsequent blocks are chained together for creating a blockchain. Including the previous block hash into the block data structure ensures that no change can be made in previous blocks without changing the current block, hence making it impossible to tamper a block without changing all previous blocks. A nonce is a random number added to hash that can be used only once and ensures the required difficulty level [2, 5]. The timestamp provides the data associated with the time of each transaction to distinguish between different records in the blockchain and to provide historical data for future reference. To handle the huge data associated with recorded transactions, a data structure is required to enable data compression and decrease the storage requirements. For this purpose, Bitcoin utilizes the Merkle tree as the data organization tool and includes the Merkle root in each block data structure. Figure 2.2 shows the Merkle tree of a blockchain with four transactions (T1, T2, T3, T4). Each transaction is encrypted using a hash. The leaves of the Merkle hash tree are the block hash values, and the non-leaf nodes are the cryptographic hash values comprising their child's hash. The combination of

Industry 4.0 Vision for the Supply of Energy and Materials: Enabling Technologies and Emerging Applications, First Edition. Edited by Mahdi Sharifzadeh.

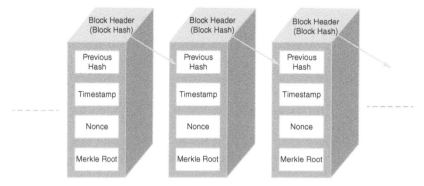

Figure 2.1 Simplified Bitcoin Blockchain. Based on S. Nakamoto, 2019.

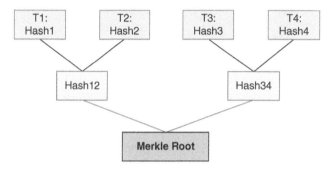

Figure 2.2 Data Organization in the Merkle Hash Tree.

hashes is stored within the Merkle root, so if any of the transactions is tampered with it can be easily detected by the block in the same chain with the advantage of data compression.

2.2 Blockchain Attributes

The decentralized nature of blockchains and utilizing consensus mechanisms for data validation and appending new blocks to the blockchain provide many advantages [1]:

1) **No need for a third-party intermediary:** In a blockchain system, the need for a central authorization organization to validate transactions is eliminated, and consensus mechanisms are used instead. As the result of peer-to-peer transactions and the elimination of an intermediary third-party such as a bank, faster transactions with a lower cost are achieved.

2) **Transparency:** The stored transactions in a public blockchain are visible to all nodes, and everyone can maintain a copy of the same public ledger and verify its correctness. Transparency of the information also improves trust among participants and increases the resiliency of the system.

3) **Security and reliability:** In a blockchain, transactions must be agreed upon by participants before recording. Then cryptography techniques are used to encrypt

approved transactions and link them as a chain. A copy of these chained transactions is distributed among all the nodes, and once a piece of data is recorded in a blockchain nobody can change it, which provides immutability and makes blockchain immune against hacks and other malicious manipulation. On the other hand, managing the data ledger is not limited locally to a certain trusted node. Multiple nodes participate in transactions, and there is no critical node to be exploited for hacking or fraud. Therefore, the blockchain improves the reliability and security of data storage.

4) **Verifiability and auditability:** Stored transactions in a blockchain are immutable and permanent. All transactions are labeled with a timestamp and a signature. Therefore, the ownership of recorded data is auditable and transactions can be traced back to find their origin. Also, transaction histories are available for authenticity verification that prevent fraud, any data disputes, and tampering.

5) **High availability:** In a blockchain, all nodes are working in a peer-to-peer network that an updated version of the data ledger is replicated on every node. Therefore, the unavailability of a single node does not stop the system to continue working [1].

2.3 Blockchain Taxonomy

Based on the data accessibility of the peers in the blockchain network, three kinds of blockchain systems could be considered, as discussed in the following:

1) **Public blockchains:** This type of blockchain is known as *permissionless* blockchain in which all recorded data are accessible and visible for all peers. Besides, every node in the blockchain network is allowed to participate in the consensus process. Therefore, all nodes in a public blockchain maintain an updated version of the recorded data, and there exist numerous distributed copies of the ledger in the blockchain network, enabling immutability. However, a large number of peers in a public network requires a high propagation time that limits transaction throughput. Bitcoin and Ethereum are popular examples of public blockchains.

   The following two types of blockchain are known as permissioned blockchains that only preselected nodes or organizations are allowed to validate transactions [1].

2) **Private blockchain:** Here, the data access of the peers is limited, and only some special nodes are permitted to participate in the consensus process. In a private blockchain, the number of nodes with a complete copy of the recorded data is limited, and the ledger could be tampered with easier. The fewer number of peers in private blockchains cause a lower propagation time that enables a higher transaction throughput.

3) **Consortium or federated blockchains:** This type of blockchain is a combination of public and private blockchains. Its main application is in the networks shared among several organizations in which preselected federated nodes are allowed to participate in the consensus process. Access to the recorded data could be public or private that is determined by the authorizing organization. As the number of peers in the consortium network is low, immutable transactions are not completely achievable. However, the lower propagation time and consequently higher transaction throughput can be achieved in a consortium blockchain.

The difference between private and consortium blockchain is that a private blockchain is a centralized system and operates by a certain organization. However, a consortium blockchain is a private blockchain that is operated by a group of organizations [1].

2.4 Blockchain Architectures

In general, the architecture of blockchains can be decomposed into four layers: data, network, consensus, and application, as shown in Figure 2.3 and discussed in the following [2]:

1 Data Layer

This layer is responsible for the data organization and storage. Stored data availability and regulation of its privacy and accuracy are very important tasks of this layer. Different blockchains have different methods for organizing and storing data. For example, while Bitcoin uses Merkle trees, Ethereum uses Merkle-Patricia to store and organize the transaction data. The other important part of this layer is data cryptography to protect the privacy and assets of users.

Cryptography in blockchains: Cryptography is a useful technique in blockchain to protect users' privacy and assets. The transparent data ledger in blockchains is accessible to all users in the network, which may cause privacy issues. On the other hand, unlike physical assets, the ownership proof of the digital assets requires digital signatures. Generally, blockchains use asymmetric encryptions and hash functions for cryptography. In asymmetric cryptography, a public key is available for all nodes. A private key is available only to a specific user and is applied to sign their launching transaction. Both public and private keys are needed for submitting a transaction. The hash function applies encryption methods to transform all different incoming messages into a message with a fixed length (Figure 2.4). Figure 2.5 offers an example for Bitcoin that uses a public key generated by a one-way hash encryption algorithm to pseudonym the users' address and protect their privacy. Also, Base58Check is an encryption protocol used to code hashed addresses for enhancing readability, compressing data, and checking for errors.

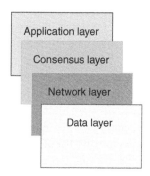

Figure 2.3 Blockchain Four-Layers Structure. Based on M. Wu, et al., 2019.

2 Network Layer

All users in blockchains are connected via an Internet-based peer-to-peer network. In this network, there is no central node or hierarchical topology. Therefore, all nodes communicate with each other in a flat topology. In blockchain networks, all nodes have equal status, with no special advantages. In this situation, the joint operation of all nodes is required to reach a successful operation for the whole network. In a blockchain, it is not required for all

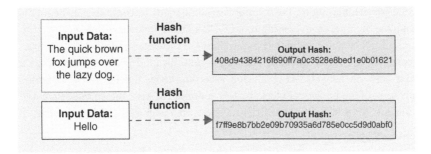

Figure 2.4 Hash Encryption.

nodes in the network to have complete functionality. In this regard, there are two kinds of nodes in the network: full and light. While the light nodes provide only a simplified payment verification (SPV), full nodes have a complete functionality, including routing, blockchain transaction storage, mining, and wallet [6].

For developing a perfect P2P network, the following characteristics are considered and analyzed:

- **Node behavior:** Different end users as nodes in blockchains may exhibit different behavior. For example, some nodes may have anomalous behavior and illegal interests, while other nodes in blockchain generally behave legally [7].
- **The size of the P2P network:** The number of nodes in the blockchain networks determines the size of that network.
- **The geographical distribution of the nodes:** Nodes in a blockchain network may be located in different places that could be distinguished via their IP addresses by utilizing an IP geolocating service. The network stability in terms of the node availability and potential interruptions

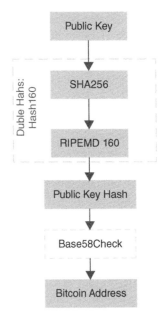

Figure 2.5 Public Key Address in Bitcoin.

- **The propagation time of the transmitted information:** All nodes in a blockchain are communicating via a communication network to send and receive information. The propagation time of this communication network constrains the transaction throughput in the blockchain [8].
- **The network effect of the BC network:** Based on Metcalfe's law, the network effect of a network is proportional to the square of the number of its nodes. The network effect means that by increasing the number of users in a network the value of the good or service in the network will be improved.
- Also, a new network model has been proposed that introduces the value of the network as the exponential of the root of the network nodes number [2].

In certain real-world applications, the operation of some industrial entities may be highly tangled requiring the interchain operation of blockchains to link these different applications and enhance the interoperability between them. In this regard, cross-chains connect various blockchain networks to enable effective communication between them. Digital asset transfer between networks is conducted according to predefined rules, while the original operation of the individual blockchain is guaranteed. Also, the cross-chain could be used to improve the scalability of blockchain systems in a huge P2P network of users [9].

3 Consensus Layer

The consensus layer provides a distributed mechanism in blockchains for reaching the agreement among all users regarding new transactions and corresponding blocks. This distributed consensus mechanism reduces or ideally removes the need for a central authority or third-party verification. Therefore, consensus mechanisms provide a distributed multi-node system without trust, which does not require costly verification of transactions by a third party.

Ideal consistency versus application consistency

Ideal consistency has three conditions:

1) Terminability: To reach a consistent result in a limited time.
2) Consensus: All nodes in the network should be able to eventually reach the same final decision.
3) Legality: The result of the decisions must comply with certain protocols.

In real applications, it is not possible to achieve these three conditions at the same time and with an acceptable cost because of the issues associated with actual distributed systems. First, there are random delays and content fault, which may render the communication systems between nodes unreliable and limit the terminability of the network. Second, there are possibilities for faults or sabotage in the node processes. Third, the simultaneous call of the different parallel processes may limit the extensibility of the system. Therefore, some of the properties of the ideal consistency should be loosened for reaching strong consistency instead of ideal consistency. However, strong consistency is not possible in engineering systems, and and a compromise needs to be made.

   While the aforementioned characteristics limit the consistency of any distributed network, additional challenges may be encountered within blockchain systems. The first issue in the blockchain is the risk of double-spending, which refers to the scenario in which the same digital coins are paid more than once. Besides, an effective blockchain system must be able to resist possible attacks such as Sybil attack, DDoS attack, and eclipse attack [10]. Therefore, a proper consensus mechanism in a blockchain must be able to handle the issues related to the consistency of the blockchain. In the following, some original and advanced consensus mechanisms are briefly introduced [2, 11–18].

Proof of work (PoW)

PoW is the original consensus mechanism in Bitcoin as one of the first applications of blockchain. In PoW, each node utilizes its computing power to compete for solving a specific puzzle problem [2, 12, 18]. The node that solves the problem first is considered as the winner and would be rewarded as the miner of the new block. In Nakamoto's proposal for Bitcoin [4], each user should pay a certain cost to get the right to kipping a copy of the blockchain ledger according to the predefined rules. Such subscription fees are paid to the winner of the mining processes when a new block is created.

Nonetheless, there is a possibility for two or more separate chains to be developed in parallel, known as blockchain forking. To avoid such conflict and reach a global consensus, the following two rules are required:

1) The rule of the longest chain: in the condition of the existence of multi-chain, the chain with the longest length is valid.
2) The rule of incentivization: only the miner who finds a valid block first would be rewarded.

Tampering the blockchain in the PoW mechanism requires at least 50% hashing power. As a result, tampering requires a high cost due to the high requirement of hashing power. In some cases, the gains from tampering can be greater than the cost; hence, malicious manipulation is not favorable [18].

Proof of stake (PoS)

The commercial large-scale application of PoW mechanisms has some disadvantages. The main drawbacks include high demand for power and other resources, low transaction throughput, and security issues against attacks. To overcome these drawbacks, instead of a high-cost mining process in PoW that requires hashing power of the nodes to find a valid nonce, in the proof of stack (PoS) algorithm the nodes with more coins are certified for generating a new block. PoS was used in PPCoin for the first time as an improved version of PoW. PoS is a proper alternative to PoW for reducing mining costs due to energy consumption in the mining process [13, 14, 19]. This mechanism is based on the assumption that the nodes owning more coins are less likely to attack the blockchain network. The limitation of this mechanism is that considering only the amount of owning coins gives more credit to richer nodes in the mining procedure and may not be fair to smaller contributors. Therefore, many solutions are considered to improve the PoS protocols [12]. A popular resolution is to consider both the amount and age of the coins to select the best mining nodes.

In the PoS mechanism, for the nodes with older coins, it is easier to mine new blocks [20]. Including coinage in the consensus algorithm reduces the difficulty for producing new blocks, but it weakens the network tolerance against double-spending attack compared with PoW. The reason should be associated with lower bifurcation costs in PoS, which produce multiple bifurcations with near to zero cost, thereby increasing the risk of double-spending attacks.

Hybrid consensus protocols

Some protocols blend the PoW and PoS mechanisms to generate new consensus algorithms to include advantages of both while compensating for their drawbacks. In this section, some of the hybrid consensus mechanisms proposed in literature are briefly discussed.

– **Delegated PoS (DPoS):** In this algorithm, some trusted nodes are selected by a voting mechanism, so the delegated nodes will mine and validate new blocks [12]. Therefore, in the DPoS algorithm, the transactions need to be confirmed only by delegated nodes and not by all the nodes in the network. As a result, transaction speed is improved [2]. DPoS is a combination of centralized and decentralized mechanisms that takes advantage of both the PoW and PoS. DPoS is faster than PoW and more democratic than PoS. On the other hand, DPoS is not a fully decentralized mechanism as PoW [13,

21], with additional scalability and transaction throughput. An DPoS algorithm implemented in Bitshares is used by industrial companies offering online services [22].

- **Proof of burn (PoB):** In this mechanism, miners do not need to consume a huge of energy and also incur the high-cost powerful hardware to mine a new block. Instead, miners must burn their virtual coins and provide proof that have paid or sent their coins to an unspendable address to be allowed for mining the new block [12, 23]. Since the burned coins are destroyed and cannot be recovered again, the risk of individuals to become too powerful to damage the blockchain network is mitigated [24]. The "burning" process is similar to fuel consumption for creating a new block, and burning more coins allows more chances for the miners to find a new block [2, 25, 26].

- **Proof of activity (PoA):** This algorithm utilizes both the PoW and PoS in parallel [13]. In the first step of PoA, all miners utilize their processing power in the PoW mechanism to solve a mathematical puzzle and find a new empty block [12]. This new block has only a header that includes the address of the previous block, the miner's public key, and a nonce [2]. In the next step, the PoS mechanism is utilized to validate and sign the new block. A committee of N nodes are randomly selected that consider the details of the new block header to validate and sign the new generated block. The new block is first, required to be validated by all validators. Only after the validation, is conducted, it is considered as a complete block, eligible for recording transactions, and is appended to the blockchain. In this method, the nodes owning more stakes are luckier to be selected as validators.

- **Practical Byzantine fault tolerance (PBFT):** This algorithm is used in an asynchronous network to prevent a Byzantine fault, which is a computer system error or malicious attack that causes Byzantine or arbitrary behavior in the network's defective nodes. In a Byzantine condition, faulty nodes can appear as functioning or faulty, and it is difficult for a fault detection system to declare a Byzantine node that [27, 28]. Therefore, a PBFT is a consensus algorithm that must be able to resist both the system errors and malicious attack. It is mathematically proven that in a BFT algorithm with n Byzantine (faulty) nodes, more than $3n + 1$ ordinary nodes are required to tolerate the Byzantine fault in the consensus procedure [29, 30]. Besides, some improved BFT algorithms such as the Byzantine consensus algorithm based on Gossip protocol (GBC), credit practical Byzantine fault tolerance (CBPFT), and Byzantine fault tolerant algorithm based on vote (vBFT) are designed to improve the PBFT algorithm in terms of energy consumption, efficiency, and scalability [31]. Five phases have been considered in the PBFT algorithm. In the "request" phase, the client sends the request to other replicas in the network. Then, in the "pre-prepare" phase, the primary replica assigns a serial number to this request to coordinate the requests' order and sends it to all other replicas. In the "prepare" phase, after receiving the request from the "pre-prepare" phase to all nodes in the network, each node checks the validity of the request. If the request and its sequence are valid, each node sends it to others. If there are n Byzantine nodes in the network, more than $2n + 1$ replicas are required to approve the request. In the "commit" phase, the serial number of the request is checked again to prevent failure in cases of primary fails. In the last step, all replicas reply to the client [2, 31].

4 Application Layer

In addition to the conventional applications of blockchains such as those in cryptocurrencies [32],, blockchains can provide the requires platform for many industrial applications in the fields of healthcare industry, financial sector, supply chains, privacy protection [33, 34], and IoTs. To enable blockchain for different applications, smart contracts are the most important extension that could be used in blockchain and are introduced in the next section. Deploying blockchain in different fields will provide many improvements and advantages in these fields [15, 35–38].

Finance

Blockchains initially were used in cryptocurrency and then were considered to provide benefits in other financial domains [39]. Using them in the finance world could eliminate the need for a trusted third party and provide a decentralized platform for transactions by verifying them and providing them with a secure registry. These two features enable blockchains for many financial activities such as [40–42] cryptocurrency, peer-to-peer transactions, stock trading, financial settlement, and insurance market.

Health care

In the health care industry, blockchains could be used for enhancing medical record processes by digitizing patients' information in the form of electronic medical records (EMRs), which allow for secure and controlled sharing among providers and eliminate the risk of altering patients' privacy [43]. Health care blockchains provide better and easier access to research results, medical innovations, and patients' information for a comprehensive analysis and proposing more effective remedial solutions [44, 45]. The other benefit of utilizing blockchains in the health care industry is to improve pharmacy supply chain management [46].

Manufacturing

In this domain, blockchains could be used to effectively reduce manufacturing cost and also to enhance anti-counterfeiting procedures, social manufacturing network, energy scheduling, and enable cloud manufacturing [47, 48].

Energy

Applying blockchains in the energy domain is mainly to manage microgrids, which are small electrical networks in distribution networks that have their own energy resources and consumptions. They also are connected to the main electrical power network and have energy exchange with the main network [49]. In a microgrid, many distributed resources are available, such as PV panels, electric vehicles, batteries, gas turbines, and responsive loads, which require smart management to provide an efficient and secure operation for the whole power system. In this regard, blockchains could be used to efficiently manage energy generation and consumption in the smart grids [42]. Utilizing a blockchain in the energy domain could enable secure energy transactions in smart microgrids, provide a suitable platform for energy scheduling and exchange for all distributed energy resources (DERs) and demand response utilization, and enhance energy loss tracking for a better cost distribution among DERs and consumptions [50–52].

Agriculture and food

In the agriculture and food industry, blockchains can be used in conjunction with information and communication technology (ICT) to enhance farming and distribution procedures that improve cost-effectiveness, enhance food quality and safety, reduce food waste, and reduce risks by reducing uncertainty. They also provide a trusted supply chain management to track food safety and prevent fraud by enhancing transparency in the food supply chain [53–57].

Entertainment

Blockchains provide many advantages in the entertainment industry, especially in online gaming and other entertainment. They give gamers a suitable and secure mechanism to better manage and protect their virtual assets for secure use in different games. Also, blockchains provide a fast payment mechanism between entertainment customers and providers and offer new business models for the entertainment business. The other gained benefit from blockchains in the entertainment domain is peer-to-peer online entertainment [58–61].

Telecommunication

Utilizing blockchains in the telecommunication industry enhances telecommunication service management, provides transparency and traceability, and makes managing telecommunication contracts more efficient [62–64].

Internet of things (IoT)

IoT refers to a new technology that links Internet networks and physical objects to facilitate human life. In this technology, Internet-based communication, computing systems, and objects cooperate to deliver the required service and economic benefits to users. In such platforms, blockchains are employed to enable IoT for many applications by improving data handling in terms of security, privacy, trust, authentication, and robustness against cyberattacks [2, 65–72].

The application of blockchains for accuracy enhancement in the industry

In industries in which accuracy is paramount, blockchains can help improve time delay, reduce human mistakes, and enable secure, fast, and efficient planning, monitoring, coordinated scheduling, and validated activities [73].

5 The application of blockchains in Smart Contracts

Smart contracts are an enabling extension for BC to make it a suitable platform for many applications. Flexible and programmable, smart contracts automates agreement procedures between contracts corresponds [74] and offer an efficient negotiation model without the need for a registered third party with lots of paperwork. Therefore, smart contracts cut administrative time and costs in the contract procedure [75–79].

References

1 Firouzi, F., Chakrabarty, K., Nassif, S., and Device, F. (2020). *Intelligent Internet of Things*.

2 Wu, M., Wang, K., Cai, X., Guo, S., Guo, M., and Rong, C. (2019). A comprehensive survey of blockchain: From theory to IoT applications and beyond. *IEEE Internet Things J.* 6 (5): 8114–8154.

3 Chahbaz, A. (2018). *An Introduction to Blockchain and Its Applications. With a Focus on Energy Management*. diplom. de.

4 Nakamoto, S. (2019). *Bitcoin: A Peer-to-Peer Electronic Cash System*. Manubot.

5 Lin, J. (2018). Analysis of blockchain-based smart contracts for peer-to-peer solar electricity transactive markets. Master Thesis, Department of Electrical and Computer Engineering, Virginia Polytechnic Institute and State University.

6 Gervais, A., Karame, G.O., Capkun, V., and Capkun, S. (2014). Is Bitcoin a decentralized currency? *IEEE Secur. Priv.* 12 (3): 54–60.

7 Huang, B., Liu, Z., Chen, J., Liu, A., Liu, Q., and He, Q. (2017). Behavior pattern clustering in blockchain networks. *Multimed. Tools Appl.* 76 (19): 20099–20110.

8 Donet, J.A.D., Pérez-Sola, C., and Herrera-Joancomart, J. (2014). The Bitcoin P2P network. In: *International Conference on Financial Cryptography and Data Security* 87–102.

9 Sun, H., Mao, H., Bai, X., Chen, Z., Hu, K., and Yu, W. (2018). Multi-blockchain model for central bank digital currency. *Parallel and Distributed Computing, Applications and Technologies, PDCAT Proceedings* 2017: 360–367.

10 Pinzón, C., and Rocha, C. (2016). Double-spend attack models with time advantange for Bitcoin. *Electron. Notes Theor. Comput. Sci.* 329: 79–103.

11 Salian, A., Shah, S., Shah, J., and Samdani, K. (2019). Review of blockchain enabled decentralized energy trading mechanisms. In: *2019 IEEE International Conference on System, Computation, Automation and Networking (ICSCAN)* 1–7.

12 Zheng, Z., Xie, S., Dai, H.-N., Chen, X., and Wang, H. (2018). Blockchain challenges and opportunities: A survey. *Int. J. Web Grid Serv.* 14 (4): 352–375.

13 Wang, W. et al.(2018). A survey on consensus mechanisms and mining management in blockchain networks. *arXiv Prepr. arXiv1805.02707.* 1–33.

14 Dinh, T.T.A., Liu, R., Zhang, M., Chen, G., Ooi, B.C., and Wang, J. (2018). Untangling blockchain: A data processing view of blockchain systems. *IEEE Trans. Knowl. Data Eng.* 30 (7): 1366–1385.

15 Casino, F., Dasaklis, T.K., and Patsakis, C. (2019). Telematics and informatics A systematic literature review of blockchain-based applications : Current status, classification and open issues. *Telemat. Informatics* 36: 55–81.

16 Yli-Huumo, J., Ko, D., Choi, S., Park, S., and Smolander, K. (2016). Where is current research on blockchain technology? – A systematic review. *PLoS One* 11 (10): e0163477.

17 Lu, Y. (2018). Blockchain and the related issues: A review of current research topics. *J. Manag. Anal.* 5 (4): 231–255.

18 Mingxiao, D., Xiaofeng, M., Zhe, Z., Xiangwei, W., and Qijun, C.(2017). A review on consensus algorithm of blockchain. In: *2017 IEEE International Conference on Systems, Man, and Cybernetics (SMC)* 2567–2572.

19 Makhdoom, I., Abolhasan, M., Abbas, H., and Ni, W. (2019). Blockchain's adoption in IoT: The challenges, and a way forward. *J. Netw. Comput. Appl.* 125: 251–279.

20 King, S. and Nadal, S. (2012). Ppcoin: Peer-to-peer crypto-currency with proof-of-stake. *Self-published Pap. August* 19.

21 Chaumont, G., Bugnot, P., Hildreth, Z., and Giraux, B.(2019). DPoPS: Delegated proof-of-private-stake, a DPoS implementation under X-Cash, a Monero based hybrid-privacy coin.

22 Schuh, F. and Larimer, D. (2015). BitShares 2.0: Financial smart contract platform. *Bitshares Financ. Platf* 12. https://medium.com/dpops-delegated-proof-of-private-stake-a-dpos/dpops-delegated-proof-of-private-stake-a-dpos-implementation-under-x-cash-a-monero-based-65a06b3b60a

23 Xu, X., et al. (2017). A taxonomy of blockchain-based systems for architecture design. In: *2017 IEEE International Conference on Software Architecture (ICSA)* 243–252.

24 Upadhyay, N. (2019). *UnBlock the Blockchain*. Springer.

25 Shahzad, B., and Crowcroft, J. (2019). Trustworthy electronic voting using adjusted blockchain technology. *IEEE Access* 7: 24477–24488.

26 Suyambu, G.T., Anand, M., and Janakirani, M. (2020). Blockchain—A most disruptive technology on the spotlight of world engineering education paradigm. *Procedia Comput. Sci.* 172: 152–158.

27 Hao, X., Yu, L., Zhiqiang, L., Zhen, L., and Dawu, G. (2018)Dynamic practical Byzantine fault tolerance. In: *2018 IEEE Conference on Communications and Network Security (CNS)* 1–8.

28 Castro, M., Liskov, B., et al. (1999). Practical Byzantine fault tolerance. *OSDI* 99 (1999): 173–186.

29 Kim, S., Lee, S., Jeong, C., and Cho, S. (2020). Byzantine fault tolerance based multi-block consensus algorithm for throughput scalability. In: *2020 International Conference on Electronics, Information, and Communication (ICEIC)* 1–3.

30 Zhang, L., and Li, Q. (2018). Research on consensus efficiency based on practical byzantine fault tolerance. In: *2018 10th International Conference on Modelling, Identification and Control (ICMIC)* 1–6.

31 Wang, H., and Guo, K. (2019). Byzantine fault tolerant algorithm based on vote. In: *2019 International Conference on Cyber-Enabled Distributed Computing and Knowledge Discovery (CyberC)*, 190–196.

32 List of cryptocurrencies – Wikipedia. https://en.wikipedia.org/wiki/List_of_cryptocurrencies.

33 Zyskind, G., Nathan, O., and Pentland, A. (2015). Decentralizing privacy: Using blockchain to protect personal data. In: *2015 IEEE Security and Privacy Workshops* 180–184.

34 Lu, Y. (2018). Blockchain: A survey on functions, applications and open issues. *J. Ind. Integr. Manag.* 3 (04): 1850015.

35 Alladi, T., Chamola, V., Parizi, R.M., and Choo, -K.-K.R. (2019). Blockchain applications for Industry 4.0 and industrial IoT: A review. *IEEE Access* 7: 176935–176951.

36 Tama, B.A., Kweka, B.J., Park, Y., and Rhee, K.-H. (2017). A critical review of blockchain and its current applications. In: *2017 International Conference on Electrical Engineering and Computer Science (ICECOS)*, 109–113.

37 Abou Jaoude, J., and Saade, R.G. (2019). Blockchain applications–usage in different domains. *IEEE Access* 7: 45360–45381.

38 Chen, W., Xu, Z., Shi, S., Zhao, Y., and Zhao, J. (2018). A survey of blockchain applications in different domains. In: *Proceedings of the 2018 International Conference on Blockchain Technology and Application* 17–21.

39 Hileman, G., and Rauchs, M. (2017). Global cryptocurrency benchmarking study. *Cambridge Cent. Altern. Financ.* 33: 33–113.

40 Suberg, W. (2017). Hyperledger blockchain "Shadows" Canadian Bank's international payments. https://cointelegraph.com/news/hyperledger-blockchain-shadows-canadian-banks-international-payments.

41 Mainelli, M., and Von Gunten, C. (2014). Chain of a lifetime: How blockchain technology might transform personal insurance.

42 Cohn, A., West, T., and Parker, C. (2017). Smart after all: Blockchain, smart contracts, parametric insurance, and smart energy grids. *Georg. Law Technol. Rev.* 1 (2): 273–304.

43 The blockchain for healthcare: Gem launches gem health network with Philips blockchain lab – Bitcoin magazine: Bitcoin news, articles, charts, and guides.

44 Peterson, K., Deeduvanu, R., Kanjamala, P., and Boles, K. (2016). A blockchain-based approach to health information exchange networks. *Proc. NIST Workshop Blockchain Healthcare* 1 (1): 1–10.

45 Kuo, -T.-T., Kim, H.-E., and Ohno-Machado, L. (2017). Blockchain distributed ledger technologies for biomedical and health care applications. *J. Am. Med. Informatics Assoc.* 24 (6): 1211–1220.

46 Benchoufi, M., and Ravaud, P. (2017). Blockchain technology for improving clinical research quality. *Trials* 18 (1): 1–5.

47 Ding, K., Jiang, P., Leng, J., and Cao, W. (Jan, 2015). Modeling and analyzing of an enterprise relationship network in the context of social manufacturing. *Proc. Inst. Mech. Eng. Part B J. Eng. Manuf.* 230 (4): 752–769.

48 Abeyratne, S.A., and Monfared, R.P. (2016). Blockchain ready manufacturing supply chain using distributed ledger. *Int. J. Res. Eng. Technol.* 5 (9): 1–10.

49 Lasseter, R.H., and Paigi, P. (2004). Microgrid: A conceptual solution. In: *2004 IEEE 35th Annual Power Electronics Specialists Conference (IEEE Cat. No. 04CH37551)* 6: 4285–4290.

50 Mengelkamp, E., Notheisen, B., Beer, C., Dauer, D., and Weinhardt, C. (2018). A blockchain-based smart grid: Towards sustainable local energy markets. *Comput. Sci. Dev.* 33 (1): 207–214.

51 Mengelkamp, E., Gärttner, J., Rock, K., Kessler, S., Orsini, L., and Weinhardt, C. (2018). Designing microgrid energy markets: a case study: The Brooklyn microgrid. *Appl. Energy* 210: 870–880.

52 Pop, C., Cioara, T., Antal, M., Anghel, I., Salomie, I., and Bertoncini, M. (2018). Blockchain based decentralized management of demand response programs in smart energy grids. *Sensors (Switzerland)* 18 (1): 162.

53 Lin, Y.-P., et al. (2017). Blockchain: The evolutionary next step for ICT E-agriculture. *Environments* 4: 3.

54 Yiannas, F. (2018). A new era of food transparency powered by blockchain. *Innov. Technol. Governance, Glob.* 12: 46–56. Jul.

55 Basnayake, B.M.A.L., and Rajapakse, C. (2019). A blockchain-based decentralized system to ensure the transparency of organic food supply chain. In: *2019 International Research Conference on Smart Computing and Systems Engineering (SCSE)* 103–107.

56 Shahid, A., Almogren, A., Javaid, N., Al-Zahrani, F.A., Zuair, M., and Alam, M. (2020). Blockchain-based agri-food supply chain: A complete solution. *IEEE Access* 8: 69230–69243.

57 Antonucci, F., Figorilli, S., Costa, C., Pallottino, F., Raso, L., and Menesatti, P. (2019). A review on blockchain applications in the agri-food sector. *J. Sci. Food Agric.* 99 (14): 6129–6138.

58 Gainsbury, S., and Blaszczynski, A. (Sep, 2017). How blockchain and cryptocurrency technology could revolutionize online gambling. *Gaming Law Rev.* 21: 482–492.

59 Kalra, S., Sanghi, R., and Dhawan, M. (2018). Blockchain-based real-time cheat prevention and robustness for multi-player online games. Proceedings of the 14th International Conference on emerging Networking EXperiments and Technologies, pp. 178–190. https://doi.org/10.1145/3281411.3281438.

60 Dutra, A., Tumasjan, A., and Welpe, I. (2018). *Blockchain Is Changing How Media and Entertainment Companies Compete.* MIT Sloan Management Review.

61 Tripathi, S. (2018). Blockchain application in media and entertainment. In: SMPTE 2018, 1–8.

62 Bohrweg, N. (2017). *Applicability of Blockchain Technology in Telecommunications Service Management.* Leiden University.

63 Dabbagh, M., Sookhak, M., and Safa, N.S. (2019). The Evolution of blockchain: A bibliometric study. *IEEE Access* 7: 19212–19221.

64 Khalaf, O.I., Abdulsahib, G.M., Kasmaei, H.D., and Ogudo, K.A. (2020). A new algorithm on application of blockchain technology in live stream video transmissions and telecommunications. *Int. J. e-Collaboration* 16 (1): 16–32.

65 Li, D., Deng, L., Cai, Z., and Souri, A. (2020). Blockchain as a service models in the Internet of things management: Systematic review. *Trans. Emerg. Telecommun. Technol.* e4139. https://onlinelibrary.wiley.com/doi/abs/10.1002/ett.4139.

66 Lao, L., Li, Z., Hou, S., Xiao, B., Guo, S., and Yang, Y. (2020). A survey of IoT applications in blockchain systems: Architecture, consensus, and traffic modeling. *ACM Comput. Surv.* 53 (1): 1–32.

67 Bhushan, B., Sahoo, C., Sinha, P., and Khamparia, A. (2021). Unification of blockchain and Internet of things (BIoT): Requirements, working model, challenges and future directions. *Wirel. Networks* 27 (1): 55–90.

68 Honar Pajooh, H., Rashid, M., Alam, F., and Demidenko, S. (2021). Multi-layer blockchain-based security architecture for Internet of things. *Sensors* 21 (3): 772.

69 Dai, H.-N., Zheng, Z., and Zhang, Y. (2019). Blockchain for Internet of things: A survey. *IEEE Internet Things J.* 6 (5): 8076–8094.

70 Viriyasitavat, W., Da Xu, L., Bi, Z., and Hoonsopon, D. (2019). Blockchain technology for applications in Internet of things – Mapping from system design perspective. *IEEE Internet Things J.* 6 (5): 8155–8168.

71 Fernández-Caramés, T.M., and Fraga-Lamas, P. (2018). A review on the use of blockchain for the Internet of things. *Ieee Access* 6: 32979–33001.

72 Wang, X. et al. (2019). Survey on blockchain for Internet of things. *Comput. Commun.* 136: 10–29.

73 Project Provenance Ltd. (2015). Blockchain: The solution for supply chain transparency | provenance. https://www.provenance.org/whitepaper.

74 Luu, L., Chu, D.-H., Olickel, H., Saxena, P., and Hobor, A. (2016). Making smart contracts smarter. In: *Proceedings of the 2016 ACM SIGSAC Conference on Computer and Communications Security*, 254–269.

75 Haiwu, H., An, Y., and Zehua, C. (2018). Survey of smart contract technology and application based on blockchain. *J. Comput. Res. Dev.* 55 (11): 2452.

76 Mohanta, B.K., Panda, S.S., and Jena, D. (2018). An overview of smart contract and use cases in blockchain technology. In: *2018 9th International Conference on Computing, Communication and Networking Technologies (ICCCNT)* 1–4.

77 Cheng, J.-C., Lee, N.-Y., Chi, C., and Chen, Y.-H. (2018). Blockchain and smart contract for digital certificate. In: *2018 IEEE international conference on applied system invention (ICASI)*, 1046–1051.

78 Macrinici, D., Cartofeanu, C., and Gao, S. (2018). Smart contract applications within blockchain technology: A systematic mapping study. *Telemat. Informatics* 35 (8): 2337–2354.

79 Karamitsos, I., Papadaki, M., and Al Barghuthi, N.B. (2018). Design of the blockchain smart contract: A use case for real estate. *J. Inf. Secur.* 9 (3): 177.

3

Robotics: A Key Driver of Industry 4.0

Mojtaba Molla-Hosseini[a], Marjan Hosseini[b], Gholamreza Vossoughi[a], and Mahdi Sharifzadeh[c],*

[a] *Mechanical Engineering Department, Sharif University of Technology, Tehran, Iran*
[b] *Civil Engineering Department, Tehran University, Tehran, Iran*
[c] *Sharif Energy, Water and Environment Institute (SEWEI), Sharif University of Technology, Tehran, Iran*
* *Corresponding author. Sharif Energy, Water and Environment Institute (SEWEI), Sharif University of Technology, Tehran, Iran*

3.1 Introduction

The concept of robotic mechanics dates back to classical times and the early 20th century [1]. Robotics is a field of study integrating computer science and engineering [1]. The aim of robotics is to help and assist humans in employing machines in the real world. Robotics consists of a diverse range of technical fields including mechanical engineering, electrical engineering, electronics, mechatronics, information engineering, computer engineering, control engineering, and software engineering.

Robots can perform many delicate tasks of such complexity that are conventionally done by humans. Moreover, robots are a lot more accurate and efficient than humans in some tasks like producing things on large scale. They are also used in situations that might be dangerous or even mortal for humans. Bomb detection and deactivation, cleaning up hazardous materials, and performing actions in environments that are impossible for a human to survive (e.g., underwater, in space, high heat) are examples of such situations.

Industry 4.0 has been defined as the automation of traditional industrial manufacturing by means of modern smart technology [2]. This revolution is marked by breakthroughs in some technologies including robotics, nanotechnology, artificial intelligence, biotechnology, the Internet of things, fifth-generation wireless technologies, and more [3]. Robotics is one of the main drivers of the Fourth Industrial Revolution (also called Industry 4.0) through its various applications referred to in the next session. Figure 3.1 shows the categories of robotic systems considered in this chapter. In the following, some categories of industrial robots are presented.

3.2 Classification of Robotic Systems

Different categories of industrial robots can be found in the literature. They can be classified according to geometry, actuators, control methods, kinematics, and more. In the following, these categories are explained in detail.

Industry 4.0 Vision for the Supply of Energy and Materials: Enabling Technologies and Emerging Applications, First Edition. Edited by Mahdi Sharifzadeh.
© 2022 John Wiley & Sons, Inc. Published 2022 by John Wiley & Sons, Inc.

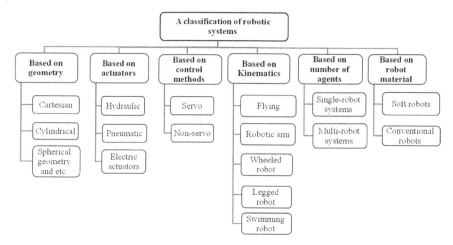

Figure 3.1 Robotic System Categories.

3.2.1 Classification Concerning Geometry

Every industrial robot can be classified in terms of both geometry and configuration. An industrial robot could have one to six degrees of freedom developed by one or several loops from the base to the end effector.

In the case in which there is only one loop, the structure is serial. In contrast, when more than one loop exists, the structure is called parallel. The main feature of this latter category is that the parallel robots benefit from their high stiffness compared with serial counterparts. On the other hand, a small workspace (i.e., space a robot could reach) is their weakness. The parallel and serial structures are also called closed-chain and open-chain manipulators.

A serial robot itself could have some subcategories with regard to both joints types and order including Cartesian geometry, cylindrical and spherical geometry, Selective Compliance Assembly Robot Arm (SCARA), articulated geometry, and the anthropomorphic (Table 3.1). Figure 3.2 indicates examples of spherical, SCARA, and parallel

Table 3.1 Classification of Robots According to Their Geometry.

Subcategory	Joints Types and Orders	Application
Cartesian geometry	Prismatic1-prismatic2-prismatic3	aterial handling and assembly
Cylindrical geometry	Revolute1-prismatic2-prismatic3	Carrying objects even of large dimensions
Spherical geometry	Revolute1-revolute2-prismatic3	Machining
SCARA	Revolute1- revolute2-prismatic3: all axes of joints are parallel	Screw tightening and assembly [5]
Articulated geometry	Revolute1-revolute2-revolute3 Joints 2 and 3 are parallel and perpendicular to joint 1	Smooth, sweeping motions, capable of reaching any point within a certain region [6]
The anthropomorphic	Revolute1-revolute2-revolute3	Packaging, welding, loading and unloading of machines and, etc.

(a)   (b)

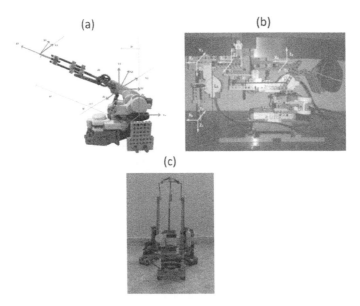

(c)

Figure 3.2   a) Spherical Robot. b) SCARA Robot. c) Parallel Robot [4], figure 02,08 (p.205,211) / with permission of JOHN WILEY & SONS, INC.

Figure 3.3   An Example of an Industrial Serial Robot for a Typical Pick and Place Task [4], figure 05 (p.209) / with permission of JOHN WILEY & SONS, INC.

robots [4]. Figure 3.3 indicates an example of an industrial serial robot that is used for typical pick and place tasks [4].

3.2.2   Classification Concerning the Actuator System

Generally, three types of manipulated variables—electric, hydraulic, and pneumatic actuators—are deployed in robots. Table 3.2 shows the different types of actuators and their pros and cons.

3.2.3   Classification Concerning the Control Strategy

Robots apply servo or non-servo control strategies. In non-servo robots, there are no closed-control loops, and only open-loop devices exist and mechanical orders are predetermined. This control method is widely used for material transfer. In contrast, in servo robots, a control loop exists and the actuation is performed with regard to online responses. Thus, the manipulation is a lot more accurate and reliable.

Table 3.2 Classification of Robots According to Their Actuator.

Actuator	Pros	Cons
Hydraulic actuators	• Suitable for heavy lifting • High-speed movement as well as rapid response time	• Exposure to leaking • Pumps are required
Pneumatic actuators	• Less leaking	• Compressibility of air leads to unsmooth movement • Air compressors are required
Electric actuators	• Pumps and air compressors are not required • There is no leaking • Comfortable designing, programming, and installing	• High initial cost, • Size of the motor may be large in some special cases, • Not safe in hazardous and flammable areas

3.2.4 Classification Concerning Kinematics (Motions)

A robot could be classified in terms of motions and movements, which can be categorized as flying robots, robots with robotic arms, wheeled robots, legged robots, swimming robots, and so on.

3.2.5 Classification Concerning the Number of Agents

Robotic systems could also be categorized in terms of the number of agents including single and multi-robot systems. In the following, a detailed discussion of multi-robot systems is presented.

3.2.5.1 Multi-Robot Systems

The robots in this category could have several agents. In this case, they are referred to as multi-robot systems or swarm robotic systems. Multi-Robot Systems (MRS) benefit from collaborations between agents.

In the following, various applications of MRS including search and rescue, foraging and flocking, formation and exploration, cooperative manipulation, team heterogeneity, and adversarial environment are discussed.

3.2.5.1.1 Search and Rescue

Patrolling and surveying indoor places performed by MRS are two examples of the search and rescue application [7]. Unmanned aerial vehicle (UAV) groups are deployed for surveillance of outdoor places such as offshore areas [8]. They proposed a scheme for a multi-robot team for the surveillance of shipwreck survivors at the sea [7].

3.2.5.1.2 Foraging and Flocking

Foraging and flocking is another application of MRS inspired by natural colonial motions such as those of ants and bees. In the literature, MRS has been applied to the task of cleaning up hazardous waste [9]. Robots are used to move waste back to the base station. In another work [10], foraging has been applied for clean-up and object collection.

3.2.5.1.3 *Formation and Exploration*
In this mission, robots must maintain a strict arrangement, avoiding obstacles in the path.

3.2.5.1.4 *Cooperative Manipulation*
An example of cooperative manipulation is a pusher–steerer system [11], which consists of a steerer robot pre-programmed with a trajectory and a pusher robot exerting a force onto the object so that the steerer robot could follow its programmed path by means of setting its heading.

3.2.5.1.5 *Team Heterogeneity*
Heterogeneity refers to a team of mobile robots handling complex tasks a lot more efficiently than a single robot owing to the diverse capabilities of its members. In a study, the robots are deployed to perform diverse tasks using speech segmented between them through their recipients [11]. Task propagation in search and rescue is another example of MRS heterogeneous applications in which the most suitable robot for a task is chosen considering the probability that a particular order given by the user is satisfied by mentioned robot capabilities [12]. For example, the command "take-off" could be performed only by an aerial robot.

3.2.5.1.6 *Adversarial Environment*
Zhang and Wang deployed a genetic algorithm to make robots learn not to enter adversarial defense regions, where they may be "killed" [13]. An example of such application could be on battlefields or mine clearance. MRS are also applied to monitor the team's actions to identify adversarial players not passing the game rules [14]. Figure 3.4 shows a multi-robot system that is forming a square structure [15].

3.2.6 Classification with Respect to Robot Materials

Robots could also be categorized according to their materials, including conventional and soft robots. Soft robots are inspired by biological systems and consist of soft materials. In contrast to conventional robots made with hard materials such as steel, titanium, or aluminum, soft robots are made from hyper-elastic materials such as polymer, silicone, rubber, and other flexible materials [16]. Due to their formable features, soft robots are limited in using conventional rigid sensors like strain gauges, encoders, or inertial measurement units.

3.2.6.1 Main Advantages of Soft Robots
Advantages of soft robots compared with conventional ones include being adaptive to wearable devices, being adaptive to curve and irregular surfaces,

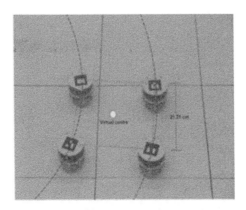

Figure 3.4 A Multi-Robot System Forming a Square Structure [15], figure 04, (p.07) / with permission of JOHN WILEY & SONS, INC.

and having safe interaction with humans, including simple gripping systems. In contrast to conventional robots, which have a limited degree of freedom of motion, soft robots have a flexible body that results in a high degree of freedom. Due to their unique features, soft robots have a varied range of applications [17]. Figure 3.5 indicates a soft robot with the capability of changing body shape [18].

3.2.6.2 Main Applications of Soft Robots

Soft robots have many application areas such as locomotion and exploration, human–machine interaction and interface, surgical and medical applications, manipulation, rehabilitation, and wearable robots [17]. More explanations about each application are provided in the following.

3.2.6.2.1 Locomotion and Exploration

Soft robots are capable of navigating and exploring unknown places due to their considerably high flexibility compared with conventional robots, which helps them move more efficiently through environments such as underwater. They mimic underwater and caterpillar-like creatures by performing motions such as jumping, rolling, and crawling [19]. For instance, smart soft composite enables them to perform motion via abdominal contraction [13]. In another study, a soft robot is developed to simulate the locomotion of earthworms, absorbing external shocks due to their soft materials [19]. Snake-like motion with wheels using soft robots is also studied [14]. In another study, advanced robots performed a more complex motion that enables them to explore indoor and outdoor environments in harsh conditions, suitable for surveillance applications [21]. Fish-like robots are developed for underwater conditions. The soft bodies of these robots enable them to perform nonlinear motions such as swimming, turning, and diving easier [22]. In another study, a robot with eight legs was developed to mimic the motion of an octopus. These legs offer the capability for traveling easily through unstructured surfaces as well as grasping [23, 24].

The design of some soft robots is not inspired by animals. A circular deformable robot that can perform crawling and jumping was developed for moving in rough areas [25].

3.2.6.2.2 Human–Machine Interaction and Interface

Applications in which there are interactions with humans usually involve the risk of injury, too. Soft robot materials are more flexible and could be safer for such applications [26].

Figure 3.5 A Mobile Soft Robot with a Flexible Body [18], figure 03 (p.02) / with permission of JOHN WILEY & SONS, INC.

3.2.6.2.3 Medical and Surgical Applications

Soft robots are also suitable for minimally invasive surgery (MIS) because of their precision, flexibility, and high degrees of freedom (DOF) that leads to more compatibility with variable work conditions and environments [27]. Their controllable stiffness leads to less damage to surrounding soft tissues as well as more accessibility inside the body [28, 29]. They are also utilized in laparoscopic surgery and endoscopy system inspired by an elephant's trunk [30, 31]. Endoscopy systems allow access via confined space and manipulation for cardiac ablation [30].

3.2.6.2.4 Applications with High Degrees of Freedom

One of the most essential features of soft robots, making them different from conventional robots, is their significantly higher DOF. Low DOF in conventional robots poses limitations for manipulation such as grasping action. Such features of soft robots also simplify the control algorithm [16].

A case in point is a tentacle-like manipulator with a pneumatic network area capable of grasping of variety of materials [32]. Another example is OctArm, which has the ability to grasp objects of varying size and shape and dynamic disturbances in the form of payload [33]. Many examples in the literature are inspired by octopus and other cephalopod limbs [34, 35].

3.2.6.2.5 Rehabilitation and Wearable Robots

Biocompatibility and bio-integrity are two other important features of soft robots, making them applicable for wearable devices and human-assistance applications with minimal possibility of injury to the robot and humans [36, 37]. Soft robotic gloves with soft actuators are another example of these applications used for hand rehabilitation in patients having grasp pathologies [37]. Some kinds of these gloves are also used for thumb rehabilitation [38] and others for training children to improve their functionality such as box and block test. Rehabilitation contains human–machine interaction [39]. Moreover, gait rehabilitation soft robots are used for spinalized rodents [40].

Light soft robots utilized as wearing clothes can assist patients by means of providing some required forces during walking. In contrast to rigid robots, light soft robots minimize the suit's unintentional disturbances due to its flexibility and light weight [41]. Other soft robots used for rehabilitation include pneumatic artificial muscle actuators for the ankle and foot [42] and in patients with mandibular motion disorders [35].

3.3 Applications of Robotics

Upon critical review of literature, a diverse range of robotics applications can be identified, including medical, space exploration, military, education, agriculture, oil and gas industry, railcar industry, and human-interactive applications as discussed in the following.

3.3.1 Medical Application

Robots have been deployed in a wide variety of medical cases, one of the most important of which is surgery, especially tumor removal [44]. Robots are used for

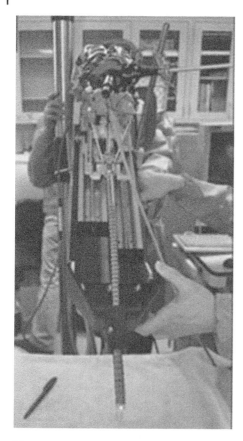

Figure 3.6 A Highly Articulated Flexible Robot for Surgery Tasks [46], figure 01 (p.02) / with permission of JOHN WILEY & SONS, INC.

monitoring medical parameters [45] or hospital management activities such as queuing systems. Trans-oral robotic surgery (TORS) is especially leveraged in surgical operations for cancer treatment [46]. Figure 3.6 shows a soft robot that is used for surgery tasks [46].

There are also ambulance robots equipped with a defibrillator to help victims during the journey to the hospital [47]. Other robots provide health care in the form of UAVs in far, congested, and inaccessible areas [48].

3.3.2 Space Exploration

The two methods of space exploration are human and robot. Robot exploration is very efficient, especially when planetary objects are distant and when the mission could pose human life at risk. Table 3.3 shows various exploration robots and missions [8].

Figure 3.7 indicates a wheeled mobile robot for space exploration tasks. Such robots should be intelligent because using instructions from Earth reach them with long delays [49].

3.3.3 Military

Robotics also has a diverse range of military applications. They can have missions such as scouts, detecting bombs, destroying mines, and medical aids [50]. Scout robots are deployed to be sent to war fields, which could be dangerous and unknown, before the major mission to gain the required information such as three-dimensional (3D) maps or taking photos and recording voices from the target areas or in the role of soldiers [50].

Unmanned robots can detect bombs and explosive disposals in lands. They can also be used as anti-tank launchers. Some of them have a gripper to pick up things and are capable of fire disrupters [50]. A major benefit of such robots is that they could move easily throw obstacles, carrying heavy things while some machines could not.

In another military application, robots are deployed to move medics all over the battlefield. Some small and fast robots go and check everywhere in the area by means of their sensors. Afterward, they spread information to the major and larger robots to carry the soldiers away from the area. Robots, communicating and sharing information, have the ability to manage the medical works faster than humans, which can save soldiers' lives.

Figure 3.8 shows an armed police robot that carries a casualty in a rescue task [51].

Table 3.3 The Characteristics of Some Robots' Missions. Based on B. V Ratnakumar and M. C. Smart, 2007.

Orbiters	Deployed for imaging planetary object's surface, monitoring daily global weather, and helping to understand their atmosphere as well.
Fly-by and sample-return missions	The robot approaches very closely to planetary object so it could perform studies of that object by means of making detailed observations.
Landed missions-Lander	Studying biology, chemical composition including organic and inorganic, seismology, meteorology, magnetic properties, physical features of planetary surfaces, appearance, and atmosphere.
Landed missions-Rovers	The robot is equipped with a relatively small payload with science instruments in order to probe the planetary object's surface or atmosphere.
Impactors and penetrators	The robot examines the planetary object's surface more closely.
Miscellaneous science missions	Optical or spectroscopic studying outside the earth's atmosphere.

Figure 3.7 A Wheeled Mobile Robot for Space Exploration Task [49], figure 01 (p.03) / with permission of JOHN WILEY & SONS, INC.

3.3.4 Education

Educational robotics is one of the applications of robots that has attracted a lot of interest in all levels of education [8]. For instance, NAO is a robot that engages children with difficulties in learning and improving the therapeutic process. Commercial robots are also used in classrooms. Children can play with robots in their playtime [52]. Educational robotics can promote classroom teaching as well [53]. They can participate in interactive learning activities and language skills. However, these types of robotics focus mostly on improving programming skills. Educational robots can be evaluated according to programming language, processor, sensors, encoder, connection, cost, and battery [52].

There are some robots called telepresence robots deployed to represent absent students, for example due to illness, who are not able to attend school programs such as classes or exams.

Figure 3.8 A Military Robot Carrying a Casualty in a Rescue Task [51], figure 02 (p.03) / with permission of JOHN WILEY & SONS, INC.

3.3.5 Agriculture

UAVs can be utilized for monitoring agricultural activities for distributing seeds, water, chemicals, and so on. They are much more cost-effective than aircraft and have less carbon footprint as well. In comparison, less pesticide is lost because of the wind, as the height of flying UAVs is only about few meters above the land [47, 54].

Other essential applications of robotics in agriculture are sowing, harvesting, and planting processes. Combining machine vision, actuators, laser technologies, and mechatronics including microsensors, computers, and electric motors have made this possible [55, 56]. In another study, a deep convolutional neural network was employed to develop a robotic automatic system to sow various kinds of grass, following certain patterns on the lawns, and avoid obstacles [56, 57].

Increasing productivity, easy handling, and requiring less labor are three main reasons to consider robotics for agricultural activities [58]. Figure 3.9 illustrates examples of robotic applications in mechanical harvesting [59].

Figure 3.9 An Industrial Robot Used for Harvesting Sweet Peppers [59], figure 05 (p.05) / JOHN WILEY & SONS, INC. / CC BY 4.0.

3.3.6 Oil and Gas Industry

Robotics has widespread applications in the gas and oil industry for inspection and monitoring of the assets structures, especially pipes and tanks [60, 61]. Remotely operated vehicles, autonomous underwater vehicles, unmanned ground vehicles, and unmanned aerial vehicles are four categories of such robots used in this field of study [60].

The pipes are usually developed in underground or underwater areas. Consequently, they are at the risk of corrosion, erosion, cracks, shocking loads, and degradation in general due to extreme weather conditions such as high and low temperatures, dust, and humidity. Manual inspection is expensive and cumbersome. There are also robots that can be inserted in the pipe and travel inside. They are equipped with cameras, X-ray, or eddy currents technology to conduct nondestructive inspection methods. Other robots may be equipped with additional instruments to measure water clarity, light penetration, and temperature, to name a few essential parameters in the natural gas industry [62].

Considering the size of petroleum and petrochemical storage tanks, human inspection is dangerous because of the risk of being poisoned with H_2S and other gases. In addition, inspection procedures impose production interruption that could be very expensive. There is a significant incentive to replace manual inspections with online robotic inspections. Such robots might be equipped with an adhesion mechanism to climb the structure. Using a magnetic field is the most common way in this regard[63]. Vacuum suction [64, 65], rails [66], or pegs and grippers/clamps are also among other methods that can be utilized for this purpose. Such robots are also equipped with high-definition cameras and other measurement devices for monitoring the tank.

Figure 3.10 shows robots that are used in the oil and gas industry. These robots must perform tasks such as detection and localization of faulty heat surfaces, pressure reading in manometers, and identification of unexpected objects in the site [67].

3.3.7 Textile

The application of robotics in the textile industry includes sewing, folding, cutting, and packing. These robots can also perform repetitive works while providing persistent accuracy. Using robotics in this industry leads to improving both the quality and efficiency of production. It can also mitigate labor costs, which can make it cost-effective.

3.3.8 Railcar Industry

A robotic arm can be employed in the railcar industry for manufacturing operations. The application of robots in the railcar industry can both increase automation and reliability as well as improve productivity. During railcar manufacturing, the robotic arm carries out gripping and handling operations as well as inspection operations and assembly. The robotic arm needs to have a high degree of dexterity, repeatability, and precision [68].

Figure 3.11 shows an industrial robot that is used for welding tasks in cars and railcars manufacturing [69]

Figure 3.10 Examples of Robots Performing a Testing Task in the Oil and Gas Industry [67], figure 08 (p.19) / JOHN WILEY & SONS, INC.

Figure 3.11 An Industrial Robot Used for Welding Tasks [69], figure 01 (p.04) / with permission of JOHN WILEY & SONS, INC.

3.3.9 Maintenance and Repair

Maintenance and repair are among the most important and vital issues in almost all industries including the nuclear industry, highways, railways, power line maintenance, aircraft servicing, underwater facilities, and coke ovens [70].

In addition to performing cost-effectively and quickly, robots could carry out some dangerous missions in which there could be even life-threatening risks if performed by a human. Two examples are releasing menacing gases during maintenance in the nuclear industry and electric shocks in power line maintenance. In the nuclear industry, there are different asset structures to be maintained, including nuclear facilities and laboratories,nuclear reactors,decommissioning and dismantling nuclear facilities,andemergency intervention.

Highways are an essential component of the transportation network. According to Table 3.4, robots could be involved in several areas in this application. Robots are commonly located in the railway maintenance shops and perform activities such as grinding, cleaning, repairing, and welding.

Power line maintenance using robots has a significant role in reducing human activities in this industry as there is the danger of electric shock and falling from high places. Therefore, electric power companies use robotics for power line maintenance extensively.

Figure 3.12 shows a robot that is inspecting the infrastructure of solar cells [71].

In aircraft servicing applications, robots can perform tasks like stripping and painting. They are also deployed as assistants for re-arming fighter aircraft.

Another application is underwater facilities. Underwater construction, cleaning the surface of the ocean (especially for the maintenance of the offshore oil industry), and repairing and inspecting communications platforms, wellheads, pipelines, and cables are some special tasks executed by robots in this field.

Table 3.4 Areas in Which Robots Could Be Involved in Highways.

Area	Application
Highway integrity management	• Pothole repair • Crack sealing
Highway marking management	• Pavement marker replacement • Paint re-striping
Highway debris management	• Litter bag pickup • On-road refuse collection • Hazardous spill clean-up • Snow removal
Highway signing management	• Sign and guide marker washing • Roadway advisory
Highway landscaping management	• Vegetation control • Irrigation control
Highway work zone management	• Automatic warning system • Lightweight movable barriers • Automatic cone placement and retrieval

Figure 3.12 A Robot Inspecting Infrastructure Solar Cells [71], figure 06 (p.08) / with permission of JOHN WILEY & SONS, INC.

Robotics is also used in coke ovens. The wall and central portion of the oven chamber are two critical assets to repair. This is a tough task owing to the high-temperature, inaccessibility of the area. As a result, components used in these robots must have high heat resistance.

3.3.10 Construction Industry

Monitoring and recording the work progress are very important in this field that can be performed by UAVs or aerial cinematography technology. This leads to better volumetric surveys and helping site energy simulation and logistic planning. It also could be used for generating conceptual 3D modeling, which leads to less manpower and cost as well as higher productivity. In some advanced cases, laser technology is added to robots so that it helps engineers have better measurements.

Information technology (IT) has helped this field, too. Some examples are building information modeling (BIM) to help improve the efficiency of construction by means of better information management, cross-disciplinary collaboration, problem-solving, construction process control, and risk management. IT has also been used in prefabrication, a very important task in construction [72]. Radio frequency identification (RFID) has also been utilized in the construction industry. Other applications in this field include virtual prototyping for construction process simulation, demolition and renovation, quantification of process emissions, accident identification, and estimation [73–76].

Robotic has also contributed to other aspects of construction such as bricklaying and welding tasks [77].

3.3.11   Environmental Issues

The first use of robotic in the environmental field involves sensors inserted in smart cities to acquire and analyze the environmental data. An example is simulators developed with 3D technology used in city mapping. These simulators consist of air pollution, humidity sensor, ultrasonic sensor, and temperature sensor [78]. The environmental data collected by camcorders then are sent to a cloud space to be used in future calculations. In addition, several environmental conditions such as UV radiation index, climate parameters, sound level, and air pollution have been monitored that could be used in defining air quality [79].

In some applications, analyzing this data facilitates evidence-based decision-making. A case in point is monitoring the wastewater during heavy rains, which could be used to alert the population in areas and avoid traffic congestion [80].

As another application with unhealthy working conditions, cleaning services, air and sediment sampling, sewage, and water systems have to be monitored. Such tasks require communication and teleoperation, digitization and mapping, and video and image capture. A visual robot structure consists of two agents having different perspectives to make a comprehensive view of the environment. This feature facilitates the inspection of sidewalk pavements, sweeping sidewalks, detecting garbage, and managing parking lots [54].

Figure 3.13 indicates a legged mobile robot deployed for performing tasks in unhealthy and difficult situations such as the case in which the area is exposed to fire and flood [71].

3.3.12   Security Services

The application of robotics for security services especially in surveillance and emergency response activities is used to reduce the crime rate and has resulted in a reduction of the cost of crime and victimization [54].

UAVs are commonly used in this field due to their fast mobility [81, 82]. Such robots can interact with cloud space. The exact location of the crimes or emergencies can be sent to the cloud as videos or reports by means of mobile applications. Another application of UAVs in this field is in mountain surveillance to find people and equipment that are buried in snow during avalanches [82].

Figure 3.13 A Legged Mobile Robot Deployed for Performing Tasks in Unhealthy and Rough Conditions [71], figure 02 (p.05) / with permission of JOHN WILEY & SONS, INC.

Similar robotic systems are deployed to control the crowd by means of patrols in charge of monitoring city locations, such as stadiums, stations, and parks in the case of emergencies. These patrols must be alert and process quickly in real time to identify any anomalies that may cause risks to the population and make the decision-making process easier in this kind of event [47, 83].

3.3.13 Social Assistance

The robotic field has also contributed to projects related to social assistance such as helping to provide the basic needs of homeless people. In a project called Raven, [47, 84], an interactive robotic object (IRO) provided social assistance by taking the physical shape of a mailbox in the street, from which people in need could retrieve basic supplies stocked by passersby wanting to help. Another example is wheelchair robots enabling elderly people to move easily. There are also horseback-riding robots used to increase patients' physical strength. A case in point is helping the elderly persons to solve motion problems such as balance difficulties and poor posture [85]. Figure 3.14 shows a social robot that can help people handle their daily tasks [86].

3.3.14 Travel Industry

The most common use of robots deployed in the field of tourism is UAVs, particularly in guiding visitors to tourist attractions. They are also used to transfer audiovisual and informative data to a tourist's mobile applications [47]. Some other advanced applications are based on passenger satisfaction consists of their perceptions and evaluations for new services. For example, a system consists of sensors and data mining techniques

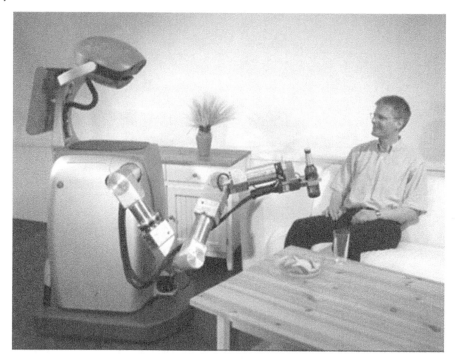

Figure 3.14 A Social Robot Helping a Person to Handle Daily Tasks [86], figure 03.20 (p.94) / with permission of JOHN WILEY & SONS, INC.

is proposed to make improvement in trips with ships in terms of security, comfort, identification, and etc. In this system, passenger satisfaction is evaluated by means of polls on each service.

Robots are also used in hotels and restaurants. For instance, in restaurants, a robot could collect prepared dishes from the kitchen and automatically serve customers.

3.3.15 Human-Interactive Applications

Human-interactive robots can be considered as another category of robotic systems. They are designed for mental therapy, communication (social activity), guidance, welfare, education, entertainment, and so on.These robots involve more interaction with humans than industrial ones. Human-interactive robots can be categorized into (a) performance, (b) teleoperated performance, (c) operation, programming, building, and control, and (d) interactive autonomous control.

3.3.15.1 Performance Robots

Performance robots are mostly developed to provide fun for people. For instance, mechanical dolls were developed in the 18th century to perform tasks like drawing a picture, writing a letter, playing the organ, and dancing. More recently, these kinds of robots have been deployed in amusement parks, museums, movies, and exhibitions [87].

3.3.15.2 Teleoperated Performance Robots

Teleoperated performance robots are controlled remotely by operators. They are more active than performance robots, which do the same job repeatedly. In contrast, teleoperated performance robots are reacting to audience actions by means of commands sent to them by the operators according to audience reactions. They are applicable in amusement parks and exhibitions [88].

3.3.15.3 Operation Robots

These robots are also mostly developed for entertainment. Some examples are a stuffed-animal game machine at amusement centers and matches between robots in football (RoboCup or robot football) and wrestling (RoboOne or robot wrestling; Robo-One Official Homepage. http://www.robo-one.com) [89].

LEGO MINDSTORMS and I-Blocks are robots applied to stimulate children's creativity. Both entertainment and education are involved to achieve these goals [89].

3.3.15.4 Interactive Autonomous Robots

Interactive autonomous robots could connect with humans through verbal or nonverbal communication. Compared with previous categories, such interactions are mostly personalized. One of these robots was developed for entertainment using nonverbal communication to attract people's interest [89]. Others could also produce conversation, showing some facial expressions. Guiding robots in exhibitions and museums [90] and mental commitment robots [88] are other examples in this category.

3.4 Challenge and Future Trends

3.4.1 General Challenges in the Robotic Field

Although robotic systems have experienced striking developments during recent years, some challenges need to be resolved. In the following, three different kinds of robotic systems—locomotion, sensors, and computer vision algorithms—potent for future developments are discussed.

3.4.1.1 Locomotion Systems

Robotic locomotion systems that are used on the ground are mainly wheel-driven. These robots are applicable in industries such as agriculture. However, these systems are strongly affected by the characteristics of environmental barriers such as rocks and branches. Besides, they could have a devastating impact on the local soil, such as soil compaction by the constant locomotion [91]. One solution for locomotion in unstructured environments is to use legged robots [92, 93]. Because legged robots have no constant contact with the ground during locomotion, they are functional in difficult-to-access environments.

3.4.1.2 Sensors

Sensors are generally high-cost modules in robotic systems. In fact, for some cases, industrial prototypes and models may be inapplicable in the real world due to their high-cost sensors. For example, in the agricultural industry, red, green, and blue (RGB; primary colors) cameras are cheaper and provide less information. On the other hand, the RGB-D (thermal, multispectral, and hyperspectral) ones provide more information such as temperature, depth, and other spectral data at a higher price. It is important to mention that the robustness of sensors is an essential issue. In this case, humidity, dust, and temperature could affect sensors' functionality. Therefore, there is a trade-off between quality and financial cost. In each application, the best solution is to choose a sensor with the minimum system requirements that could operate properly. The development of low financial cost sensors to have higher ingress protection and the ability to function properly is a challenge to be addressed [91].

3.4.1.3 Computer Vision Algorithms

Machine learning algorithms such as MLP, RBF, SVM, CNN, and R-CNN [91] and more advanced techniques such as deep learning-based and reinforcement learning algorithms in robotics are important subjects being widely investigated.

Efficient real-time algorithms with low response time have significant potential for industrial application. In agricultural cases, machine learning algorithms are applicable in disease identification, classification of ripeness (ripe–unripe), weed detection, yield estimation, and so forth. The short-term and long-term variations of the crop may have negative impacts on the computer vision algorithm performance. As a result, new computer vision algorithms that adapt to these changes and require devices with low processing power and cost are recommended.

3.4.2 Challenge and Future Trends in Multi-Robot Systems

MRS require inherently complex control algorithms. They are capable of being deployed in many industrial and non-industrial applications due to their outstanding features referred already. Many topics, challenges, and future trends in this field have not been solved yet and include the following.

Most of the control algorithms in multi-robot systems are based on swarm intelligence systems in which it is hard to predict future behaviors of the system. There is the possibility of not achieving desired configurations, or there might be even the risk of instability. A set of parameters (numeric and structural) affects a system's behavior and could lead to phase transitions [94–96] and chaotic behaviors [96, 97]. Modeling these systems is also a challenge due to many unknown disturbances in the dynamic environment [97]. Very few of these challenges, which have limited the use of these complex systems in life-critical applications such as airplane control, have been addressed [97]. There have been several efforts to prove convergence [97, 98], while they often consider its asymptotic properties [98]. However, applicable algorithms with more difficult aspects such as speed of convergence are a future framework in this field.

Another important challenge is that different motions presented in multi-robot systems have diverse parameters, algorithms, and convergence analysis. Each scenario may have a specific cyber-physical system that is suitable for that mission. Some research has considered performing an analysis of typical classes of swarm behaviors, among them foraging [94–96], coverage [97–100], aggregation and pattern formation [100–103], and cooperative tracking [97, 98, 104]. On the other hand, a comprehensive algorithm, which should be the combination of each scenario and its associated cyber-physical system, does not exist. This can be a serious and vital gap in this field because missions often are a combination of different tasks.

Several task-allocation mechanisms capable of being used in multi-robot systems were introduced in [104–107]. However, all these algorithms have been validated in relatively simple scenarios as well as in systems including homogeneous swarms, so more developed and complex algorithms that can be used in heterogeneous swarm systems are needed.

The majority of the works that have been deployed and tested on physical and real scenarios are restricted to the use of simple and small robots. These robots have accessories such as sensors and processors and the small amount of data exchanged between them (e.g., [108–110]). As a result, they are not suitable for complex real-world tasks. On the other hand, some cyber-physical systems such as self-driving cars [111] are too complex in terms of algorithm, mechatronics, and behavioral features to be used in swarm robotic systems. In summary, there is a trade-off between design simplicity and the ability to perform complex tasks reliably and effectively.

In real-world tasks, an important further challenge when multi-robot systems are deployed is system maintenance, which seems to be a more difficult task compared with conventional systems. This task might become easier if the same multiple components are used to monitor the faults and fixing them when needed. Therefore, fault detection and isolation in these systems have become a vital requirement. There have been intensive research efforts in this field [112], though applicable algorithms that are capable of being used in the real world are an outstanding challenge. In this regard, several future trends could be identified.

Animals have been an important source of inspiration in robotic systems, e.g., swarm intelligence widely used in multi-robot systems. Other examples include honeybees [113, 114], fish [115–117], cockroaches [118], and cows [119] in autonomous robotic systems. This is an emerging field due to its specific and outstanding features. There is also recent studies on programmable bacteria whose foraging and collective decision-making enables them to spread in difficult environments [120, 121]. One application of this study is to let a swarm of bacteria transfer drugs to tumors interorganically [122].

Another forthcoming subject is using molecular networks to design complex interactions among agents in a multi-robot system. They could have a magnificent contribution in this field due to their outstanding features. Though complex and enormous, these networks could have several desired features in multi-robot systems such as scalability, resilience, and robustness. There are already several studies in the field of nanonetwork in which non-traditional communication types such aselectromagnetic or molecular communication [123] are used. Swarm robots could get inspired by the molecular communication between cells using mechanisms such as transformation, synthesis, reception, propagation, and emission of molecules [124].

Unfortunately, little work is done to use humans as inspiration for swarm intelligence [111, 125], despite several relatable and rich features like having natural language and developing the most advanced and complex societies and cultures.

3.5 Summary and Conclusion

Robotics has made a great contribution to human development in the 21st century, especially thanks to its magnificent impacts on industrial applications. In this chapter, different classes of robots were categorized in terms of geometry, actuators, control methods, kinematic and motion, number of agents, and robot materials. Multi-robot systems and soft robots have been introduced as advanced topics in the robotic field with their applications. Then, several industrial and nonindustrial robotic applications including the medical industry, space exploration, military, education, agriculture, oil and gas industry, textile, railcar industry, maintenance and repair, construction industry, environmental issues, security service, social assistance, travel industry, and human-interactive applications that have great impacts on Industry 4.0 were reviewed. Finally, several challenges and future trends in the robotic field were reported.

References

1 Nocks, L. (2007). *The Robot: The Life Story of a Technology*. Greenwood Publishing Group.

2 Library, G.N. Robotics-Wikipedia. https://en.wikipedia.org/wiki/Robotics#cite_note-1 (accessed 23 August 2021).

3 Needham, J., Wang, L., Métailie, G., and Huang, H.T. (1954). *Science and Civilisation in China*. Cambridge University Press.

4 Berenguel, M., Rodríguez, F., Moreno, J.C., Guzmán, J.L., and González, R. (2016). Tools and methodologies for teaching robotics in computer science & engineering studies. *Comput. Appl. Eng. Educ.* 24 (2): 202–214.

5 Yamaha Motor Company. SCARA robots (Application examples) – Industrial Robots | Yamaha Motor Co., Ltd. https://global.yamaha-motor.com/business/robot/lineup/application/ykxg (accessed 23 August 2021).

6 Academi, R. Articulated Geometry-21008-Robotpark Academy. http://www.robotpark.com/academy/articulated-geometry-21007 (accessed 23 August 2021).

7 Stancovici, A., Micea, M.V., and Cretu, V. (2016). Cooperative positioning system for indoor surveillance applications. In: *2016 International Conference on Indoor Positioning and Indoor Navigation (IPIN)*, 1–7.

8 Ratnakumar, B.V. and Smart, M.C. (2007). Aerospace applications. II. Planetary exploration missions (orbiters, landers, rovers and probes). In: *Industrial Applications of Batteries*, 327–393. Elsevier.

9 Parker, L.E. (1998). Alliance: An architecture for fault tolerant multirobot cooperation. *IEEE Trans. Robot. Autom.* 14 (2): 220–240.

10 Schneider-Fontan, M., and Mataric, M.J. (1998). Territorial multi-robot task division. *IEEE Trans. Robot. Autom.* 14 (5): 815–822.

11 Brown, R.G. and Jennings, J.S. (1995). A pusher/steerer model for strongly cooperative mobile robot manipulation. In: *Proceedings 1995 IEEE/RSJ International Conference on Intelligent Robots and Systems. Human Robot Interaction and Cooperative Robots*, vol. 3, 562–568.

12 Rossi, A., Staffa, M., and Rossi, S. (2016). Supervisory control of multiple robots through group communication. *IEEE Trans. Cogn. Dev. Syst.* 9 (1): 56–67.

13 Zhang, D., and Wang, L. (2007). Target topology based task assignment for multiple mobile robots in adversarial environments. In: *2007 46th IEEE Conference on Decision and Control*, 5323–5328.

14 Weigel, T., Gutmann, J.-S., Dietl, M., Kleiner, A., and Nebel, B. (2002). CS Freiburg: Coordinating robots for successful soccer playing. *IEEE Trans. Robot. Autom.* 18 (5): 685–699.

15 Gutiérrez, H., Morales, A., and Nijmeijer, H. (2017). Synchronization control for a swarm of unicycle robots: Analysis of different controller topologies. *Asian J. Control* 19 (5): 1822–1833.

16 Rus, D., and Tolley, M.T. (2015). Design, fabrication and control of soft robots. *Nature* 521 (7553): 467–475.

17 Lee, C., et al. (2017). Soft robot review. *Int. J. Control. Autom. Syst.* 15 (1): 3–15.

18 Nishikawa, Y., and Matsumoto, M. (2018). Lightweight indestructible soft robot. *IEEJ Trans. Electr. Electron. Eng.* 13 (4): 652–653.

19 Lin, H.-T., Leisk, G.G., and Trimmer, B. (2011). GoQBot: a caterpillar-inspired soft-bodied rolling robot. *Bioinspir. Biomim.* 6 (2): 26007.

20 Onal, C.D., and Rus, D. (2013). Autonomous undulatory serpentine locomotion utilizing body dynamics of a fluidic soft robot. *Bioinspir. Biomim.* 8 (2): 26003.

21 Tolley, M.T., et al. (2014). A resilient, untethered soft robot. *Soft Robot.* 1 (3): 213–223.

22 Katzschmann, R.K., Marchese, A.D., and Rus, D. (2016). Hydraulic autonomous soft robotic fish for 3D swimming. *Experimental Robotics*, 405–420.

23 Cianchetti, M., Calisti, M., Margheri, L., Kuba, M., and Laschi, C. (2015). Bioinspired locomotion and grasping in water: The soft eight-arm OCTOPUS robot. *Bioinspir. Biomim.* 10 (3): 35003.

24 Calisti, M., et al. (2012). Design and development of a soft robot with crawling and grasping capabilities. In: *2012 IEEE International Conference on Robotics and Automation*, 4950–4955.

25 Sugiyama, Y., and Hirai, S. (2006). Crawling and jumping by a deformable robot. *Int. J. Rob. Res.* 25 (5–6): 603–620.

26 Wang, W., Lee, J.-Y., Rodrigue, H., Song, S.-H., Chu, W.-S., and Ahn, S.-H. (2014). Locomotion of inchworm-inspired robot made of smart soft composite (SSC). *Bioinspir. Biomim.* 9 (4): 46006.

27 Lee, C., et al. (2014). Pneumatic-type surgical robot end-effector for laparoscopic surgical-operation-by-wire. *Biomed. Eng. Online* 13 (1): 1–19.

28 Cianchetti, M., Ranzani, T., Gerboni, G., De Falco, I., Laschi, C., and Menciassi, A. (2013). STIFF-FLOP surgical manipulator: mechanical design and experimental characterization of the single module. In: *2013 IEEE/RSJ International Conference on Intelligent Robots and Systems*, 3576–3581.

29 Cianchetti, M., et al. (2014). Soft robotics technologies to address shortcomings in today's minimally invasive surgery: the STIFF-FLOP approach. *Soft Robot.* 1 (2): 122–131.

30 Deng, T., Wang, H., Chen, W., Wang, X., and Pfeifer, R. (2013). Development of a new cable-driven soft robot for cardiac ablation. In: *2013 IEEE International Conference on Robotics and Biomimetics (ROBIO)*. 728–733.

31 Chou, C.-P., and Hannaford, B. (1996). Measurement and modeling of McKibben pneumatic artificial muscles. *IEEE Trans. Robot. Autom.* 12 (1): 90–102.

32 Martinez, R.V., et al. (2013). Robotic tentacles with three-dimensional mobility based on flexible elastomers. *Adv. Mater.* 25 (2): 205–212.

33 McMahan, W., et al. (2006). Field trials and testing of the OctArm continuum manipulator. In: *Proceedings 2006 IEEE International Conference on Robotics and Automation, 2006. ICRA 2006*, 2336–2341.

34 Calisti, M., et al. (2011). An octopus-bioinspired solution to movement and manipulation for soft robots. *Bioinspir. Biomim.* 6 (3): 36002.

35 Mazzolai, B., Margheri, L., Cianchetti, M., Dario, P., and Laschi, C. (2012). Soft-robotic arm inspired by the octopus: II. From artificial requirements to innovative technological solutions. *Bioinspir. Biomim.* 7 (2): 25005.

36 Majidi, C. (2014). Soft robotics: A perspective—current trends and prospects for the future. *Soft Robot.* 1 (1): 5–11.

37 Polygerinos, P., Wang, Z., Galloway, K.C., Wood, R.J., and Walsh, C.J. (2015). Soft robotic glove for combined assistance and at-home rehabilitation. *Rob. Auton. Syst.* 73: 135–143.

38 Maeder-York, P., et al. (2014). Biologically inspired soft robot for thumb rehabilitation. *J. Med. Device.* 8 (2).

39 Polygerinos, P., Galloway, K.C., Savage, E., Herman, M., O'Donnell, K., and Walsh, C.J. (2015). Soft robotic glove for hand rehabilitation and task specific training. In: *2015 IEEE International Conference on Robotics and Automation (ICRA)*, pp. 2913–2919.

40 Song, Y.S. et al. (2013). Soft robot for gait rehabilitation of spinalized rodents. In: *2013 IEEE/RSJ International Conference on Intelligent Robots and Systems*, 971–976.

41 Asbeck, A.T., Dyer, R.J., Larusson, A.F., and Walsh, C.J. (2013). Biologically-inspired soft exosuit. In: *2013 IEEE 13th International Conference on Rehabilitation Robotics (ICORR)*, 1–8.

42 Park, Y.-L., et al. (2014). Design and control of a bio-inspired soft wearable robotic device for ankle–foot rehabilitation. *Bioinspir. Biomim.* 9 (1): 16007.

43 Sun, Y., Lim, C.M., Tan, H.H., and Ren, H. (2015). Soft oral interventional rehabilitation robot based on low-profile soft pneumatic actuator. In: *2015 IEEE International Conference on Robotics and Automation (ICRA)*, 2907–2912.

44 Cammaroto, G. et al. (2020). Alternative applications of trans-oral robotic surgery (TORS): A systematic review. *J. Clin. Med.* 9 (1): 201.

45 Oueida, S., Kotb, Y., Aloqaily, M., Jararweh, Y., and Baker, T. (2018). An edge computing based smart healthcare framework for resource management. *Sensors* 18 (12): 4307.

46 Rivera-Serrano, C.M., et al. (2012). A transoral highly flexible robot: Novel technology and application. *Laryngoscope* 122 (5): 1067–1071.

47 Mohamed, N., Al-Jaroodi, J., Jawhar, I., Idries, A., and Mohammed, F. (2020). Unmanned aerial vehicles applications in future smart cities. *Technol. Forecast. Soc. Change* 153: 119293.

48 Arias Fisteus, J., Sánchez Fernández, L., Corcoba Magaña, V., Muñoz Organero, M., Fernández-Rodríguez, J.Y., and Álvarez-García, J.A. (2017). A scalable data streaming infrastructure for smart cities. *En JARCA 2016 : XVIII Jornadas de Sistemas Cualitativos y sus Aplicaciones en Diagnosis, Robótica e Inteligencia Ambiental Almería: CEUR Workshop Proceedings.*

49 Roehr, T.M., Cordes, F., and Kirchner, F. (2014). Reconfigurable integrated multirobot exploration system (RIMRES): Heterogeneous modular reconfigurable robots for space exploration. *J. F. Robot.* 31 (1): 3–34.

50 Voth, D. (2004). A new generation of military robots. *IEEE Intell. Syst.* 19 (4): 2–3.

51 Choi, B., Lee, W., Park, G., Lee, Y., Min, J., and Hong, S. (2019). Development and control of a military rescue robot for casualty extraction task. *J. F. Robot.* 36 (4): 656–676.

52 Pachidis, T., Vrochidou, E., Kaburlasos, V.G., Kostova, S., Bonković, M., and Papić, V. (2018). Social robotics in education: state-of-the-art and directions. In: *International Conference on Robotics in Alpe-Adria Danube Region*, 689–700.

53 Papastergiou, M. (2009). Exploring the potential of computer and video games for health and physical education: a literature review. *Comput. Educ.* 53 (3): 603–622.

54 Abbasi, M.H., Majidi, B., and Manzuri, M.T. (2018). Deep cross altitude visual interpretation for service robotic agents in smart city. In: *2018 6th Iranian Joint Congress on Fuzzy and Intelligent Systems (CFIS)*, 79–82.

55 Ampatzidis, Y., De Bellis, L., and Luvisi, A. (2017). iPathology: robotic applications and management of plants and plant diseases. *Sustainability* 9 (6): 1010.

56 Rivera, R., Amorim, M., and Reis, J. (2020). Robotic services in smart cities: An exploratory literature review. In: *2020 15th Iberian Conference on Information Systems and Technologies (CISTI)*, 1–7.

57 Piran, S.J., Majidi, B., and Manzuri, M.T. (2019). Deep interpretation of Parkland environment for autonomous landscaping robot for the green smart city. In: *2019 4th International Conference on Pattern Recognition and Image Analysis (IPRIA)*, 185–189.

58 Suprem, A., Mahalik, N., and Kim, K. (2013). A review on application of technology systems, standards and interfaces for agriculture and food sector. *Comput. Stand. Interfaces* 35 (4): 355–364.

59 Arad, B., et al. (2020). Development of a sweet pepper harvesting robot. *J. F. Robot.* 37 (6): 1027–1039.

60 Yu, L., et al. (2019). Inspection robots in oil and gas industry: A review of current solutions and future trends. In: *2019 25th International Conference on Automation and Computing (ICAC)*, 1–6.

61 Shukla, A., and Karki, H. (2013). A review of robotics in onshore oil-gas industry. In: *2013 IEEE International Conference on Mechatronics and Automation*, 1153–1160.

62 Chen, H., Stavinoha, S., Walker, M., Zhang, B., and Fuhlbrigge, T. (2014). Opportunities and challenges of robotics and automation in offshore oil & gas industry. *Intell. Control Autom.* 5 (3): 136–145.

63 Kawaguchi, Y., Yoshida, I., Kurumatani, H., Kikuta, T., and Yamada, Y. (1995). Internal pipe inspection robot. In: *Proceedings of 1995 IEEE International Conference on Robotics and Automation*, vol. 1, 857–862.

64 Xiao, J., Sadegh, A., Elliot, M., Calle, A., Persad, A., and Chiu, H.M. (2005). Design of mobile robots with wall climbing capability. In: *Proc. IEEE AIM, Monterey, CA*, 438–443.

65 Longo, D., and Muscato, G. (2006). The Alicia/sup 3/climbing robot: A three-module robot for automatic wall inspection. *IEEE Robot. Autom. Mag.* 13 (1): 42–50.

66 Yamamoto, S. (1992). Development of inspection robot for nuclear power plant. In: *Proceedings 1992 IEEE International Conference on Robotics and Automation*, 1559–1560.

67 Merriaux, P., Dupuis, Y., Boutteau, R., Vasseur, P., and Savatier, X. (2018). Robust robot localization in a complex oil and gas industrial environment. *J. F. Robot.* 35 (2): 213–230.

68 Daniyan, I., Mpofu, K., Ramatsetse, B., and Adeodu, A. (2020). Design and simulation of a robotic arm for manufacturing operations in the railcar industry. *Procedia Manuf.* 51: 67–72.

69 Banga, H.K., Kalra, P., Kumar, R., Singh, S., and Pruncu, C.I. (2021). Optimization of the cycle time of robotics resistance spot welding for automotive applications. *J. Adv. Manuf. Process.* e10084.

70 Parker, L.E., and Draper, J.V. (1998). Robotics applications in maintenance and repair. *Handb. Ind. Robot.* 2: 1023–1036.

71 Delmerico, J., et al. (2019). The current state and future outlook of rescue robotics. *J. F. Robot.* 36 (7): 1171–1191.

72 Wong, J., Zhang, J., and Lee, J. (2015). A vision of the future construction industry of Hong Kong. *ISARC. Proceedings of the International Symposium on Automation and Robotics in Construction*, vol. 32, 1.

73 Cheng, J.C.P., and Ma, L.Y.H. (2013). A BIM-based system for demolition and renovation waste estimation and planning. *Waste Manag.* 33 (6): 1539–1551.

74 Das, M., Cheng, J.C.P., and Shiv Kumar, S. (2014). BIMCloud: a distributed cloud-based social BIM framework for project collaboration. *The International conference on Computing in Civil and Building Engineering (ICCCBE)*, 41–48.

75 Wong, J.K.W., Li, H., Chan, G., Wang, H., Huang, T., and Luo, E. (2014). An integrated 5D tool for quantification of construction process emissions and accident identification. *ISARC. Proceedings of the International Symposium on Automation and Robotics in Construction*, vol. 31, 1.

76 Huang, T., Kong, C.W., Guo, H., Baldwin, A., and Li, H. (2007). A virtual prototyping system for simulating construction processes. *Autom. Constr.* 16 (5): 576–585.

77 Holt, M. (2021) Website: P. Robots in UK Construction Industry – Planning and Consulting http://pandcltd.co.uk/robots-in-uk-construction-industry.

78 Adil, S.H., Abd Aziz, A., Akber, T., Ebrahim, M., Ali, S.S.A., and Raza, K. (2017). 3D smart city simulator. In: *2017 IEEE 3rd International Symposium in Robotics and Manufacturing* Automation *(ROMA)*, 1–5.

79 Alexandru, P., Andrei, M., Cristina-Mădălina, S., and Stan, O. (2018). Smart environmental monitoring beacon. In: *2018 IEEE International Conference on Automation, Quality and Testing, Robotics (AQTR)*, 1–4.

80 Righetti, F., Vallati, C., and Anastasi, G. (2018). IoT applications in smart cities: A perspective into social and ethical issues. In: *2018 IEEE International Conference on Smart Computing (SMARTCOMP)*, 387–392.

81 Ermacora, G., et al. (2014). A cloud based service for management and planning of autonomous UAV missions in smart city scenarios. In: *International Workshop on Modelling and Simulation for Autonomous Systems*, 20–26.

82 Silvagni, M., Chiaberge, M., Sanguedolce, C., and Dara, G. (2017). A modular cloud robotics architecture for data management and mission handling of unmanned robotic services. In: *International Conference on Robotics in Alpe-Adria Danube Region*, 528–538.

83 Rahman, A., Jin, J., Cricenti, A., Rahman, A., Palaniswami, M., and Luo, T. (2016). Cloud-enhanced robotic system for smart city crowd control. *J. Sens. Actuator Networks* 5 (4): 20.

84 Tan, H., Yang, F., Tummalapalli, N.T., Bhatia, C., and Barde, K. (2017). Raven: a street robot to address homelessness. In: *Proceedings of the Companion of the 2017 ACM/IEEE International Conference on Human-Robot Interaction*, 397–398.

85 Chen G. L., et al. (2002). Biofeedback control of horseback riding simulator. In Proceedings: *International Conference on Machine Learning and Cybernetics*, vol. 4, 1905–1908.

86 Mann, W.C. (2005). *Smart Technology for Aging, Disability, and Independence: The State of the Science*. John Wiley & Sons.

87 Hirai, K. (1998). Humanoid robot and its applications. In: *IARP 1st Int. Workshop on Humanoid and Human Friendly Robotics, Tsukuba*, V–1.

88 Shibata, T., and Wada, K. (2011). Robot therapy: a new approach for mental healthcare of the elderly–a mini-review. *Gerontology* 57 (4): 378–386.

89 Kitano, H. (1998). *RoboCup-97: Robot Soccer World Cup I*, Vol. 1395. Springer Science & Business Media.

90 Bischoff, R., and Graefe, V. (2004). HERMES-a versatile personal robotic assistant. *Proc. IEEE* 92 (11): 1759–1779.

91 Oliveira, L.F.P., Moreira, A.P., and Silva, M.F. (2021). Advances in agriculture robotics: A state-of-the-art review and challenges ahead. *Robotics* 10 (2): 52.

92 Birrell, S., Hughes, J., Cai, J.Y., and Iida, F. (2020). A field-tested robotic harvesting system for iceberg lettuce. *J. F. Robot.* 37 (2): 225–245.

93 Ge, Y., Xiong, Y., Tenorio, G.L., and From, P.J. (2019). Fruit localization and environment perception for strawberry harvesting robots. *IEEE Access* 7: 147642–147652.

94 Castello, E., et al. (2016). Adaptive foraging for simulated and real robotic swarms: The dynamical response threshold approach. *Swarm Intell.* 10 (1): 1–31.

95 Gazi, V., and Passino, K.M. (2004). Stability analysis of social foraging swarms. *IEEE Trans. Syst. Man, Cybern. Part B* 34 (1): 539–557.

96 Ducatelle, F., et al. (2014). Cooperative navigation in robotic swarms. *Swarm Intell.* 8 (1): 1–33.

97 Hoff, N.R., Sagoff, A., Wood, R.J., and Nagpal, R. (2010). Two foraging algorithms for robot swarms using only local communication. In: *2010 IEEE International Conference on Robotics and Biomimetics*, 123–130.

98 Zhang, F., Bertozzi, A.L., Elamvazhuthi, K., and Berman, S. (2017). Performance bounds on spatial coverage tasks by stochastic robotic swarms. *IEEE Trans. Automat. Contr.* 63 (6): 1563–1578.

99 Schroeder, A.M., and Kumar, M. (2016). Design of decentralized chemotactic control law for area coverage using swarm of mobile robots. In: *2016 American Control Conference (ACC)*, 4317–4322.

100 Mahadev, A., Krupke, D., Fekete, S.P., and Becker, A.T. (2017). Mapping and coverage with a particle swarm controlled by uniform inputs. In: *2017 IEEE/RSJ International Conference on Intelligent Robots and Systems (IROS)*, 1097–1104.

101 Correll, N., and Martinoli, A. (2007). Robust distributed coverage using a swarm of miniature robots. In: *Proceedings 2007 IEEE International Conference on Robotics and Automation*, 379–384.

102 Gasparri, A., Priolo, A., and Ulivi, G. (2012). A swarm aggregation algorithm for multi-robot systems based on local interaction. In: *2012 IEEE International Conference on Control Applications*, 1497–1502.

103 Antonelli, G., Arrichiello, F., and Chiaverini, S. (2010). Flocking for multi-robot systems via the null-space-based behavioral control. *Swarm Intell.* 4 (1): 37–56.

104 Turgut, A.E., Çelikkanat, H., Gökçe, F., and Şahin, E. (2008). Self-organized flocking in mobile robot swarms. *Swarm Intell.* 2 (2): 97–120.

105 Givigi, S.N., Jr., and Schwartz, H.M. (2006). A game theoretic approach to swarm robotics. *Appl. Bionics Biomech.* 3 (3): 131–142.

106 Senanayake, M., Senthooran, I., Barca, J.C., Chung, H., Kamruzzaman, J., and Murshed, M. (2016). Search and tracking algorithms for swarms of robots: a survey. *Rob. Auton. Syst.* 75: 422–434.

107 Brambilla, M., Ferrante, E., Birattari, M., and Dorigo, M. (2013). Swarm robotics: A review from the swarm engineering perspective. *Swarm Intell.* 7 (1): 1–41.

108 Rubenstein, M., Ahler, C., Hoff, N., Cabrera, A., and Nagpal, R. (2014). Kilobot: A low cost robot with scalable operations designed for collective behaviors. *Rob. Auton. Syst.* 62 (7): 966–975.

109 Jdeed, M., Zhevzhyk, S., Steinkellner, F., and Elmenreich, W. (2017). Spiderino-a low-cost robot for swarm research and educational purposes. In: *2017 13th Workshop on Intelligent Solutions In Embedded Systems (WISES)*, 35–39.

110 Arvin, F., Murray, J., Zhang, C., and Yue, S. (2014). Colias: An autonomous micro robot for swarm robotic applications. *Int. J. Adv. Robot. Syst.* 11 (7): 113.

111 Schranz, M., et al. (2021). Swarm intelligence and cyber-physical systems: concepts, challenges and future trends. *Swarm Evol. Comput.* 60: 100762.

112 Christensen, A.L., OGrady, R., and Dorigo, M. (2009). From fireflies to fault-tolerant swarms of robots. *IEEE Trans. Evol. Comput.* 13 (4): 754–766.

113 Polic, M., Salem, Z., Griparic, K., Bogdan, S., and Schmickl, T. (2017). Estimation of moving agents density in 2D space based on LSTM neural network. In: *2017 Evolving and Adaptive Intelligent Systems (EAIS)*, 1–8.

114 Stefanec, M., Szopek, M., Schmickl, T., and Mills, R. (2017). Governing the swarm: Controlling a bio-hybrid society of bees & robots with computational feedback loops. In: *2017 IEEE Symposium Series on Computational Intelligence (SSCI)*, 1–8.

115 Bonnet, F., et al. (2019). Robots mediating interactions between animals for interspecies collective behaviors. *Sci. Robot.* 4 (28): 1–8.

116 Bonnet, F., Kato, Y., Halloy, J., and Mondada, F. (2016). Infiltrating the zebrafish swarm: Design, implementation and experimental tests of a miniature robotic fish lure for fish–robot interaction studies. *Artif. Life Robot.* 21 (3): 239–246.

117 Landgraf, T., Bierbach, D., Nguyen, H., Muggelberg, N., Romanczuk, P., and Krause, J. (2016). RoboFish: increased acceptance of interactive robotic fish with realistic eyes and natural motion patterns by live Trinidadian guppies. *Bioinspir. Biomim.* 11 (1): 15001.

118 Halloy, J., et al. (2007). Social integration of robots into groups of cockroaches to control self-organized choices. *Science (80)* 318 (5853): 1155–1158.

119 Correll, N., Schwager, M., and Rus, D. (2008). Social control of herd animals by integration of artificially controlled congeners. In: *International Conference on Simulation of Adaptive Behavior*, 437–446.

120 Shklarsh, A., Ariel, G., Schneidman, E., and Ben-Jacob, E. (2011). Smart swarms of bacteria-inspired agents with performance adaptable interactions. *PLoS Comput. Biol.* 7 (9): e1002177.

121 Jinek, M., Chylinski, K., Fonfara, I., Hauer, M., Doudna, J.A., and Charpentier, E. (2012). A programmable dual-RNA–guided DNA endonuclease in adaptive bacterial immunity. *Science (80)* 337 (6096): 816–821.

122 Felfoul, O., et al. (2016). Magneto-aerotactic bacteria deliver drug-containing nanoliposomes to tumour hypoxic regions. *Nat. Nanotechnol.* 11 (11): 941–947.

123 Akyildiz, I.F., Jornet, J.M., and Pierobon, M. (2011). Nanonetworks: A new frontier in communications. *Commun. ACM* 54 (11): 84–89.

124 Akyildiz, I.F., Pierobon, M., Balasubramaniam, S., and Koucheryavy, Y. (2015). The Internet of bio-nano things. *IEEE Commun. Mag.* 53 (3): 32–40.

125 Parpinelli, R.S., and Lopes, H.S. (2011). New inspirations in swarm intelligence: A survey. *Int. J. Bio-Inspired Comput.* 3 (1): 1–16.

4

Cloud Computing and Its Impact on Industry 4.0: An Overview

Mahdi Sharifzadeh[a,], Hossein Malekpour[a], and Ehsan Shoja[b]*

[a] Sharif Energy, Water and Environment Institute (SEWEI), Sharif University of Technology, Tehran, Iran
[b] Computer Engineering Department, Sharif University of Technology, Tehran, Iran
* Corresponding author. Sharif Energy, Water and Environment Institute (SEWEI), Sharif University of Technology, Tehran, Iran

4.1 Introduction

There are various cloud-based applications that we may use every day without thinking about them. Every time you share a file on your Google Drive, download a file, use a mobile app, or play an online game, actually you are benefiting from cloud services. All these services are from the servers and some digital spaces that are probably away from the users around the world.

When you use a hard drive or a USB stick for storing your information, you can access it only when your hard drive or USB stick is available, but keeping your information on Google Drive, OneDrive, or AirDrop is much different. In the latter case, you can access it from the Internet with any computer that is connected at any time and in any place. Now, what is cloud computing?

In this chapter, we try to collect some useful information about cloud computing and some related topics such as *grid* and *edge computing*. The contents of this chapter are organized into six sections. At first, *cloud computing* is introduced, some definitions of it are presented, and a historical view of cloud computing evolution is described. Next, the architecture and types of cloud services—infrastructure as a service (IaaS), platform as a service (PaaS), and software as a service (SaaS)—are discussed. The next section represents the common features of the cloud. Then, different deployment models of cloud computing are given, and next some advantages and disadvantages of cloud computing technology are discussed. Finally, a brief review of grid, fog, and edge computing is given.

4.1.1 What Is Cloud Computing?

At first let's look at how analyst firms, IT companies, and academics define *cloud computing*. They attempt to answer what it is exactly—in other words, its main characteristics. Here are some examples:

> Cloud computing is a style of computing in which scalable and elastic IT-enabled capabilities are delivered as a service using Internet technologies [1].

Digital transformation is helping companies rapidly drive efficiency, agility, and connectivity as they use technology to transform their business processes into something easier, faster, and more secure, flexible and profitable. Cloud computing technology is a cornerstone to digital transformation [2].

The idea of delivering personal (e.g., email, word processing, presentations.) and business productivity applications (e.g., sales force automation, customer service, accounting) from centralized servers [3].

The main feature of these definitions is the usage of IT infrastructure and applications as a service for customers.

The following definition by Berkeley RAD Lab shows a different perspective of clouds—from the provider side:

Cloud Computing refers to both the applications delivered as services over the Internet and the hardware and systems software in the datacenters that provide those services. The services themselves have long been referred to as Software as a Service (SaaS). The data center hardware and software is what we will call a cloud. When a cloud is made available in a pay-as-you-go manner to the general public, we call it a *public cloud*; the service being sold is utility computing. We use the term *private cloud* to refer to internal datacenters of a business or other organization, not made available to the general public. Thus, cloud computing is the sum of SaaS and utility computing but does not include private clouds. People can be users or providers of SaaS, or users or providers of utility computing [4].

The National Institute of Standards and Technology (NIST) first defined cloud computing as:

...A pay-per-use model for enabling available, convenient, on-demand network access to a shared pool of configurable computing resources (e.g., networks, servers, storage, applications, services) that can be rapidly provisioned and released with minimal management effort or service provider interaction [5].

After further review and more input on the matter, they revised their definition to the following:

Cloud computing is a model for enabling ubiquitous, convenient, on-demand network access to a shared pool of configurable computing resources (e.g., networks, servers, storage, applications, and services) that can be rapidly provisioned and released with minimal management effort or service provider interaction. This cloud model is composed of five essential characteristics, three service models, and four deployment models [5].

Here is one last concise definition:

Cloud computing is a specialized form of distributed computing that introduces utilization models for remotely provisioning scalable and measured resources [6].

As a general definition, all the infrastructure, hardware, software, and systems designed for scalable remote access to hardware and software services are called *cloud*. In cloud computing, scalable and often virtualized resources are offered as a processing service through local area networks and the Internet. This model focuses on providing some services to the user by demand, in which the user doesn't need special equipment for processing and the location of this processing is not important to them. These services can be compared to a power grid in which the consumer is provided with the necessary energy to use in their electrical devices only by connecting through a socket, without the need to know how electricity is generated and the exact place of its production [7].

According to the previous discussion, a summary of cloud computing features is as follows:

- Virtualization and dynamic scalability on demand are the main features of clouds.
- Cloud computing describes a model for on-demand delivery of computing power based on pay-per-use business models.
- Software, platform, and infrastructure are provided in as a service manner.

Historically, cloud computing was first offered by commercial providers, such as Amazon, Google, and Microsoft. It presents a model on which computing services and resources are viewed as a cloud. The main rule of this model is offering software, platform, and infrastructure *as a service* [8].

In Figure 4.1, a brief historical view of cloud computing evolution is described. In 1999 Salesforce developed the concept of software as a service by delivering services to enterprises by their website. In 2002, Amazon launched Amazon Web Services (AWS), a suite that includes storage, computation, and other services. It can offer an organization tools such as compute power, database storage, and content delivery services. Two years later, it launched Elastic Compute Cloud (EC2) to let some small concerns and

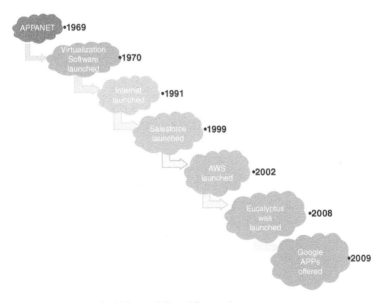

Figure 4.1 Historical Views of Cloud Computing.

users run their computer software in the cloud. In 2008, the first open-source AWS API compatible platform, Eucalyptus was launched and can develop private clouds. And in 2009, Google introduced new application browsers such as Google Apps that provide cloud services.Salesforce.com

The purpose of technologies such as electricity grids, and now cloud computing, is to allow access to large amounts of computing power virtually. These technologies aim to deliver computing as a utility. Utility computing describes a business model for delivering computing power on demand, such that consumers pay providers based on usage (pay-as-you-use). Similarly, when we use some services such as water, electricity, and gas from traditional public utility services, we must pay them based on our usage [9].

Figure 4.2 shows the correlation of technologies that create cloud computing. According to this figure, cloud computing is the result of the advancement in hardware, distributed computing, Internet technologies, and systems management [8].

4.1.2 Concepts and Terminologies

This section briefly presents basic terms and notions of cloud computing and provides an essential understanding of these key concepts.

4.1.2.1 IT Resource

IT resource or simply *resource* in the context of cloud computing refers to any IT-related hardware or software such as storage and network devices, physical and virtual servers, software programs, and services.

4.1.2.2 Cloud

As also stated in the previous section, *cloud* is a term that refers to an ecosystem of remotely accessible IT resources owned and governed by an organization and configured to support rapid provision and measured services.

4.1.2.3 Cloud Provider and Cloud Consumer

Cloud provider and *cloud consumer* are the terms that respectively refer to the owner and the user of the cloud and its resources.

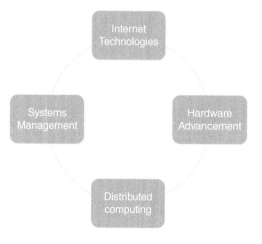

Figure 4.2 Cloud Computing—the Collaboration of Some Technologies. Based on W. Voorsluys et al., 2001.

4.1.2.4 Cloud Service

Cloud services refer to any IT resources (e.g., infrastructures, software, services) hosted by cloud providers and made remotely accessible to users on demand via the Internet. Not all the IT resources within the cloud are required to be remotely accessible.

4.1.2.5 Scaling

The term *scaling* refers to the addition or extraction of IT resources to meet the user's changing demands. There are two types of scaling: horizontal and vertical. Horizontal scaling happens when new nodes or machines are added to the system. In contrast, vertical scaling only strengthens or upgrades the existing nodes. The terms *scaling out* or *scaling in* are respectively used in horizontal scaling to denote the increase or decrease of the nodes. Scaling up or scaling down also refers to the increase or decrease of the resources through vertical scaling, respectively.

4.1.2.6 Virtualization

Virtualization is the process of running several virtual instances of resources over physical machines. Virtualization enables cloud providers to serve the user demands with fewer resources and hence to efficiently utilize their available resources. It also allows cloud consumers to pay only for the number of resources they need.

4.2 Architecture and Types of Services in Cloud Computing

Various architectures for cloud computing are proposed by researchers. Menken recognized seven core parts for cloud computing: application, client, infrastructure, platform, service, storage, and processing power [10]. Miller [11] examined the different ways a company can benefit from cloud computing for the development of its business applications. He considered four types of cloud service development: software as a service, platform as a service, web services, and on-demand computing [11]. Yousef et al. introduced five layers of cloud computing, namely, cloud application, cloud software environment, cloud software infrastructure, software kernel, and firmware/hardware [12].

At first glance, it seems that these classifications are different in some respects. However, in essence, they describe the same phenomena and have a very common origin. In general, cloud computing provides three types of capabilities as a service and delivers three model services: (1) infrastructure as a service; (2) platform as a service; and (3) software as a service.

As shown in Figure 4.3, each layer provides a segment of the cloud's services, but they work together seamlessly. These cloud capabilities are also known as the *cloud layers*. End users can use the IT capabilities of cloud computing by accessing these layers. We adopt this classification *and* will discuss each service in more detail in the next sections. First, the three layers of cloud computing—IaaS, PaaS, and SaaS—and their interconnection are discussed a followed by a brief review of a new service of clouds called *serverless*.

4.2.1 Infrastructure as a Service

As shown in Figure 4.3, the Infrastructure services are considered in the bottom layer of cloud computing systems. In this layer, basic services (server, processor, storage, and other resources) are provided as needed. IaaS can be considered the most straightforward type of cloud-based service. This type of service is used for companies that need a high degree of personalization. The most important advantage of IaaS is its additional capacity, which is available to the users on demand. IaaS enables businesses to provide cost-effective resources as well as quality infrastructure for their development.

Typically, virtualized infrastructure as a service is offered by IaaS providers instead of selling raw hardware infrastructure. PaaS and SaaS providers can access IaaS services based on standardized interfaces [13]. With IaaS, the third party will host infrastructures such as networking equipment, servers, and data storage. However, users will usually have their operating system and interface. A business developer may use an IaaS provider to create an experimental environment before using internal software.

A good example of IaaS is Amazon Web Services, which uses Amazon Simple Storage Service (S3) to store data and EC2 for processing. Joyent is applied to run webpages and rich Web applications [2]. As another example, Sun introduced the Network. Application Catalog in March 2007. It allows software developers and communities to immediately run their applications online.Network.com

4.2.2 Platform as a Service

Platform as a service is the higher-level layer in cloud architecture, above the infrastructure layer. It is a type of cloud computing service in which the service provider offers a platform service to customers and enables them to create and maintain infrastructure such as software, software development, administration, and management without the need to provide commercial services. These services provide more programmability features for customers. IaaS virtualize raw computing and storage services and a platform is a bridge that connects software applications to the infrastructure. Therefore, PaaS is an abstraction layer between IaaS and PaaS. Software developers can write their application codes and upload them to a platform without a need to consider the underlying hardware infrastructure (IaaS) [13].

These services allow developers to create their applications using the tools provided by the cloud provider. They can include preconfigured features that users can subscribe to and use. PaaS is typically delivered in a workplace data center by providing a private cloud, as software or a device in the client firewall. PaaS Hybrid Cloud offers a combination of these two types of cloud services to customers.

An interesting example of PaaS is Nebula. It provides an improved alternative for NASA researchers to share their scientific data sets with other people. Nebula is an open-source cloud computing platform [14]. Another example of PaaS is Map Reduce

Figure 4.3 Three Layers of Cloud Computing.

(MR). This framework offers two functionalities, namely, map and reduction functions. Other well-known examples of PaaS are Google App Engine, Force.com, Windows Azure, and Amazon Elastic.

4.2.3 Software as a Service

Probably the most critical part of cloud computing is software. As shown in Figure 4.3, SaaS is the most visible of the three layers of cloud computing for end users. It is about the actual services that are accessible to users. SaaS is simply a software distribution model in which the service provider hosts customer applications and provides them to customers via the Internet. Many of these end users access cloud services through the Internet using Web browsers or an intranet using individual application programming interfaces (APIs).

This type of cloud computing does not require the download and installation of software on each user's personal computer and saves technical staff time. Also, services such as repairs, maintenance, and troubleshooting are performed entirely by the service provider. In this method, the user can use the software with an account on the Internet. It is also the responsibility of the service provider to update and maintain the software tool. Typically, users of a SaaS service have no knowledge or control over basic infrastructure, either being a software platform based on PaaS or other hardware infrastructures [13]. Examples of this type of service include Google apps such as Google Docs, Google Play, and Gmail.

4.2.4 Serverless Cloud Computing

In addition, Google App Engine introduced another type of cloud computing service called serverless [15], a cloud computing model in which the servers are in a cloud service and dynamically manage resource allocation. For instance, nowadays, every person, company, or organization needs to use a website. On the other hand, hosting Web services requires knowledge, expertise, and the support of experts. Serverless processing is the use of cloud Web services without using a traditional server and wasting all system resources. The administrator of a website will have no worries about site hosting settings, maintenance, support, and uptime of their server. It should be noted that the word *serverless* does not mean that there is no server, but it does mean that the user will not need to provide the server and manage and maintain it.

Serverless computing enables developers to build applications faster because they don't need infrastructure management. Cloud service providers can use serverless applications to automatically manage, scale, and provision the infrastructure needed to run the code. The server is still executing code, but the tasks related to infrastructure configuration and management are invisible to developers. Serverless computing allows developers to focus on main works and helps teams to promote their productivity. It also allows organizations to optimize their resources and focus on their innovations [2]. Serverless computing has various benefits including:

- No infrastructure management: Users don't need to know how to manage servers or apply complex settings; they can simply pay a monthly fee.
- Faster time to market: Helps teams promote productivity and increase sales.

- Cost savings: Users pay for the service they use during the month. For example, a website with low monthly traffic does not need to rent a physical server with as many features as possible.
- Dynamic scalability: The infrastructure dynamically scales up and down seamlessly based on workload demands.
- More efficient use of resources: Organizations can optimize resources and stay focused on innovation.

4.2.4.1 Serverless Application Patterns

Serverless applications are built using various patterns by developers, including [2]:

- Serverless workflows
- Serverless application environments
- Serverless functions
- Serverless Kubernetes
- Serverless API gateway

4.3 Main Characteristics

The most common features of the majority of cloud environments are (1) multitenancy and resource pooling; (2) on-demand usage; (3) elasticity; (4) broad network access; (5) measured usage; and (6) resiliency.

Note that resiliency is not included in the NIST definitions as one of the key characteristics of cloud computing environments [6, 7]. Therefore, in some sources, only five characteristics are discussed. Cloud platforms can be measured or compared based on their value offerings for each of these features. Figure 4.4 depicts the six aforementioned characteristics of clouds, and they are discussed in the following.

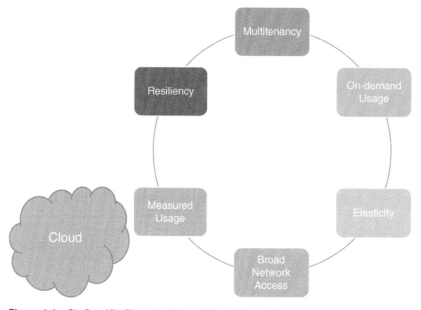

Figure 4.4 Six Specific Characteristics of Cloud Computing.

4.3.1 Multitenancy and Resource Pooling

A software operation mode where a single program serves different customers (tenants) is called multitenancy. Thus, in a multitenant environment, several different customers can access and share the same program. By using multitenancy models, cloud providers can pool their IT resources to serve multiple consumers. Resource pooling allows cloud providers to dynamically assign and reassign resources such as storage, processing, memory, and network bandwidth based on the consumer's demand.

4.3.2 On-Demand Usage

On-demand usage, also known as *on-demand self-service*, refers to a service granted by cloud providers to cloud consumers to unilaterally (i.e., without requiring asking for permission) self-provision the IT resources on demand and whenever required. Once configured, these can be done automatically without requiring any further human interaction with the cloud consumer (service provider) or cloud provider.

4.3.3 Elasticity

The ability of the cloud to rapidly (and in most cases automatically) adapt to the workload changes by scaling out or in the IT resources based on the demand or other predetermined conditions is called *elasticity*. This feature gives the illusion of infinite resource availability to the consumers and is one of the distinctive characteristics of cloud computing that separates it from older computing paradigms such as grid computing. Cloud elasticity also enables a pay-as-you-go cost model where the customer pays only for the resources they consume.

4.3.4 Broad Network Access

Ubiquity is another important characteristic of cloud environments. This feature represents the ability of the cloud service to be widely accessible. This means that resources should be accessible through standard mechanisms and protocols that are available on various thin or thick client platforms.

4.3.5 Measured Usage

The ability of a cloud to automatically keep track of the usage of its resources by an abstract metering capability is a characteristic denoted by *measured usage* or *measured service*. Based on this feature, the cloud provider can monitor, control, measure, and report the usage of its provisioned services. This will give transparency and provide both the consumer and the provider with an account of what has been used. Hence, the cloud provider would be able to charge the consumer only for the actual consummation of the resources, taking into account the duration that the resources were granted. Note that the measurements are not necessarily used for billing purposes. Measured usage can also be used for the sake of general monitoring, related usage reporting, or predictive planning.

4.3.6 Resiliency

Resiliency refers to the ability of a system to provide continuous operation, operate under adverse conditions, and rapidly recover in the face of failures. In cloud computing, resiliency is the use of redundant implementations of resources as a form of failover across different physical locations (spread within the same cloud or across multiple clouds). This feature can also increase the availability and reliability of a cloud service.

4.4 Cloud Deployment Models

4.4.1 Types of Clouds

In this section a brief description and review of different types of cloud computing is presented.

4.4.1.1 Public Clouds

Public clouds have the most diverse functionalities; however, they are cost-effective, and there is no obligation to pay the prerequisites after the work is done. A public cloud is for Web development systems and is managed by a third party that stores computing resources such as servers and storage space and accesses this data using a Web browser. Public clouds are paid for by people who receive services from them.

Public clouds offer benefits such as easy scalability, high reliability, easy management, and cost-effectiveness, but they are not the safest option for sensitive data [2]. Generally, they are used for (1) scalable and flexible IaaS for storage and compute services and (2) powerful PaaS for developing and implementing cloud-based applications. Google is the most popular public cloud, and it offers a wide range of cloud computing services to the general public.

4.4.1.2 Private Clouds

Private clouds offer privacy and are not designed to be shared with everyone. A private cloud provides the same general cloud advantages, but it uses its proprietary hardware. The private cloud means using cloud infrastructure only by a client or organization [16]. Privacy and security in the private cloud are so that it offers full control on the accessibilities. This type of cloud belongs exclusively to one person or one company and can be physically located on the company site [17].

Private clouds provide a self-service interface that controls services, allowing IT employees to quickly provide the resources they need, allocate, and provide them on demand. They also provide fully automated resource pool management from computing to storage, analysis, and middleware. In a private cloud, complex security is designed for specific company conditions. [18]

There are some advantages of using a private cloud for clients or organizations, such as high-level security, more control on the server, and adjustability [2]. The disadvantages of private clouds include the high costs of installation and maintenance and limited remote access. Hewlett Packard Enterprise (HPE), VMware, Microsoft, and IBM are the major private cloud vendors [18].

4.4.1.3 Hybrid Clouds

The architecture in which multiple clouds communicate (community, private, or public) is called a hybrid cloud. This type of cloud infrastructure is a combination of some information technology architectures, which includes a degree of management capability and workload in two or more environments. This type of cloud is the best because the cloud space can be both private and public. It is very convenient when you want to transfer data. Hybrid clouds are limited by the technology that allows applications and support data to move between private and public clouds, giving businesses more flexibility [13].

Cloud hybrids are used when cloud burst occurs, meaning that resources are limited and cannot be developed so that public resources must be used by connecting the organization's private cloud to the public cloud. More sensitive applications are placed on the private cloud, and the rest are placed on the public cloud to reduce costs [2].

Hybrid clouds are also employed in distributed networks to access information remotely and share data. Hybrid technology enables employing the computational capacity of the other servers that are connected to the network. For example, an organization may use a private cloud as the main website server, and for other applications like sharing multimedia files and working with heavy data it could employ a hybrid cloud. Some cloud providers can also connect to develop a hybrid cloud [17]. These providers are separate organizations but have limited connections due to a data transfer technology.

The top features of a hybrid cloud include allowing companies to store important applications and sensitive data in traditional data centers or private clouds. In addition, a hybrid cloud can provide benefits from public cloud resources such as SaaS for the latest applications and IaaS for flexible virtual resources. It also facilitates the portability of data, applications, and services, and provides more options for implementation models.

The deployment of a hybrid cloud has some advantages [2] include a high degree of flexibility and scalability, affordability, and advanced security, but network-level communications may conflict because it uses both private and public cloud technology.

Hybrid cloud service providers such as Google and Microsoft make it easy for users to connect local resources to the public–private cloud. This allows the so-called hypervisor layer to be created, where virtual machine mechanisms are created [19].

4.4.1.4 Community Clouds (Federations)

This type of cloud enables sharing information between multiple organizations with common interests and minimizes costs as costs are shared between consumer companies. For example, this concept aims to allow multiple people in an organization to work on a joint program or a project. In a federation architecture, a centralized cloud infrastructure solves the specific problems of the business sector by combining the services provided.

An example of utilization of a community cloud is federal agencies in the United States that share similar requirements using a common network [20].

4.4.2 Public vs. Private Clouds

A common feature of both public and private clouds is that one needs to pay for the use of the feature. Data centers provide hardware foundations and are usually associated with cloud drift. For this reason, companies virtualize them to increase flexibility and deployment. Through data virtualization, companies can increase their server usage by up to 35% [13].

The difference is that a public cloud (e.g., Google services) is not limited to a single user, where they are exposed to consumers as public payment. However, a private cloud could be dedicated only to a person or one company that has complete control over it. The advantage of a public cloud is that it achieves all the benefits of virtualization while maintaining complete control over the infrastructure

4.4.3 Hybrid Clouds vs. Cloud Federations

Single clouds can be combined. For example, in a hybrid cloud a combination of public and private clouds allows an organization to run some data on public infrastructure and some on private infrastructure. A mixed cloud space monitors the distribution of internal and external infrastructure and their security in the distribution of applications in different environments. The term *cloud federation* means cooperation between two public clouds, even if a private cloud is involved. Infrastructure providers should provide extensive and comparable computing resources. A super federation is a collection of individual clouds that can work together through standard interfaces according to basic principles [13].

4.5 Points, Risks, and Challenges

Cloud computing is an emerging technology that is used by almost all existing businesses. Whether the technology is public, private, or hybrid, *cloud computing* has become a key driver of competition between companies. In this section, the advantages and risks associated with cloud computing are discussed.

4.5.1 Advantages

4.5.1.1 Cost Savings
Perhaps the most important advantage of cloud computing is its IT cost savings. No matter the type or size of a business, when it comes to minimizing capital and operating costs, it will be profitable to utilize a cloud computing infrastructure.

For companies, especially start-ups, cost control is crucial for survival in competitive markets. In the past, only large companies could physically create, build, and install software and hardware, especially at a very high cost. Thus, small and medium-sized companies that did not have such possibility automatically would withdraw from the competition. However, the benefits of cloud computing have made it accessible even for small companies to remotely access their software needs. So they pay both for their consumption and do not need to spend money to build infrastructure [19].

By minimizing up-front financial investments, cloud computing allows organizations to redirect these expenses to their core business or make additional investments in special features only. Such cost reduction is not limited just to the procurement of resources namely hardware, software, or networking infrastructure. Cloud computing also offers the ability to pay only for the capacity needed and eliminates infrastructure running, housing, purchasing, maintenance, and management costs [20].

Another cost-saving situation can be linked to the availability of batch processing in the cloud platforms. For example, consider a situation where a company has to complete some batch-centric tasks or perform parallel processing. The company should either heavily invest in the acquisition of its resources or accept the cost of delays and more processing hours. On the other hand, the company could use a cloud platform and do the jobs on the cloud's almost "unlimited" resources in a fraction of time.

In a nutshell, the cost benefits of cloud computing can be listed as follows:

- Lower initial capital investment
- Higher degrees of utilization and the pay-as-you-go cost model
- Lower maintenance and staff costs
- Lower energy costs and carbon footprint
- Higher degrees of resiliency without expensive investment in redundancy
- Fewer time delays

4.5.1.2 Scalability and Elasticity

Cloud scalability and elasticity avoid lacking or wasting resources. As stated earlier, scalability is the ability of a system to increase or decrease its resources, and elasticity is the ability of the system to take advantage of its scalability.

The ability of the cloud to automatically increase or decrease scalability to meet user demands also can save businesses from unpredictable demands during peak times. Another advantage of scalability and elasticity is the flexibility that it brings to the organizations. Normally, it is not easy to handle business growth and adapt to shifting requirements and unexpected spikes in demands. Cloud scalability and elasticity would allow the companies to easily and rapidly adjust to changing demands.

Software tools made available on the cloud are highly customizable and increase the power and storage space requirements. Adopting the bandwidth to the user requirements in cloud-based software is very convenient. Clouds allow a user to automatically change workloads at the right time according to traffic and workload, without compromising performance [21].

4.5.1.3 Accessibility

Cloud computing allows users to access the information they need at any time and place using any device with an Internet connection. Using a cloud infrastructure increases the efficiency and productivity of different businesses. It can also facilitate collaboration and sharing between users in different locations. The importance of this feature was especially underlined during the COVID-19 pandemic. When everyone was in quarantine, cloud-based solutions became more and more popular due to their broad accessibility.

4.5.1.4 Agility

By clouds, a server can be prepared to be set up in a few minutes, whereas buying and using the same server in one place can take weeks or months [21]. This agility offers opportunities to try out new ideas and transform businesses in the dynamic market.

4.5.1.5 Reliability

Users need to have control over situations where there is a possibility of sudden events. Care must always be taken in maintaining the data so that it can be retrieved in the event of a problem. The use of cloud services reduces the occurrence of such situations because the information is stored in the cloud and backups are made regularly. Data recovery is also easy. Thus, data backup, disaster recovery, and business continuity are made easier with clouds [2].

By using a platform as a managed service, cloud computing can be much more secure and compatible than an internal IT infrastructure. Most cloud service providers offer a 99.99% trust level service in a service-level agreement (SLA) to their users.

4.5.1.6 Security

In 2017, Gartner predicted the security improvement of public clouds: "Through 2020, public cloud infrastructure as a service (IaaS) workloads will suffer at least 60% fewer security incidents than those in traditional data centers" [22]. The information stored on each server is highly confidential, which is why many companies upload their confidential information to the cloud. Cloud service providers should provision special sets of policies and techniques for enhancing the security of the user's systems and protecting their data, software, and hardware from potential threats [2].

4.5.2 Risks and Challenges

Despite its high speed, efficiency, and creativity, cloud computing also has disadvantages. Some of the risks and challenges associated with cloud computing are now discussed.

4.5.2.1 Security Issues

Cloud service providers implement the best security information to store important information; however, all the sensitive information of organizations is sent to a third party, that is, the provider of cloud computing services. When data is sending over the cloud, there is a possibility that the information will be hacked or stolen by hackers.

Users often have concerns about the security of these spaces, especially when it comes to their medical records or financial background. There are rules for cloud computing that must ensure the security of users. The Federal Risk and Authorization Management Program (FedRAMP) provides a standardized approach to security authorizations for cloud service Offerings. For more information see the FedRAMP webpage [20].

4.5.2.2 Less Control

Although cloud computing companies will grant access for organizations to view CPU and RAM performance status and to personalize the space, users' accessibility and control are still very limited. If there is a problem or error in the programming code or there is a problem in the hardware, it will be difficult to detect and diagnose it. Administrative tasks, such as server access, update, and middleware management, may also not be allowed.

4.5.2.3 Migration

Another main challenge for companies using cloud services is the process of transferring existing applications to new servers [23]. Because of some new conditions that might occur, companies and users that use cloud services always are concerned about how they can change their service provider to a new supplier with the minimum amount of time, charge, and by no affection on their data loss.

4.5.2.4 Vendor Lock-In

Open standards are necessary for all practices of using the Web as a "pervasive computer." With the increase in the number of cloud service providers, portability has become even more important. If a company is dissatisfied with one provider's service or if the service provider withdraws from the business, transferring may not be easy or cost-effective and the company should reformat the data and applications and convert them. Transferring to a new provider is complex and time-consuming, and it might be needed to hire qualified staff to bring services into the company. In addition, users have become increasingly dependent on the Web and its providers, so when service providers change their terms of use or operating methods over time, their users will feel trapped and helpless [24]. Examples include imposing new restrictions on the use of a feature or disabling it for several months to improve it. Providers may be also keen to remove a feature that has been offered for free when the demand is high.

4.5.2.5 Accessibility Compromised

Since cloud service providers offer service every day to many users, unpredicted demands may cause sporadic problems or, worse, power outages. These factors can lead to the temporary suspension of users' business processes. In addition, to use the cloud computing service, access to a permanent connection to the Internet is a basic requirement, and it is a limitation for the user of the cloud service.

4.5.2.6 Legal Issues

Normally, cloud consumers are not aware of the whereabouts and the physical location of their IT resources. Cloud providers may establish their data centers and server farms in affordable locations with cheap energy and infrastructure. There is a valid concern that a user or a company's data may end up in a country with different regulations and policies, such as weaker data protection policies than its origin country.

4.6 Grid, Edge/Fog, and Cloud Computing

4.6.1 Grid Computing

In the 1990s, the Cloud Computing Association introduced grid computing. Such architecture emerges when strong computers are connected by fast data communication to support scientific, computational, and big data applications, and its aim is to leverage existing computing resources available for complex computing by geographically distributed sites [25].

In simple terms, grid computing offers the ability to use networked systems to enhance computational capabilities. Here, it is possible to create a strong and large "virtual" device using the services and features of the systems to perform large computing operations that one device or system is unable to handle on its own. Grid computing is a method for processing large computational tasks. Here, the system can perform its massive processing by distributed computers around the world that are connected to this network. This network uses computer resources in times of unemployment. It is the technology of sharing network resources of different and heterogeneous domains based on receivable services and is considered as a middle layer that allows secure and efficient calculations and data transfer, which is broadly accepted by the industry. In grid computing, instead of having expensive dedicated resources, most of which will usually be unemployed, low-cost resources are provided from distributed resources over a network.

Foster and Kesselman defined grid computing as "a computational grid is a hardware and software infrastructure that provides dependable, consistent, pervasive, and inexpensive access to high-end computational capabilities" [26]. IBM also defines grid computing as "...The ability, using a set of open standards and protocols, to gain access to applications and data, processing power, storage capacity and a vast array of other computing resources over the Internet" [25]. Insight Research defines it as "a form of distributed system wherein computing resources are shared across networks" [27], and with extensive calculations capabilities to integrate computer systems, on-demand, shrinking the data and functionality that is not available to an individual or group of machines.

Grid computing provides benefits and opportunities at both the IT management and business levels [13]. Processing resources, data storage, applications, scientific tools, and communications can be shared in a grid [28]. This architecture offers direct access to computers and software. Such resource sharing may be relevant to other areas of application, so it should be presented in a general way and not just for specific applications only for the sake of high performance.

In general, three approaches are presented: (1) improving the efficiency of applications; (2) data access; and (3) improving the services available in the grid for the design of grid architecture. With this view, grid systems are divided into three parts:

- Computational grid models
- Data grids
- Service grids

A resource-focused view classifies grids into the following categories [29]:

- Compute grids, which focus on computing resource sharing
- Application grids, which focus on management and access to remote software and libraries
- Data grids, which focus on data accessibility and storage management
- Service grids, which support the sharing of services efficiently

Here, resources are shared through cluster grids (e.g., Linux cluster of diaDexus in 2003), enterprise grids (e.g., grid of the pharmaceutical company Novartis in 2003), community and partner grids (e.g., White Rose Grid in 2009), and utility grid services (e.g., Sun Grid Compute Utility in 2006) [13].

4.6.2 Relationship between Grid and Cloud Computing

Cloud computing has indeed taken some of its features from *grid computing*, but the two concepts should not be confused. The most important difference is that in grid computing, the computer and other user's devices also participate in the processing operations. However, in cloud computing a user does not need to do any processing work, requiring only a very simple hardware interface to meet all its needs. In fact, in cloud computing all services are fully managed by the server, and the user does the minimum work just to connect with the cloud environment. Given that all processes in cloud computing are performed by the server, the user sees only the result. Table 4.1 shows some differences between grid and cloud computing.

Cloud and grid computing can complement each other. Cloud evolved from the grid and relies on the grid [13].

4.6.3 Edge Computing

In the past, cloud computing was considered because of its flexibility and cost-effectiveness, in which network management was focused. Today, with the development of the Internet, cloud computing addresses new challenges.

The reason for the delay of remote program servers is due to the additional load of the radio access networks (RANs) [30]. In addition, Internet of things (IoT) devices are increasingly being used. Things like capacity constraints, intermittent connectivity, security, and precise latency in cloud computing can't be fixed properly [31]. Large-scale network of things and device and network-level heterogeneity have created new challenges in IoT [32, 33]. To meet these challenges, an advanced cloud computing model is needed that transcends centralized architecture and reduces constraints.

To integrate IoT with cloud capabilities, data is generated by IoT sensors. Big data is used to determine reactions to events in the analysis. However, sending all files and data over the cloud requires a high-bandwidth network. A resolution is the so-called network edge capability to support IoT systems [34].

In edge computing, massive amounts of data don't transmit to the centralized cloud infrastructure, and they can be processed at the network edges to reduce energy consumption and better bandwidth usage. Edge computing is faster and has better response quality than cloud computing, so edge computing can be considered for IoT infrastructures [35, 36].

Cisco defined edge computing as using some new technologies, resources and applications such as 5G and IoT under the precise and very fast networked system [37].

Emerging technologies in computing and communications—software defined networking (SDN) and network function virtualization (NFV) [38]—transfer centralized cloud computing to the edge networks [34]. NFV enables edge devices to provide computing services and network operations by creating multiple virtual machines (VMs) [39].

Table 4.1 Comparing Grid and Cloud Computing.

Grid Computing	Cloud Computing
Assigns multiple servers to work	Virtualizes servers to compute multiple tasks simultaneously
Used to run a program for a limited time	Can provide a higher level of details
Tool for unifying resources and computations	Growing trend in the external deployment of IT resources

Significant growth has been observed in the investment of industrial networks and in network-focused researches [39]. For instance, to reduce latency, IoT providers such as mobile operators are prone to deploying the applications at the edge of the network [39]. A new architectural element called cloudlets should be used to support low latency requirements [40]. In partnership with Carnegie Mellon University, Vodafone, Intel, and Huawei launched Open Edge Computing (OEC) in June 2015 to accelerate the development of cloudlets [39].

4.6.3.1   Edge Computing vs. Cloud Computing

The computational process of edge computing is different from cloud computing. In cloud computing, the information must transfer to the central data center very quickly, such that the decision-making procedure is not disrupted, but in edge computing the data is processed at the device level. Edge computing is preferable to cloud computing for organizations because it is better for real-time decision-making systems. Cloud computing is used when data processing is not time-driven. In addition to latency, for distributed networks and places where centralized data center connectivity is limited or impossible, edge computing is preferred to cloud computing.

Different types of edge computing technology are fog computing, multi-access edge computing (MEC), emergency response units, cloudlets, and microdata centers [37].

4.6.4   Fog Computing

Fog computing is another technology that is very close to edge computing. It was introduced by Cisco in 2012 [41], and in November 2015 ARM, Cisco, Dell, Intel, Microsoft, and Princeton University founded the OpenFog Consortium to accelerate its progress [42].

As shown in Figure 4.5, cloud computing can be related to the IoT devices by the fog network. These devices are called fog nodes, and they can be located anywhere with a network connection. A node can be any computing device with storage and network connection. For example, industrial controllers, routers, switches, and embedded servers are fog nodes [39].

Similar to edge computing, in fog computing IoT data is also analyzed close to the source of data where it is collected and thus minimizes latency. It unloads a lot of network traffic from the main network and holds sensitive data inside the network. It also monitors sensitive information by analyzing specific IoT information [43].

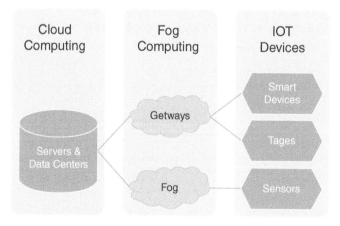

Figure 4.5   Real-Time Computing Done by Fog While Intense Computing Is Processed in the Cloud.

4.6.4.1 Relationship between Edge Computing and Fog Computing

Both fog and edge computing push data toward analytics operating systems that are either where the data is located or close to where it is located. Both technologies help organizations not only to use the cloud to analyze data but make decisions very fast and effective. The main difference between edge computing and fog computing is the placement of the data processing [44–46].

Jessica Califano, head of marketing and communications at Temboo, says, "Fog computing and edge computing are effectively the same things. Both are concerned with leveraging the computing capabilities within a local network to carry out computation tasks that would ordinarily have been carried out in the cloud" [47].

Cisco describes the relationship between fog and edge computing this way: "Fogging enables repeatable structures in the edge computing concept so that enterprises can easily push compute power away from their centralized systems or clouds to improve scalability and performance" [37].

Paul Butterworth, co-founder at Vantiq declares the relation between fog and edge computing very well. He uses the term edge computing when the computing process is physically close to the sensor and devices and considers fog computing as a machine that may be physically more distant from the sensors and actuators and transport the edge computing activities to processors that are connected to the LAN or into the LAN hardware [47].

4.7 Summary

This chapter presents an overview of cloud computing. Anyone who uses the Internet has profited from cloud computing facilities. Some new technologies such as 5G, IoT, smart home, smart cities, and Industry 4.0 are based on cloud computing and related techniques, including grid, fog, and edge computing. The focus of cloud computing is to provide capabilities and facilities as a service on-demand to customers. In general, cloud computing can provide three model services, namely, infrastructure as a service (IaaS), platform as a service (PaaS), and software as a service (SaaS). The main characteristics of all these model are dynamic scalability, on-demand usage, and virtualization. Cloud computing is deployed in four types: public, private, hybrid, and community. Each of them has advantages and disadvantages and is used for a special purpose based on their corresponding properties. The concept of cloud computing is recently extended to the grid, fog, and edge computing. Cloud and grid computing can complement each other. Cloud evolved from the grid and relies on the grid. Edge computing is useful when it is needed to avoid delays while using a communicating medium. Additionally, Edge computing is preferred to cloud computing for distributed systems.

References

1 Gartner Glossary. (2021). Cloud computing. https://www.gartner.com/en/information-technology/glossary/cloud-computing is a style, a service using Internet technologies.

2 Microsoft. (2021). What is cloud computing? https://azure.microsoft.com/en-us/overview/what-is-cloudcomputing/.

3 Rangan, K., Cooke, A., Post, J., & Schindler, N. (2008). The Cloud Wars : 100 + billion at stake. Analyst (May), 1–90.

4 Armbrust, M., et al. (2009). Above the clouds: A Berkeley view of cloud computing.

5 Mell, P.M., and Grance, T. (2011). *SP 800-145. The NIST Definition of Cloud Computing*. National Institute of Standards & Technology.

6 Erl, T., Puttini, R., and Mahmood, Z. (2013). *Cloud Computing: Concepts, Technology, & Architecture*. Pearson Education.

7 Armbrust, M., et al. (2010). A view of cloud computing. *Comm. ACM* 53 (4): 50–58.

8 Voorsluys, W., Broberg, J., and Buyya, R. (2011) Introduction to cloud computing. In: *Cloud Computing: Principles and Paradigms* (ed. R. Buyya, J. Broberg, and A. Goscinski), 1–41. John Wiley & Sons Ltd.

9 Buyya, R., Yeo, C. S., Venugopal, S., Broberg, J., and Brandic, I. (2009). Cloud computing and emerging IT platforms: Vision, hype, and reality for delivering computing as the 5th utility. *Futur. Gener. Comput. Syst* 25 (6): 599–616.

10 Blokdijk, G., and Menken, I. (2009). Cloud Computing - The Complete Cornerstone Guide to Cloud Computing Best Practices: Concepts, Terms, and Techniques for Successfully Planning, Implementing. Cloud Computing Technology - Second Edition.

11 Miller, M. (2009). Cloud computing : Web-based applications that change the way you work and collaborate online. *Que Publ.* (1): 45–49.

12 Youseff, L., Butrico, M., and Da Silva, D.(2008). Toward a unified ontology of cloud computing.

13 Stanoevska-Slabeva, K., and Wozniak, T. (2010). Cloud basics – An introduction to cloud computing. In: *Grid and Cloud Computing: A Business Perspective on Technology and Applications* (ed. K. Stanoevska-Slabeva, T. Wozniak, and S. Ristol), 47–61. Springer Berlin Heidelberg.

14 Riso, A.M., and Curtis, G. (2010). NASA cloud computing platform: Nebula. https://www.nasa.gov/offices/ocio/ittalk/06-2010_cloud_computing.html#.

15 Evans, J. (2015). Whatever happened to PaaS? TechCrunch.

16 Cisco. (n.d.). What is cloud computing? https://www.cisco.com/c/en/us/solutions/cloud/what-is-cloud-computing.html.

17 Dobran, B. (2018). What is cloud computing in simple terms? Definition & examples.

18 JavaTPoint. (2011–2021). Types of cloud. https://www.javatpoint.com/private-cloud.

19 Google. (n.d.) What is cloud computing? https://cloud.google.com/learn/what-is-cloud-computing.

20 fedRAMP. (n.d.). Securing cloud services for the federal government. https://www.fedramp.gov.

21 IBM. (n.d.). Cloud computing: A complete guide. https://www.ibm.com/cloud/learn/cloud-computing-gbl#:~:text=Cloud computing and IBM Cloud,%2C IoT%2C blockchain and more.

22 Gartner. (2017). *Cloud Strategy Leadership*. Gartner, Inc. https://www.gartner.com/imagesrv/books/cloud/cloud_strategy_leadership.pdf, p.14.

23 Durcevic, S. (2019). Cloud computing risks, challenges & problems businesses are facing. https://www.datapine.com/blog/cloud-computing-risks-and-challenges/.

24 Cloudflare. (2021). What does "vendor lock-in" mean? https://www.cloudflare.com/learning/cloud/what-is-vendor-lock-in.

25 PartnerWorld for Developers. (n.d.). IBM Solutions Grid for Business Partners: Helping IBM Business Partners to Grid-enable applications for the next phase of e-business on demand. ibm.com/partnerworld/developer, p.2.

26 Foster, I., and Kesselman, C. (2001). Computational grids. In: *Vector and Parallel Processing – VECPAR 2000*, 3–37.

27 Anonymous. (2006). *Grid Computing: A Vertical Market Perspective 2006–2011*. Insight Research Corporation.

28 Stanoevska-Slabeva, K., and Wozniak, T. (2010). Grid basics. In: *Grid and Cloud Computing: A Business Perspective on Technology and Applications* (ed. K. Stanoevska-Slabeva, T. Wozniak, and S. Ristol), 23–45. Springer Berlin Heidelberg.

29 Baker, M., Buyya, R., and Laforenza, D. (2002). Grids and grid technologies for wide-area distributed computing. *Softw. – Pract. Exp.* 32 (15): 3–5.

30 Peng, M., and Zhang, K. (2016). Recent advances in fog radio access networks: Performance analysis and radio resource allocation. *IEEE Access*: 5003–5009.

31 Ganz, F., Puschmann, D., Barnaghi, P., and Carrez, F. (2015). A practical evaluation of information processing and abstraction techniques for the Internet of things. *IEEE Internet Things J* 2 (4): 340–354.

32 Brogi, A., and Forti, S. (2017). QoS-aware deployment of IoT applications through the fog. *IEEE Internet Things J* 4 (5): 1185–1192.

33 Shi, W., Cao, J., Zhang, Q., Li, Y., and Xu, L. (2016). Edge computing: Vision and challenges. *IEEE Internet Things J* 3 (5): 637–646.

34 Satyanarayanan, M. (2017). The emergence of edge computing. *Computer (Long. Beach. Calif).* 50 (1): 30–39.

35 Lin, J., Yu, W., Zhang, N., Yang, X., Zhang, H., and Zhao, W. (2017). A survey on Internet of things: Architecture, enabling technologies, security and privacy, and applications. *IEEE Internet Things J* 4 (5): 1125–1142.

36 Zhang, T., Li, Y., and Philip Chen, C.L. (2021). Edge computing and its role in industrial Internet: Methodologies, applications, and future directions. *Inf. Sci. (Ny)* 557: 34–65.

37 Cisco. (n.d.). What is edge computing? https://www.cisco.com/c/en/us/solutions/computing/what-is-edge-computing.html.

38 Bizanis, N., and Kuipers, F.A. (2016). SDN and virtualization solutions for the Internet of things: A survey. *IEEE Access* 4: 5591–5606.

39 Ai, Y., Peng, M., and Zhang, K. (2018). Edge computing technologies for Internet of things: A primer. *Digit. Commun. Networks* 4 (2): 77–86.

40 Satyanarayanan, M., Bahl, P., Cáceres, R., and Davies, N. (2009). The case for VM-based cloudlets in mobile computing. *IEEE Pervasive Comput* 8 (4): 14–23.

41 Bonomi, F., Milito, R., Zhu, J., and Addepalli, S. (2012). Fog computing and its role in the nternet of things. In: *MCC'12 – Proceedings of the 1st ACM Mobile Cloud Computing Workshop*, 13–15.

42 Yousefpour, A., et al. (2019). All one needs to know about fog computing and related edge computing paradigms: A complete survey. *Journal of Systems Architecture* 98: 289–330.

43 Caprolu, M., Di Pietro, R., Lombardi, F., and Raponi, S. (2019). Edge computing perspectives: Architectures, technologies, and open security issues. In: *2019 IEEE International Conference on Edge Computing (EDGE)*, 116–123.

44 Mouradian, C., Naboulsi, D., Yangui, S., Glitho, R. H., Morrow, M. J., and Polakos, P. A. (2018). A comprehensive survey on fog computing: State-of-the-art and research challenges. *IEEE Communications Surveys and Tutorials* 20 (1): 416–464.

45 Aazam, M., and Huh, E. N. (2016). Fog computing: The cloud-IoT/IoE middleware paradigm. *IEEE Potentials* 35 (3): 40–44.

46 Mukherjee, M., Shu, L., and Wang, D. (Jul, 2018). Survey of fog computing: Fundamental, network applications, and research challenges. *IEEE Commun. Surv. Tutorials* 20 (3): 1826–1857.

47 Ismail, K. (2018). Edge computing vs. fog computing: What's the difference? https://www.cmswire.com/information-management/edge-computing-vs-fog-computing-whats-the-difference.

5

Applications of Artificial Intelligence and Big Data in Industry 4.0 Technologies

Reza Rezazadegan and Mahdi Sharifzadeh[*]

Sharif Energy, Water and Environment Institute (SEWEI), Sharif University of Technology, Tehran, Iran
* Corresponding author. Sharif Energy, Water and Environment Institute (SEWEI), Sharif University of Technology, Tehran, Iran

5.1 Introducing Artificial Intelligence (AI)

AI can be described as enabling machines to do tasks that humans can perform [12]. Unlike natural sciences or mathematics in which problems are solved based on fundamental laws or axioms, in machine learning we attempt to solve problems using heuristic and probabilistic algorithms. There are different ways to classify AI methods. On a first note, AI can be divided into symbolic AI and machine learning (ML). Symbolic AI, which belongs to the first wave of artificial intelligence, tries to encode human knowledge into machines using symbolic logic. One tries to emulate intelligence by mechanistic manipulation of the symbols. This way, AI is regarded as modeling the relationship between symbolic variables and structures [13]. Expert systems, which mimic the thinking of a human expert such as a doctor, are an example of symbolic AI. On the other hand, ML, which is responsible for the current AI and data science boom, tries to learn numerically from a pool of labeled or unlabeled data.

ML is already widely used in industry [7]. Its applications include fault detection, maintenance, decision support, and customer relationship management [8]. ML has also been used in predictive maintenance [9], in supporting decision-making in machining processes [10], and quality improvement [11]. ML methods can generally be classified as supervised, unsupervised, or reinforcement learning. In supervised learning, we have a set of labeled data (called training data), and we want to be able to predict the labels for data points, not in our training set. The labels can be either discrete (e.g., categorical) or continuous. In unsupervised learning, we want to obtain insights on unlabeled data based on a rule or heuristic. Reinforcement learning does not use labeled data and does not try to obtain information from unlabeled data. It instead aims at finding the optimal action plan for an agent, based on a state–action–reward model of the interaction of the agent with the environment.

According to Pedro Domingos, we have a taxonomy of AI algorithms as follows [14].

Connectionist

These algorithms mimic the working of the brain by trying to obtain a hierarchical understanding of data. Neural networks and deep learning are the main examples of connectionist AI [15].

Evolutionary or Bio-Inspired

Inspired by the theory of evolution, these methods, commonly known as genetic algorithms and evolutionary optimization, aim at finding the answer to a problem by searching the space (*landscape*) of all possibilities while trying to maximize a fitness function. Evolutionary methods are suitable for problems with too many degrees of freedom (independent variables) or highly irregular (non-smooth) fitness functions (the so-called rugged landscapes) [16]. Though inspired by biology, neural networks are excluded from this category.

Bayesian

As opposed to deterministic methods, which find exact values for model parameters and thus for the answer, in Bayesian methods we assume that model parameters come from a probability distribution, called the *prior* [17]. The cornerstone of Bayesian statistics is the *Bayes rule*, which we write as follows:

$$P\big(f(x)\,|\,x\big) = \frac{P(x, f(x))}{P(x)} = \frac{P(x\,|\,f(x))P\big(f(x)\big)}{P(x)} \tag{5.1}$$

Here $f(x)$ is the function (such as a classifier or a quantity dependent on x) that we want to estimate. On the right hand $P\big(x\,|\,f(x)\big)$ can be approximated from the available training data. The Bayes rule then tells us how to use this information to approximate $f(x)$ for unobserved data. Not just that, but it also tells us how probable various possible values of $f(x)$ are. The probability $P\big(f(x)\big)$ is the aforementioned prior, which governs the distribution of the parameters of the model. Its choice is aided by the underlying scientific facts about the problem. Since $P(x)$ does not depend on the model, it can be omitted.

By Analogy or Instance-Based Learning

This posits that the label of an item can be conferred from the labels of items similar to it—hence, the name *instance-based*. K-nearest neighbors (KNNs) and support vector machines (SVMs) are examples of learning by analogy.

Inductive Logic Programming

This is a subfield of symbolic AI that uses logic programming for representing knowledge and hypotheses. Logic programming is a computer programming paradigm based on the mathematical field of formal logic [18].

On a different note, AI methods can be either *deterministic* (determining a definite answer in each case) or *probabilistic*, that is, giving a probability distribution of answers for each problem. Note that from a probabilistic model P one can obtain a deterministic one by taking the most likely answer in P. However, this would omit the other possible solutions that may be nearly as probable. This, together with the fact that probabilistic methods can handle uncertainty (noise) in data much better, is among the reasons that probabilistic methods such as Bayesian learning or probabilistic time series forecasting are increasingly more adopted in AI.

Supervised Machine Learning

These methods can roughly be described as estimating (inter- or extrapolating) a uni- or multi-variable function, some of whose values are known to us. Such a function f can be thought of as a function on a Euclidean space R^n of dimension n, that is, $f : R^n \rightarrow R^N$. The elements of such a space are called n-dimensional *vectors*. We follow the convention of writing vectors (but not their components) in boldface whenever there is a possibility of ambiguity. Whereas quantities such as length, temperature, time, and blood pressure are already given numerically, things such as categories and text can be converted to numbers and vectors as well [19]. Digital images are also given by the array of red, green, blue (RGB) or grayscale values of their pixels. This means that 100×100 grayscale images have 10^4 raw features, whereas RGB images of the same size have up to 3×10^4 features.

In practice, we are given the values of f for N different vectors called the *training set*

$$f(x_1) = v_1, f(x_2) = v_2, \ldots, f(x_n) = v_N \tag{5.2}$$

and we want to be able to approximate $f(x)$ for general x. The components of the vectors $x = (x_1, x_2, \ldots, x_n)$ are called the *features* of the data and the v_i are the *labels*. If the values of f are discrete, called categories (or classes), then our problem is called a *classification problem*. If it instead takes continuous values then we have a *regression problem*. (Note that even though an unsupervised learning problem such as clustering is conceptually related to classification, in AI literature, classification describes only a supervised learning task.) A general method [20] for estimating $f(x)$ is to consider a set of basis functions $\varphi_1(x), \varphi_2(x), \ldots, \varphi_N(x)$ and assume that $f(x)$ is a linear combination of them:

$$f(x) = \sum_1^N w_i \phi_i(x) = \langle w, \phi(x) \rangle \tag{5.3}$$

where $\mathbf{w} = (w_1, w_2, \ldots, w_N)$ and $\Phi(x) = \left(\varphi_{1(x)}, \varphi_{2(x)}, \ldots, \varphi_{N(x)} \right)$ can be regarded as a mapping from R^n to R^N. The symbol $\langle a, b \rangle$ denotes the inner product of a and b. Training then involves finding the values of the weights w_i that "best" fit the available data. The exact meaning of "best" fit is decided by the choice of the model. As such, many classification methods have a regression counterpart as well, and vice versa.

Transfer Learning

This refers to using a model trained on a data set in a different but related problem. The importance of transfer learning stems from the computational power and time needed for training many AI models such as deep neural networks. For example, obtaining word embeddings (i.e., vector representations of words) for words in the general language involve training a model on a large corpus of texts such as Wikipedia or Google News articles [19]. This is a time-consuming procedure and requires a lot of computational power. However, the computed word embeddings are now available for download for several models.

Even when a trained model is used for a different but related problem, it can cause the new training to converge faster [21, 22]. Pre-trained models are now available from different sources such as Google's TensorFlow (a deep learning library) or Open AI's Generative Pre-trained Transformer (GPT), which can produce meaningful text [23].

5.2 Classification Methods

Typical classification problems include spam filtering, character or facial recognition, and image classification. These methods can also be used, for example, in predicting defects in disk drive manufacturing [24]. Typical classification methods include decision trees (DTs), Bayesian classifiers, k-nearest neighbors, support vector machines, relevance vector machines (RVMs), ensemble methods, and logistic regression.

Decision Trees

DTs involve a hierarchical set of if-then-else decisions. They are trained based on the information-theoretic notion of entropy, which is a measure of the mixture (chaos) of a given set. More precisely, given a probability (or frequency) distribution $\{P(x)\}_{x \in X}$, its *Shannon entropy* is given by

$$H = \sum_{x \in X} P(x) log_2 \big(P(x) \big).$$ (5.4)

When X contains only one class x_0 then $P(x_0) = 1$ and hence H is zero. On the contrary, the more classes we have in our set, the higher the entropy.

Starting with a labeled training data set at the root of the tree, the training algorithm (such as iterative dichotomiser 3, or ID3) tries to reduce the entropy of the set S_v at the vertex v by using the given features as clues [25] (Figure 5.1). The algorithm is greedy in the sense that at each node v, it uses the feature that divides the set S_v in such a way that entropy is reduced most.

A set with zero entropy has only one label and thus is fully classified. Therefore, if the sets at the leaves of a decision tree have zero entropy, we have achieved our goal of classification. Decision trees are easy to train and intuitive to understand. Additionally, they do not make assumptions about the distribution of data. They however suffer from low accuracy in the sense that they do not generalize well to data, not in the training set (overfitting) [26]. Ensemble methods can remedy this problem.

An example of a decision tree is depicted in Figure 5.1. In this example, we classify a set of iris flowers into three different species (setosa, versicolor, and virginica) based on two features: petal length and petal width. As you can see, moving from the root of the decision tree toward the leaves, entropy (class mixture) decreases. This example is generated using Python libraries Scikit-Learn and dtreeviz.

The (Naive) Bayesian Classifier

This probabilistic classifier uses the estimated distribution of the features in each class to confer the likelihood of a member belonging to a class. We continue the previous discussion on Bayesian learning.

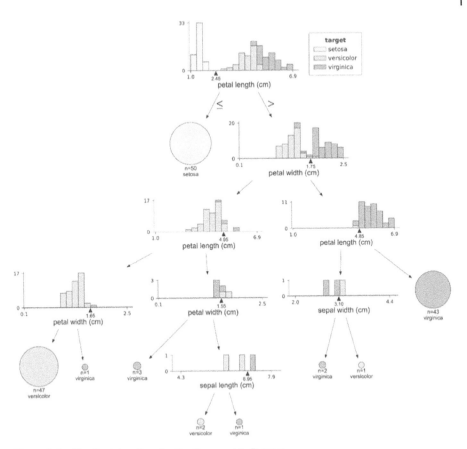

Figure 5.1 The Decision Tree for the Famous Iris Dataset.

Let $x_1,...,x_n$ be the features. One can compute the probability $P(C_k \mid x)$ of an element $x = (x_1, x_2,..., x_n)$ to belong to the class C_k using the Bayes rule:

$$P(C_k \mid x) = \frac{P(x \mid C_k) P(C_k)}{P(x)} \tag{5.5}$$

For example, if we have two classes, C_0 being "man", C_1 being woman and x represents weight, then, setting $k = 0$, the previous equation reads: the probability of a person with weight x being a man equals the probability of a man having weight x multiplied by the proportion of men in the society, divided by the probability of having weight x in general.

Assuming conditional independence, we have

$$P(x \mid C_k) = P(x_1 \mid C_k) P(x_2 \mid C_k) \cdots P(x_n \mid C_k) \tag{5.6}$$

To be able to compute the probabilities on the right, one assumes each one to come from a parametric distribution such as the normal (Gaussian) or Bernoulli distributions. One can then approximate the mean and standard deviation of each feature x_i in the class C_k based on training data and use the formula of the distribution being used, to compute the conditional probabilities $P(x_i \mid C_k)$. The probability $P(C_k)$ in (5.5) is given by the size of C_k in the sample, and $P(x)$ can be dropped as the value of x is known and is independent of C_k.

Even though the conditional independence assumption (5.6) is quite a strong assumption and does not hold in general, naive Bayes is quite a strong classifier [27].

K-Nearest Neighbors

Given a labeled training set X and an unobserved instance x, the algorithm assigns to x the most frequent class of the k elements of X that are closest to x (i.e., its neighbors). The number k is a hyperparameter to be chosen. To measure nearness we need a distance function, which is typically the Euclidean distance for continuous data or edit (Hamming) distance for discrete data [28]. For example, if $k = 6$ and the labels of the KNNs of x are {0, 0, 1, 0, 2, 1} then the label 0 is assigned to x.

KNN is used, for example, in facial recognition in conjunction with a method of dimensionality reduction applied to images [28, 29]. It can be used in determining the position of control rods in a nuclear reactor [30] and is also used in recommender systems [31]. KNN is very sensitive to noise in data in the sense that adding noise to the data can dramatically change its results [28]. This sensitivity is higher when k is smaller.

Linear Discriminant Analysis (LDA)

Imagine having two sets of data D_1 and D_2 with n_1 and n_2 elements, respectively. For a nonzero vector w the inner product

$$\langle x, w \rangle = \sum_{i=1}^{n} x_i w_i \tag{5.7}$$

gives the component (projection) of x in the direction of w. We want to find a w such that the projection $\langle x, w \rangle$ can "best" separate the two classes. To make the notion of "best" more precise, let $m_i = \sum_{x \in D_i} x / n_i$ be the mean of each class and $m_i = \dfrac{1}{n_i} \sum_{x \in D_i} \langle x, w \rangle$ be the mean of projected components of each class.

Let $s_i^2 = \sum_{x \in D_i} \left(\langle x, w \rangle - m_i \right)^2$ be the so-called *scatter* of each class which is a measure of deviation from the mean. We want a w that maximizes

$$\frac{\mid m_1 - m_2 \mid^2}{s_1^2 + s_2^2} \tag{5.8}$$

that is, that maximizes the difference between the projected means while keeping the scatter of each class in check. The optimal w can be shown to be $s_w^{-1}(m_1 - m_2)$ where

$$S_w = \sum_{i=1}^{2} \sum_{x \in Di} (x - m_i)(x - m_i)^t \tag{5.9}$$

In case the distribution of features in each class is normal with mean vector μ_i for class D_i and the same covariance matrix Σ for both classes, then w is also be given by $\Sigma^{-1}(\mu_1 - \mu_2)$.

LDA can easily be extended to the case of multiple classes by defining within-class and between-class scatter matrices [32]. The optimal direction is then the one whose projection maximizes the ratio of between-class to within-class scatter. LDA can also be used as a method of dimensionality reduction by projecting the data to the subspace generated by the means of the classes. LDA has been used in face recognition to reduce the dimension of the feature vectors (image arrays in this case) and then to use KNN for classification [33]. It has also been used for bankruptcy prediction [34].

Support Vector Machines and Kernel Methods

The SVM algorithm tries to find the hyperplane in the feature space R^n that provides the widest margin between the two categories (Figure 5.2). Remember that any hyperplane in Euclidean space is given by an equation of the form $\langle w, x \rangle + w_0 = 0$ for some vector w and scalar constant w_0. If our data points are $x_1, x_2, \ldots, x_n$ then, as in Logistic Regression, we denote the class of x_i by $y_i \in \{1, -1\}$. Being on a side of the hyperplane is decided by the sign of $\langle w, x \rangle + w_0$. We, however, want to have a margin between the hypersurface and the points and therefore we require

$$y_i\left(\langle w, x \rangle + w_0\right) \geq 1 \tag{5.10}$$

for each i. The margin between the two cases when this inequality is equality, equals $\dfrac{1}{\|w\|} + \dfrac{1}{\|w\|}$ and therefore in SVM, we want to minimize $\|w\|$ subject to the constraints in (5.10). In Figure 5.2 we see an example of a SVM applied to the iris data set. The two species are Setosa (yellow) and Versicolor (red). The dividing hyperplane (in this case

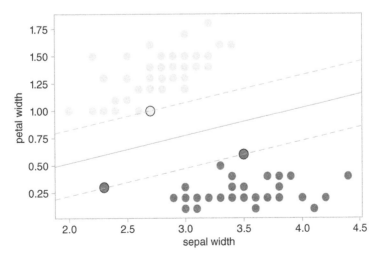

Figure 5.2 Applying SVM to the Iris Data Set with Two Features: Sepal Width and Petal Width.

a line) is depicted as a solid line. The support vectors (i.e., the data points closest to the dividing hyperplane) are marked with a black boundary.

In case the two categories are not linearly separable, kernel methods can be used. Kernel methods are tantamount to using a different inner product (and therefore distance function) than the Euclidean inner product hx, yi. As the Euclidean distance of two vectors is given by $\|x - y\|^2 = \langle x - y, x - y \rangle$, one can use a different inner product such as $<x, y>_K = K(x, y)$ to measure distances and angles. Here K is a function called the kernel. Common examples of kernels include the quadratic kernel $K(x, y) = <x, y>^2$ used commonly in natural language processing and the Gaussian kernel $K(x, y) = exp\left(-\|x - y\|^2 / \sigma^2\right)$, where σ is a hyperparameter. Note also that using a kernel is generally equivalent to mapping the data to a higher-dimensional space so that the data becomes more separable. If Φ is such a transformation then we would have $<x, y>_K = K(x, y) = \langle \Phi(x), \Phi(y) \rangle$. For example, for $K(x, y) = <x, y>^2$, $\Phi(x)$ is the vector whose components are binomials $x_i y_i$ for different indices i, j. Using the kernel is, however, more computationally efficient. Figure 5.3 depicts an example of kernel SVM. To be able to separate the two classes in this picture with SVM, we used the RBF kernel given by $K(u, v) = exp\left(-\|u - v\|^2\right)$. This is equivalent to introducing a new coordinate $r = exp\left(\|u\|^2\right)$ and applying linear SVM in the 3-dimensional space given by coordinates (x, y, r) in which the two classes are linearly separable.

SVM provides high classification accuracy [35] but has poor uncertainty management and needs cross-validation to determine the hyperparameters [36]. SVM is used in cancer research [37] and drug design [38] and is also applied in image recognition such as handwriting classification. This enables it to be used to detect defective products [39, 40]. In industrial applications, SVM can be used in monitoring the conditions of machines and fault monitoring [39, 41, 42]. It has applications in quality control as well [43].

Relevance Vector Machines

Even though SVM is a powerful classification method, it is not probabilistic; therefore Microsoft researcher M. Tipping attempted a Bayesian variant of SVM [44]. In this method, as is common in Bayesian statistics, we regard the weights in (5.10) as samples from probability distributions, which we assume to be normal with mean zero and with standard deviations that are taken as hyperparameters.

The conditional probability of the observed data given the (hyper-)parameters (including the parameters) can be computed by assuming its distribution to be a multivariable Gaussian of the form:

$$p(x \mid y) = 2\pi\sigma^2 \, exp\left(-\frac{1}{2\sigma^2}\|y - \Phi w\|\right) \tag{5.11}$$

where $w = (w_1, w_2, ..., w_N)$ and $\Phi_{i,j} = K(x_i, x_j)$. One can then use the Bayes rule to compute the conditional probability of the parameters given the data which in turn gives us a probabilistic solution to the classification problem

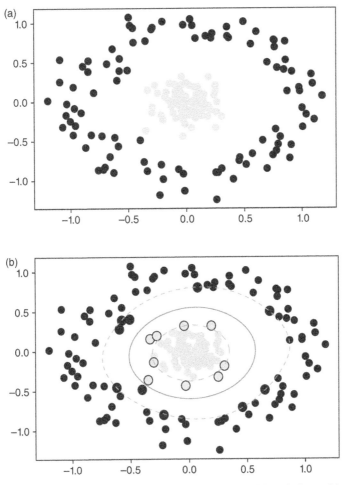

Figure 5.3 (a) Example of Two Classes That Are Not Linearly Separable. (b) Decision Boundary and Margin for SVM with This Kernel.

RVM is sparse in the sense that a significant number of the weights in (5.10) are zero for each x. This makes it easy to see which basis functions are relevant to the classification of each instance x. RVM are, however, prone to falling into local minima in the optimization (solution) process and are computationally costly [45].

Ensemble Methods

In these methods, a probabilistic ensemble of weak classifiers is produced to obtain a strong classifier. There are several different ensemble methods. In *bootstrapping (bagging)* [46] the same classifier is trained on different samples of the data (which can possibly have overlap). The classification of individual elements is then decided by the majority vote of these classifiers. In case the classifier model is a decision tree, the resulting ensemble classifier is called a *random forest*.

In *boosting* methods, one starts with all the data having the same weight, and a sample of the data is fed to a classifier. In the second iteration, the misclassified points of the first iteration are given a higher weight, and then a sample is drawn from all the points and fed to the classifier again. The higher weight means that the misclassified points are more likely to be among the sampled points. AdaBoost and XGBoost [47] are two of the most important boosting methods which compete with neural networks in their widespread applications and accuracy. Teams using these methods have won many Kaggle competitions [47].

Logistic Regression

If our classification involves only two categories, e.g., $x_1,...,x_k$ in the first category, and $x_{k+1},x_{k+2},...,x_N$ in the second, then we can represent our data set by points

$$Z = \left\{ (x_1,0),...,(x_k,0),(x_{k+1},1),...,(x_N,1) \right\} \tag{5.12}$$

We can then find the parameters for the logistic function

$$F(x) = \frac{1}{1+e^{-(ax+b)}} \tag{5.13}$$

which best separates the two categories. This is equivalent to finding a,b such that $F(x)$ best approximates the data points in Z (Figure 5.4). The method for finding a,b is similar to that of linear regression discussed in the next section. Figure 5.4 shows logistic regression for the iris data set to separate the Setosa and Versicolor species (same as in Figure 5.2). The difference here is that we are using only one feature (i.e., sepal width). The two species are depicted on the y = 0 and y = 1 lines respectively. As you see, there is an overlap between the sepal width values of the two classes. The inverted-S-shaped curve is the logistic function fitted to best separate the two sets of data points.

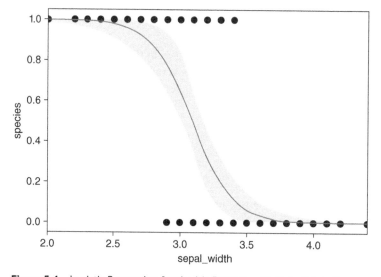

Figure 5.4 Logistic Regression for the Iris Dataset, with Data Points Corresponding to the Setosa Species on the *y* = 1 Line and the Ones Corresponding to Versicolor on the *y* = 0 Line.

5.3 Regression Methods

As mentioned before, in regression we have a set of arguments and values

$$(x_1, y_1), (x_2, y_2), \ldots, (x_n, y_n) \tag{5.14}$$

and want to approximate a function that would give us $f(x_1) = y_1, f(x_2) = y_2, \ldots, f(x_n) = y_n$. We also mentioned that the difference between regression and classification depends on whether the labels y_i are discrete and continuous; therefore, classification methods usually have a regression counterpart as well and vice versa. For example, *KNN regression* is a regression by averaging the values of the k nearest neighbors of a point, weighted by the inverse of their distances to the point. Typical regression methods include least squares regression, neural network regression, support vector regression, gaussian process regression, thin plate spline, and the Taylor polynomial.

Ordinary Least Squares (OLS) Regression

This regression finds a parametric function such as a linear function $f(x) = \langle x, a \rangle + b$, a polynomial $P(x) = a_0 + a_1 x + a_2 x^2 + \cdots + a_n x^n$, or the aforementioned logistic function that minimizes the cost (error)

$$E = \sum_i (f(x_i) - y_i)^2.$$

The answer to the minimization problem can be found using differential calculus, that is, taking the derivative of E with respect to the parameters. In more complicated situations one would use the method of gradient descent, which involves following the direction of $-\nabla E$ to reach a minimum of E. Figure 5.5 shows an example of using regression to estimate the amount of tip based on the total bill price. The shaded region in the picture depicts the 95% confidence interval. (We used Python's Seaborn library to generate this figure.)

When we want to find a linear function $f(x) = \langle x, a \rangle + b$ then the answer is given as follows. Note that when the independent variable x is multidimensional, this method is called *multiple linear regression*. However, the procedure is similar to the case of scalar x. If X is the matrix, called the *design matrix*, whose columns are the training points $\{x_i\}_{i=1}^N$, $\mathbf{y}$ the vector whose entries are the associated values $\{y_i\}_{i=1}^N$ and $\mathbf{a} = (a_1, a_2, \ldots, a_n)^t$ then we want to find a vector $\mathbf{a}$ that minimizes

$$\|X\mathbf{a} - \mathbf{y}\| \tag{5.15}$$

It is given by

$$\mathbf{a} = (X^t X)^{-1} X^t \mathbf{y}. \tag{5.16}$$

OLS is a very classic method going back to K. F. Gauss in the early 19th century. He used this method to forecast the position of the newly discovered asteroid Ceres, after

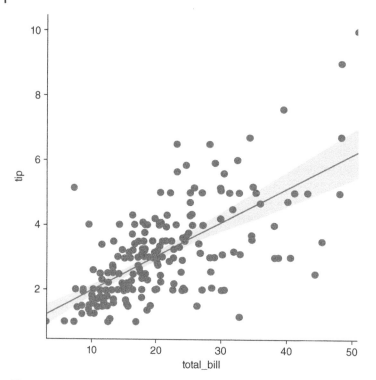

Figure 5.5 Using Linear Regression to Estimate the Tip a Restaurant Customer Pays (y-Axis) Based on Their Total Bill (x-Axis).

reappearing from behind the sun. OLS is the first method choice for regression in various scientific and industrial problems. However, it makes a number of assumptions such as the variance of the errors being independent of the independent variables (called *homoscedasticity*). Moreover, the features of the data need to be uncorrelated.

Ridge and Lasso Regression

If there is linear dependence or correlation between the features, then the columns of the design matrix X will not be linearly independent and thus the inverse in (5.16) will not exist. The remedy is to add diagonal terms to $X^t X$ (called *Tikhonov regularization* [48]), which is equivalent to adding such terms to the cost function. In ridge regression the cost function is given by

$$\left\| X\mathbf{a} - \mathbf{a} \right\|^2 + \left\| \lambda\mathbf{a} \right\|^2 \tag{5.17}$$

where λ is a hyperparameter. The solution is given by

$$\mathbf{a} = \left(X^t X + \lambda^2 I \right)^{-1} X^t \, \mathbf{y} \tag{5.18}$$

where I is the identity matrix. Lasso regression is similar to ridge, with the difference that cost function is given by $\|X\mathbf{a} - \mathbf{y}\|^2 + \|\lambda\mathbf{a}\|$.

Support Vector Regression

This regression penalizes only errors that are more than a given margin. More precisely the cost function is zero when the $|f(x) - y| < \epsilon$ and equals $|f(x) - y| - \epsilon$ otherwise. This type of cost function makes the model less sensitive to noise. It also avoids unnecessary fluctuations of the solution [49].

What makes SVR even more powerful is the ability to use kernels to enable nonlinear regression (*kernel SVR*). SVR has been used for forecasting the annual load of electricity grids [50], predicting short-term traffic patterns [51], and energy optimization for carbon fiber production [52], among others.

Gaussian Process Regression (GPR)

Also known as Kriging, GPR assigns probabilities to all the functions that can fit the training data [53]. If D is the domain of the data, one takes a discrete set of N points $x_1, x_2, ..., x_N$ in D, including the training (labeled) points. One then takes the N-dimensional multivariate Gaussian distribution $N(0, K(y, y_0))$ with mean zero and covariance matrix given by a kernel function K. Here $y = (y_1, y_2, ..., y_N)$ and each y_i corresponds to the value of the desired function on the corresponding point x_i. Training this model is as simple as requiring the y_i for the training points to coincide with the associated labels. However, since this is restrictive and can lead to overfitting, one introduces a (Gaussian) error term. Sampling from this distribution gives us a function that fits the data. We can take the mean of the distribution at each point to get the answer to the regression or classification problem. Moreover, we can use the distribution to obtain confidence intervals for the prediction; something that OLS regression is not capable of.

The choice of the Kernel function K determines which functions are more likely to be sampled. There are various kernel functions to choose from such as:

- Linear kernel, $K(y, y') = a^2 + b^2(y - c)(y' - c)$

- Periodic kernel, $K(y, y') = \sigma^2 \exp(-\dfrac{2\sin^2(\pi(y - y')/p)}{l^2})$

- Radial basis function (RBF) kernel, $K(y, y') = \sigma^2 \exp(-\dfrac{\|y - y'\|^2}{2l^2})$

They result in linear, periodic, and mixtures of Gaussians to be sampled, respectively. Different kernels can also be combined using multiplication.

GPR is a powerful nonlinear and probabilistic regression method. However, it has a high computational cost, and therefore approximation methods have been developed [53]. GPR has been used for the prediction of both cyclic and calendar aging of lithium-ion batteries [54, 55], among others.

Thin Plate Spline

This method finds a smooth function that at the same time best approximates the data (14) and has minimal variation. If the data points $\{x_i\}_{i=1}^{N}$ are two-dimensional, this is tantamount to finding the smoothest surface that best approximates the data points.

This is achieved by minimizing the sum of squared errors plus the integral of the second derivatives of the function:

$$E(f) = \sum_{N}^{i=1} \|f(x_i) - y_i\|^2 + \lambda \int \sum_{i=1}^{n} \sum_{j=1}^{n} \left(\partial_{x^i} \partial_{x^j} f(x)\right)^2 dx^1 dx^2 \cdots dx^n. \tag{5.19}$$

Here $x^1, x^2, \ldots, x^n$ are the components of x and λ is a hyperparameter. It can be shown that there is a unique f minimizing the above error [56]. The minimizing function can be computed by using a basis of radial functions of the form

$\varphi(\|x - x_i\|)$ in (5.3) where $\varphi(r) = r^2 \log r$.

TPS is used as a transformation for extraction of features and thereby comparison of biological and medical images, and using them is diagnosis [57]. It has also been used for the optimization of the complex thermal processing of carbon fiber [58].

5.4 Unsupervised Learning

As mentioned in the introduction, unsupervised learning pertains to obtaining information from unlabeled data. In this section, we review the main unsupervised learning methods.

Clustering

In dimensions up to 3, we have an intuitive notion of separating unlabeled data points into clusters whose points are close to each other but less so to the points of other clusters. Clustering is the algorithmic method of such a subdivision for higher-dimensional, and more generally, complicated data. There are various clustering methods available of which we mention a few widely-used ones. Note that all clustering methods are inherently heuristic. However, there are criteria for evaluating the quality of the resulting clusters, such as the sum of squared error, the silhouette coefficient, and the Rand index [59].

K-Means Clustering

One of the most well-known clustering methods in machine learning, k-means clustering aims at minimizing within-cluster variances. In this method, one specifies k, the number of clusters, beforehand. Initially, k random points are chosen in the data set, considered as clusters. At each step, each data point p is assigned to the cluster whose mean is closest to p. Then the means of the resulting new clusters are computed and in the next step, the assignment of points is done anew. This process is repeated until assignments are not changed anymore. Even though it is an efficient clustering algorithm, k-means does not give satisfactory results if the clusters involved are not separable with spheres. For example, the two sets in Figure 5.3 cannot be separated by k-means clustering because the centroid of the blue set lies in the span of the yellow set.

k-means is used for vector quantization, such as for reducing the number of colors of an image to a fixed number k, by clustering the RGB values of the colors in the image and then replacing each color with the mean of the cluster it belongs to [25].

Hierarchical Clustering

These are the methods that give us a hierarchy of clusters of data. In *agglomerative clustering*, in the beginning, each data point is a cluster. As the distance threshold increases, clusters start to merge until finally we are left with a single cluster. This can be illustrated by a *dendrogram* that depicts the dynamics of the merger of clusters. The desired clusters can be obtained by choosing a distance threshold, which is equivalent to cutting the dendrogram at a specific height. This choice can be done either by specifying the number of desired clusters or by studying the histogram of pairwise distances of the data points. The gaps (minima) in this histogram correspond to thresholds at which significant changes in the distance structure of the data set occur. Whether two clusters A, B should be merged at distance threshold can be decided by the following:

- If a point of A is closer than ϵ to a point of B (*single linkage*)
- If all points of A are closer than ϵ to all points of B (*complete linkage*)
- If the mean of the pairwise distances between points of A and B is less than ϵ (*average linkage*)
- In *ward linkage* the distance between two clusters is defined as the difference between the variance of their union and the sum of variances of the two. In this linkage method, at each step clusters are chosen to be merged which minimizes the increase in the sum of variances of clusters.

An example of Agglomerative Clustering is depicted in Figure 5.6. The data used, consists of about 1800 8×8 black and white images of hand-written digits. As such each of the data points can be regarded as a point in the 64-dimensional Euclidean space. Hence, one can use Euclidean distance to measure the distances between hard-written digits and therefore cluster them. Agglomerative clustering with Ward linkage is used. Each branch point in the dendrogram corresponds to a distance threshold where two clusters are

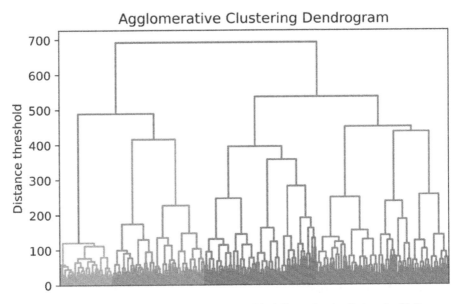

Figure 5.6 Dendrogram for Agglomerative Hierarchical Clustering Applied to the Digits Dataset.

merged together (going from the bottom to the top). In such clustering methods, one has to choose the distance threshold based on the available evidence. For example in this case we know that there are exactly 10 digits and hence there should be near 10 clusters.

Agglomerative clustering can be used for identification using biometrics of the hand [60], caching in device-to-device networks by clustering user preferences [61], and even deciding the similarity of two HIV strands in a criminal case [62], among many other industrial and scientific applications.

Density-Based Spatial Clustering of Applications with Noise (DBSCAN)

DBSCAN is one of the main density-based clustering methods that takes into account not only the proximity of the points but also their spatial density [63]. In other words, DBSCAN permits linkage only through points that have a dense enough neighborhood. Note that in a method such as single linkage, two dense clusters may be merged if there is only one link between a point of one and the other. DBSCAN, however, does not allow such weak links.

To achieve this, DBSCAN uses a distance threshold as well as a minimum number of neighbors m as parameters. For a point q to be reachable from a point p (and hence be in the same cluster), there must be a sequence of points in between, such that each point is closer than to the next and all the points (except possibly the peripheral points p,q) must have at least m neighbors within *the* distance of them (including the point itself). The latter points are called core points. A point that is not reachable from any core point is considered noise.

Even though computationally more costly, DBSCAN is capable of separating clusters of arbitrary shapes, even when one cluster is surrounded by another [63], such as in Figure 5.3. DBSCAN is used by Netflix to detect unhealthy servers [64]. Other notable applications of DBSCAN include constructing phylogenetic trees from DNA sequences [65].

Dimensionality Reduction

High dimensional data has been on the rise in machine learning and AI [66]. Typical sources of high dimensional data include digital images (recorded as arrays of their pixel values) or text converted to vectors using latent representations. One can add any data with a high number of features (such as health statistics of patients) to the list. High dimensional data is difficult (if not impossible) to visualize and high dimensionality increases the computational cost of ML algorithms dramatically [66]. Moreover, in high dimensions, the notion of distance is not as meaningful as in low dimensions. If we sample vectors by sampling their coordinates from a multivariable Gaussian with mean zero and variance 1, then any two such vectors are orthogonal to each other with a probability that quickly converges to 1 as the dimension increases [67].

Dimensionality reduction (DR) refers to methods for projecting high dimensional data to low (typically 2- or 3-) dimensional space to make visual interpretation possible and enhance training performance. DR can be regarded as a method of feature extraction from data. Note that clustering can also be regarded as an extreme form of dimensionality reduction.

Principal Component Analysis

Going back to 1901, PCA is one of the main methods of linear dimensionality reduction. Given a data set, it gives us a linear map (change of coordinates) such that the new coordinates are ordered by the variance of data along them. Coordinates with low variance can be regarded as noise and hence dropped. This way, one can reduce the dimension of the data to the most important aspects which (in the case of PCA) are called the *principal components*. Note that each principal component is a linear combination of the original coordinates (features) whose weights can be obtained from the inverse of the aforementioned linear map.

PCA is done by applying singular value decomposition (a generalization of diagonal-ization/eigendecomposition to non-square matrices) to the covariance matrix of the data [67]. PCA can be used, together with a classification method, in facial recognition [25], simplifying the output of molecular dynamics simulations [68] and portfolio opti-mization [69].

Linear discriminant analysis can be used as a dimensionality reduction method as well, when the data comes divided into two or more classes.

Nonlinear Dimensionality Reduction

NDR involves nonlinear methods of reducing data dimension, either by obtaining a projection map or by giving a low dimensional representation of the data. Note that linear DR methods such as PCA are not capable of finding nonlinear correlations in data such as nonlinear transformations of images. NDR is closely related to *manifold learning*, that is, trying to find a nonlinear manifold (of lower dimension than the fea-ture space) that fits the data [70]. Manifold can be thought of as generalizations of surfaces to higher dimensions.

Self-organizing maps and *autoencoders* are two methods of NDR that use neural networks.

Laplacian Eigenmaps Method

This method borrows ideas from differential geometry to use in dimensionality reduction [71]. To a data set, we can associate its proximity (neighborhood) graph G which has a node for each data point and two nodes are connected by an edge if their distance is less than a threshold ϵ. The Laplace operator ∇ associated to G is a linear map defined on the vector space V consisting of linear combinations of vertices of G. Let n denote the dimension of V which equals the number of vertices of G which we denote by n. Let D be the degree matrix such that $D_{v,v}$ is the degree of the vertex v and $D_{v,w} = 0$ if $v \, 6 = w$. Let W be the adjacency matrix of G, that is, $W_{u,v} = 1$ if u,v are connected in G and zero otherwise. ($W_{u,v}$ can also be weighted by the distance of the data points x,y represented by vertices u,v, that is, $W_{u,v} = \exp(-\|x - y\|^2)/\sigma$ with σ being a parameter.) We set

$$\nabla = D - W \tag{5.20}$$

and call it the *Laplace operator* of G. It can be thought of as the discrete analogue of the Laplace-Beltrami operator on Riemannian manifolds [72]. The eigenfunctions of ∇ are functions f in V defined by $\nabla f = \lambda f$ for some scalar constant λ.

Since ∇ is a self-adjoint matrix, it is diagonalizable; that is, it has a basis of eigenfunc-tions $f_0, f_1, \ldots, f_n$ where f_0 corresponds to the eigenvalue 0. Manifold regularization sends the datapoint corresponding to vertex v to the vector

$$\mathbf{f}(v) = \left(f_{1(v)}, f_{2(v)}, \ldots, f_{n(v)} \right).$$

(5.21)

Note that the map $f(v)$ is defined for general graphs G, and it can be shown [71] that it gives a "best" embedding in the sense that vertices close to each other in the graph are mapped to points close to each other in the Euclidean space. More precisely, this map minimizes the following cost function.

$$\frac{1}{2} \sum_{v,w} (f(v) - f(w))^2 w_{ij}$$

(5.22)

In addition, note that this method depends only on the pairwise distances between data points and not directly on their coordinates. Laplacian Eigenmaps can be used to obtain low dimensional representations of predictors in medical data for diagnostic purposes [73] or obtaining distinctive phonetic dimensions from speech recordings [71].

Kernel PCA is one of the most widely used manifold learning methods [74]. As mentioned before, in linear PCA we take the matrix X whose columns are our data vectors and then take the singular value decomposition (SVD) of XX^t. In kernel PCA one first applies a mapping Φ to each column and then takes SVD. The mapping Φ comes from a kernel K in the sense that $\langle \Phi(x), \Phi(y) \rangle = K(x, y)$. Kernel PCA can be used, for example, in novelty detection [75], that is deciding if the test data is different from the training (normal) data when there is not enough of the abnormal data to lend itself to a new class.

Topological Data Analysis

This relatively new branch of machine learning uses the pure-mathematical discipline of topology [76, 77]. TDA can be described as an enhanced method of clustering that takes into account the relations between data points as well. It can also be regarded as a dimensionality reduction method that, unlike the aforementioned ones, gives an abstract (non-Euclidean) representation of data. More specifically, consider a (high dimensional) data set P equipped with a distance function. For each distance threshold, we consider the *proximity graph* $G = G_\epsilon(P)$ whose vertices are data points and in which any two points whose distance is less than are connected. When $\epsilon = 0$ we have a set of vertices with no edges and when ϵ is large enough then $G_\epsilon(P)$ is a complete graph. (Figure 5.7.) Note that the connected components of $G_\epsilon(P)$ are the same as the clusters of P in single-linkage clustering with distance threshold ϵ.

An important aspect of TDA is that it takes into account not only the first-order relations between data points (i.e., the edges of $G_\epsilon(P)$) but also the higher-order ones involving multiple elements. Formally, this means that one takes the *clique complex* of $G_\epsilon(P)$, which is a hypergraph in which any clique (complete subgraph) in G is considered as a hyper-edge (technically called a simplex) which can intuitively be regarded as being "filled in." (There are other methods for obtaining a simplicial complex from a data set as well such as the alpha complex or witness complex.) This hypergraph can be regarded as an abstract space (meaning that it is not embedded in Euclidean space), and one then uses the algebraic-topological notion of homology [78] to find the voids (empty areas) in it. For example, if G is a polygon graph (cyclic graph) then it obviously has a void in its middle. The same is true for a cube-shaped graph. Complete graphs on

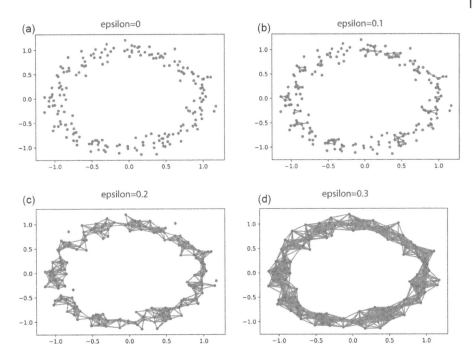

Figure 5.7 Demonstration of TDA for a Data Set with the Shape of an Annulus. As the Distance Threshold Is Increased, More Points Are Connected by Edges and Therefore More Cliques (Complete Subgraphs, Triangles in This Case) Are Formed. Finally, for a Suitably Large Value of the Distance Threshold, the Associated Simplicial Complex Takes the Form of an Annulus.

the other hand have no voids in them. The voids, roughly speaking, correspond to areas in which our data lacks instances and their study gives a good approximation of the topological shape of the data under study. An important advantage of TDA is the fact that it is robust under small perturbations of data [79], and therefore its results are more qualitative and reliable than NDR and clustering.

Another aspect of TDA is that, instead of trying to find an optimal value of the distance threshold, it takes into account all the possible values of ϵ and studies the evolution of the topological shape of the data set as ϵ varies. This way, to each data set, TDA associates a set of birth and death pairs (given by their corresponding ϵ value), for voids of different dimensions. Due to its high computational cost, TDA is usually computed only for 1-dimensional voids

Figure 5.8 shows the persistent diagram corresponding to the same data as in Figure 5.7. The red and blue dots correspond to 1 and 2-dimensional components of the complex respectively. Each such component is marked with the ε threshold it comes to existence at (on x-axis), and the threshold at which it dies (on y-axis). In dimension zero, when $\varepsilon = 0$ each data point gives us a zero-dimensional connected component. As ε increases, these components start to merge and for large enough ε (marked as infinity), only one connected component is left. In dimension 1, we have several components that are born and then die shortly after. They correspond to the cycles of length > 3 in the proximity graph G_ε. As ε grows large enough, only one such component is left (the red dot at infinity) which corresponds to the hole in the middle of the annulus in Figure 5.8. Computation of persistent homology and plotting of this figure is done using the Gudhi [80] library for Python.

There is a simpler variant of TDA, which involves using filtration of the data to divide the data set into overlapping bins $B_1, B_2, \ldots, B_k$. One then clusters the points of each bin separately and a cluster in B_i is connected to a cluster in B_{i+1} if the two share an element [81]. This way one obtains a graph-valued clustering of the data that, to an extent, captures the "shape" of the data.

Since its introduction, TDA has found applications in many scientific and applied problems such as coverage problems in sensor networks [82], complex networks [83–85], remote sensing [86], the study of viral evolution [87] and classifying agents based on their behavior in agent-based models [88]. It has even led to the formation of a dedicated company called Ayasdi, for using TDA in industrial and business problems.

5.5 Semi-supervised Learning and Active Learning

Most of the data in today's world is unlabeled and labeling data, such as texts or images, is expensive [89]. For this reason, we are looking for methods that can incorporate a large amount of unlabeled data, in addition to the small amount of labeled data to improve classification or regression results. The way the unlabeled data helps is by giving an estimate of the geometry (as in manifold learning) or the frequency distribution of data. The former is used in graph-based semi-supervised learning and the latter in generative models.

Graph-Based Models

Using this model, one adds a term reflecting the proximity of the points, to the cost function of the classifier. This reflects the idea that points close to each other in the

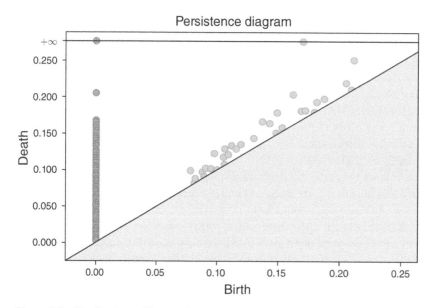

Figure 5.8 The Persistent Diagram Corresponding to the Annulus Data Set.

feature space tend to have similar classifications. If f is the function to be interpolated (as in classification or regression) then the additional term is of the form

$$\lambda_1 \|f\| + \lambda_2 \sum_{i,j} W_{i,j} \big(f(x_i) - f(x_j)\big)^2 \tag{5.23}$$

where W is the adjacency matrix of the proximity graph of the data, and λ_1, λ_2 are hyperparameters. The sum is over the unlabeled data.

Generative Models

With this model, one uses a similar model as in the naive Bayes classifier with the difference of incorporating the unlabeled data. Let $x_1, \ldots, x_k$ be labeled data with labels $y_1, \ldots, y_k$ and $x_{k+1}, \ldots, x_l$ unlabeled. We want to compute $P(C|x, \theta)$, where θ denotes the parameters of the probability distribution, x is a data point and C is a class. We have $P(C|x, \theta) = P(x, C|\theta)/P(x|\theta)$. For each value of θ, maximizing this probability gives us a classifier. We postulate that the best classifier is the one corresponding to a θ which best fits the unlabeled data, that is, maximizes $P\big(\{x_i\}_{i=k+1}^l \mid \theta\big)$. Therefore the most suitable classifier that fits both labeled and unlabeled data best is the one maximizing the following [89]:

$$\log\Big(\sum_{i=1}^k p(x_i, y_i \mid \theta)\Big) + \lambda \log\Big(\sum_{i=k+1}^l p(x_i \mid \theta)\Big) \tag{5.24}$$

Active Learning

This method involves querying users by the learning algorithm to label specific data points. This is usually done for points with high classification uncertainty in the model such as points close to the separating hyperplane of an SVM [90].

5.6 Neural Networks and Deep Learning

Neural networks (NN) are a learning method inspired by the working of biological brains and they have been used in most machine learning and AI problems including both supervised and unsupervised learning [15].

Remember that in OLS regression, one wants to approximate the available data using a parametric function. Training then involves finding the optimal values of the parameters using derivative or (equivalently) the method of gradient descent. NNs basically involve a similar problem with the main difference that our parametric function is given by the composition of artificial neurons (AN). (Composition here is meant in the mathematical sense; that is, the composition of a function f with g is given by $f\big(g(x)\big)$.)

An *artificial neuron or perceptron* is a function of the form $h\big(\varphi(x)\big)$ where $\varphi(x) = w_1 x_1 + w_2 x_2 + \cdots + w_n x_n + b$ is a multilinear function and h is nonlinear and called the *activation function*. The w_i are called the *weights* of the AN. Examples of activation functions include the *Heaviside step function*, ($h(x) = 0$ for $x < 0$ and

$h(x)=1$ for $x \geq 0$), the *logistic function*, $1/(1+e^{-x})$ and the *hyperbolic tangent*, $tanh(x)$. AN is based on biological neurons that can receive electrical input from their dendrites (modeled by the x_i) and would fire if the processed input exceeds a given amount (Figure 5.9).

Similar to biological neurons, AN can be composed in a variety of ways. One generic method is to consider several *layers* of ANs each consisting of N artificial neurons. The number N is called the *width* of the layer. If φ is an AN in layer k then it is post-composed to all the ANs in layer $k-1$. This means that if the ANs in layer $k-1$ are $\psi_1,...,\psi_N$ then we have the contribution $\varphi(\psi_1,\psi_2,...,\psi_N)$ in layer k (Figure 5.10). This would be the case for all the ANs in all the d layers except for the first layer. The first layer is called the *input layer* and is fed the data directly. Similarly, the last layer is the *output layer*. The layers in the middle are called *hidden layers*. Such a system is called an *Artificial Neural Network (NN)*. The *depth* of the NN is the number of its hidden layers. An NN is called deep if its depth is greater than 1. An NN typically has many parameters involved in it which are the weights of the AN in it. This large number of parameters may be the reason for the success of NN in solving complicated AI problems. However, the layered nature of NN enable them to confer information from data in a hierarchical manner.

A schematic depiction of an NN is provided in Figure 5.10. Each node, except for the ones in the input layer, consists of a perceptron. The input layer has 4 nodes, meaning the input data has 4 features. For each data point, the features are fed to all the perceptrons in the top hidden layer. The output of these artificial neurons is then fed to the perceptrons in the second layer, which in turn are fed to the output layer consisting of a single artificial neuron. Training such an ANN on a labeled data set involves finding weights for all the involved artificial neurons that minimize error, using the backpropagation algorithm.

The universal approximation theorem [91, 92] states that any function (satisfying very mild conditions) can be approximated by an NN of either enough depth with bounded width or enough depth with bounded width. This is a significant result as our function can even be an assignment such as "a function that takes an image and tells us if it is an image of a dog, a cat or neither."

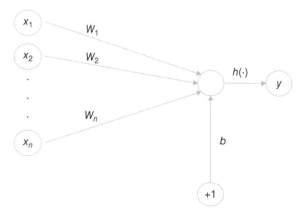

Figure 5.9 Schematic Picture of a Perceptron (Artificial Neuron). The Input Variables Are $x_1, x_2,...,x_n$ the w_i Are the Weights.

Input Layer

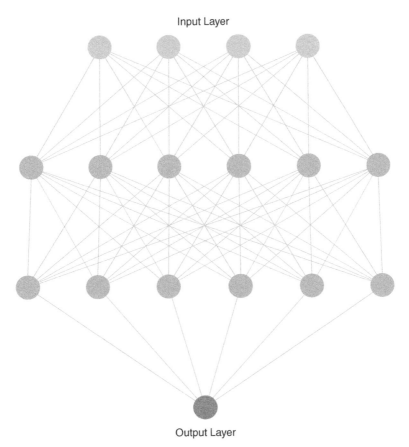

Output Layer

Figure 5.10   An Example of a Feed-Forward Neural Network with Two Hidden Layers (in Blue).

The training of an NN is based on the method of gradient descent. Starting with random numbers as the network weights, one takes the error of the output of the NN with that of the function to be approximated. The error is measured according to a chosen *loss function*. Starting from the output layer, one computes the contribution of each AN to the error and this error is recursively propagated back to the preceding layers to compute the contribution of each AN in the network to the error. If an artificial neuron A is connected with neurons $A_{k_1}, A_{k_2}, \ldots, A_{k_n}$ in the proceeding layer and the weights of these connections are $w_{n_1}, w_{n_2}, \ldots, w_{n_N}$ then the error $E(A)$ of this neuron is given by

$$h'(A)^{\circ} \cdot \sum_{i=1}^{N} w_{n_i} E\left(A_{k_i}\right) \tag{5.25}$$

Here h' is the derivative of h. This method of recursively computing the contribution of the neurons to the error is called *backpropagation*.

Then starting from the first hidden layer, one updates the weights of each node based on their error. This means that to a weight corresponding to an edge in the network, is added the product of the value of its incoming node with the error of its outgoing one. Each time the weights are adjusted in this way is called an *epoch*. One either specifies the number of epochs or runs the algorithm until the error is reduced to a satisfactory level.

Convolutional Neural Networks (CNN)

Images are two-dimensional arrays, and therefore one does not get satisfactory results if they are fed directly to a typical NN whose input is a one-dimensional array. To remedy this problem, one introduces convolution layers in which each neuron represents a pixel and, unlike fully connected NN, the neuron corresponding to a pixel is connected only to the neurons corresponding to neighboring pixels in the next layer [93]. What these convolutional layers do to the data is similar to the action of traditional image processing filters: they average pixels in a small neighborhood into a new pixel in the next layer. This operation is traditionally called a convolution. This averaging is weighted with weights that are given by the aforementioned w_i. In traditional image processing, one uses simple convolutions for edge detection, sharpening, Gaussian blur, and so forth. However, unlike traditional filters, the weights of CNN convolutions are learned, depending on the task at hand. Unlike fully connected NN which learns about global features of data, CNN finds local patterns.

The convolution layers reduce the size of the image as well. The size is further reduced using a max-pooling layer which takes the maximum of pixel values in (say, 3 × 3) neighborhoods. Max-pooling denoises the data as well. After the pooling, the image is flattened into a one-dimensional array and fed into a fully connected NN for tasks such as classification.

CNN use the *ReLU activation function* which is zero for $x < 0$ and equals x for $x \geq 0$. CNN are used in the detection of disorders from radiological images [94] and classification of images of texts [95], among others.

Recurrent Neural Networks

The NNs described so far are called feed-forward NNs because the flow of information in them is always in one direction. However in the study of time series, text, speech, and other sequential forms of data it is helpful to use NNs that involve a loop. Imagine we have a sequence of data such as $z_0, z_1, z_2, \ldots, z_n$. An RNN [96] has a main unit that is run n times, receiving z_t as input at iteration t. In addition to the input, an RNN has another ingredient called the state. The output of each iteration of the RNN is fed to the state of the next iteration. The next iteration then uses the state it received together with the input z_t to produce its own output. RNN have been used, for example, in speech recognition [97].

Long Short-Term Memory (LSTM)

This special type of RNN was introduced to deal with a problem in RNN called *vanishing gradient* [98]. This refers to the fact that information fed to an NN is lost after passing through too many layers. Similarly, an RNN cannot keep track of the dependencies between z_t and z_{t+T} if T is large. In addition to the state, LSTM have another component called *carry*, which is tasked with carrying information from older data

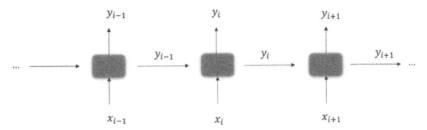

Figure 5.11 Schematic Depiction of a Recurrent Neural Network.

points z_t to the newer iterations. The output of each LSTM iteration t is based on the input of its last iteration, z_t and the carry c_t. The next carry c_{t+1} is computed from c_t as well as the input and state at time t.

LSTM have achieved milestones in speech and handwriting recognition [97, 99] and have been adopted by tech companies such as Google, Apple [100], and Facebook for such tasks as well as machine translation [101].

Transformers

Transformers [102] are similar to RNN with the notable difference that data does not have to be fed to them in its sequential order. This allows for parallel processing of data and makes it possible to train data on very large databases which were not previously possible to use before. Transformers are one of the most advanced tools used in natural language processing [103, 104].

Graph Neural Networks

Whereas traditional data analysis studies data as isolated data points, modern data science takes into account the relations among data points as well [105]. Such binary relations can be conceptualized as edges (links) in a network. Examples of networked data include social networks, citation networks, protein-protein interaction networks, metabolic networks, neuronal networks (not to be confused with artificial NN!), gene regulatory networks, RNA secondary structure, molecules and knowledge graphs. In addition, any data set equipped with a notion of distance can be converted to a network by choosing a distance threshold and connecting the points whose pairwise distances are less than.

As the input layer of a CNN is a 2D grid, GNN can be regarded as a generalization of CNN with an arbitrary graph G as the input layer [106, 107]. More precisely a GNN is a generalized RNN with multiple recurrent units, each corresponding to a vertex of the graph G. The output of each iteration is fed to the next iteration based on the edges of the graph.

Given labels (vectors) for some of the vertices of the graph, the purpose of the GNN is to compute the label for each vertex. The computed labels give us an embedding of the graph into Euclidean space.

GNN have been used in traffic prediction [108], graph representation [109] and recommender systems [110]. GNN can also be used to generate graphs based on the input data [111, 112] and are expected to help AI researchers with modeling high-level human reasoning, which involves conferring relations between concepts from real-world experiences [107].

Generative Adversarial Networks (GAN)

GANs [113] belong to the domain of *generative AI* (i.e., the art and science of producing content—text, imagery, speech, or music) using AI. GAN consist of a generator NN, which produces synthetic images, and a discriminator, which compares the output of the generator to a database of real images and scores them according to their similarity to the real images. GAN are capable of producing realistic images of persons who do not really exist. GAN can be used to generate fashion photos without the need to hire a model[1] and modeling jet formations [114].

Autoencoders

This class of neural networks is used for nonlinear dimensionality reduction [115]. They typically consist of a layer (coder) that maps the data into a lower-dimensional space and another layer (decoder) that maps the data back to the original space. The loss function measures the difference of the output with the input data (hence the name autoencoder). This way the lower-dimensional output of the coder can be regarded as the gist of the data and can be used in dimensionality reduction. There are also *denoising autoencoders* that compare data with a partially corrupted version of it. They are trained to recover the original data from the corrupted one.

Self-Organizing Maps

This type of NN tires to fit a grid to the data [116]. The grid consists of a layer of neurons whose state is a vector, initiated randomly. At each step, the state vectors are compared with a point in the data set (using Euclidean distance), and the best matching unit (BMU) is chosen. The states of the vectors in the grid are then updated to bring them closer to that of the BMU. The magnitude of this change is dampened over distance from BMU and over time (iteration). Unlike other types of NN, SOM use *competitive learning*, which means the neurons are compared with each other to measure their closeness to the target.

5.7 Bio-Inspired Methods

AI and ML problems often involve optimizing an objective or loss function f. When f is differentiable, as, in the case of OLS regression, one can use derivative to find its optima (assumed here to be minima). If the equation for the zeros of the derivative is too complicated to solve, one can use the method of gradient descent to find the minimum, although this method has the potential to fall into a local minimum of f.

If f is not differentiable, for example when its domain is discrete, then one may have to explore the whole domain of the function to find the optima. Since the domain

---

1 https://www.cdotrends.com/story/14300/rise-ai-supermodels

of the function is often prohibitively large, algorithms have been developed to use the symmetries of the function f or heuristics to make the search process more efficient [117]. One such method is *dynamic programming* (DP), which can be used when the optimization problem can be divided into overlapping subproblems and the optimum of the main problem can be constructed from the optima of the subproblems.

The simplest example of using DP is to compute the factorial of a number N using the relation $k! = k \cdot (k-1)!$ instead of computing $k!$ for each $1 \le k \le N$ separately. More generally DP is used when the domain of the problem has a tree structure and the value or optimum corresponding to a node can be computed from that of its children. One then starts computing from the leaves of the tree till the sought-after value at the root is reached. DP is used in many algorithms such as dynamical time warping for computing the distance between time series of different lengths [118], computing the minimum free energy structure of an RNA (ribonucleic acid) molecule [119], among others.

However, when f is not continuous and there is no such recursion, we have to use a search method that combines *exploration* of the domain of f and *exploitation* of the fact that f can have a form of regularity in small neighborhoods, at least in some directions. Such is the case in formal formulations of the natural process of evolution: we have a population of organisms, each with a given *fitness* associated to them that reflects their ability to survive in the current environment. The organisms reproduce and genetic mutations occur in the process, causing diversification of the organisms across generations. It is the more fit individuals that have a more likelihood of reproducing (*selection*). In evolutionary optimization (EO), of which genetic algorithms are a subfield, one uses similar principles to solve optimization problems that are impossible to solve using other methods because of the irregularity of f or the complexity of its domain [120]. There is however a notable difference between EO and natural evolution in that the latter is an open-ended procedure that, unlike the former, does not have an end goal.

One represents the elements of the domain of f by strings of bits, numbers, strings or symbols. Real valued or tree-based representations are also possible. One starts the simulation with an initial population of the individuals, that is, elements of the domain of f, which can be chosen at random. The function f is now called the *fitness function*. Elements of the population are sampled (selected) randomly in such a way that more fit members are given priority. For each sampled individual, a mutated copy (offspring) of it is added to the population. Mutation occurs based on a chosen mutation rate and it replaces a symbol in the representative string (of the individual selected to reproduce) with another one. Members of the population die based on a rate independent of their fitness. This way the overall fitness of the population is expected to increase over time. -

EO has been used in the design of electronic circuits [120], for computer-aided design of molecules [121], for designing a satellite antenna for NASA [122] and for designing controllers for robots, helicopters and a supply ship [123].

5.8 Reinforcement Learning

RL is similar to evolutionary optimization in the sense that it involves large state spaces. It however differs in two key aspects: RL is concerned with cumulative returns over time instead of just an optimal end goal, and the optimization method used in RL differs from EO.

We have a state-space S consisting of the states the agent and the environment can take, as well as an action space A. For each state s and action a we have a probability of transitioning to the state s^0. Moreover, each state–action–state triplet (s,a,s') results in a *reward* $r_a(s,s')$. The *Markov property* states that this probability is independent of the earlier states. A *policy* π is a rule that given a state s, tells us which action to take. Policies can be either probabilistic or deterministic. A deterministic policy is given by a map from S to A. The reward process can be regarded as a result of the interaction of an agent with its environment and thereby learning to optimize its policy. We want to find a policy, called optimal policy, that maximizes the *cumulative reward*

$$R = E\left[\sum_{i=0}^{\infty} \gamma^i r_{at}\left(s_t, s_t + 1\right)\right] \tag{5.26}$$

Here γ is a constant in $[0,1]$ called the discount rate, signifying the fact that the effect of the rewards diminishes over time. The action a_t is chosen based on the policy π and the current state s_t and therefore we may write $a_t = \pi(s_t)$. The previous mean is given with respect to the probability $P_{a_t}(s_t, s_{t+1})$ that the current action applied to state s_t, results in state s_{t+1}.

The value function of a policy $V^\pi(s)$ tells us how beneficial is for the agent to be at the state s, formally defined as the expected value of R, given the policy π and the initial state s. The Bellman equation [124] tells us that

$$V(s) = P_{\pi(s)}(s,s')\left(r_\pi(s)(s,s') + \gamma V(s')\right). \tag{5.27}$$

Similarly, the policy $\pi(s)$ can be expressed recursively in terms of $V(s')$ and thus, $V(s)$ and $\pi(s)$ can be computed using dynamic programming.

What we described thus far pertains to the general framework of Markov decision models. In RL, the transition probabilities and the rewards are not known a priori and are obtained using simulations, for example, by playing a computer game such as Go over and over again. One also uses the value of state–action pairs, $Q(s,a)$, instead of $V(s)$. Since it is not feasible to use dynamic programming to find the optimal policy for most practical RL problems, sampling and functional approximations are used instead. In the Monte Carlo method, for a given policy π, for an initial state, trajectories are sampled using the probability distribution of π to be able to approximate $Q(s,a)$. The policy is then updated in a greedy way by requiring that the updated policy imply an action that maximized $Q(s,\cdot)$.

Deep RL combines RL with deep learning [125]. It has triumphed in beating humans in the games of Atari [126] and Go [127]. It has potential applications in supporting decision-making in manufacturing ramp-up processes [11], self-driving cars [128], smart grids [129], and robotics [130–132].

5.9 Time Series Analysis

A time series is a time-dependent quantity $\{z_t\}$ whose values are given over a discrete set of time moments $t \in \{t_0, t_1, \ldots, t_N\}$. Time series analysis [133, 134], involves problems such as forecasting (estimating z_t for $t > t_N$), anomaly detection and time series clustering. Because of the ubiquity of time dependent quantities, time series analysis (TSA) is widely used in science, engineering, finance and business. In industry, TSA is used for forecasting manufacturing time series [135].

To analyze time series statistically, one assumes that the values z_t are samples from a time dependent probability distribution Z_t. In other words, the incorporation of probability distributions in modeling time series enables us to take the effect of uncertainties, noise and disturbance (also called residue or error) into account. A general strategy for forecasting time series is as follows.

- Determining p such that Z_t is determined by $Z_{t-1}, Z_{t-2}, \ldots, Z_{t-p}$.
- Using regression to model the dependence of Z_t on $Z_{t-1}, Z_{t-2}, \ldots, Z_{t-p}$ and extrapolating it for $t > n$.

For a given time series, the number p can be guessed by looking at scatter plots of (z_t, z_{t-k}), called *lag k plots*. For small k we can see correlation between z_t and z_{t+1}, unless our time series is noise. As k increases this correlation decreases and if p is such that correlation is not observed in lag k plots for $k > p$ it is our desired parameter.

Note that in general the dependence of Z_t on $Z_{t-1}, Z_{t-2}, \ldots, Z_{t-p}$ can be nonlinear and complicated. For this reason one considers simple linear models such as the autoregressive model, moving average, and their combination ARMA, which are capable of modeling a lot of time series. If the aforementioned dependence can be assumed to be linear, and the residue is normally distributed, we have an autoregressive model $AR(p)$:

$$z_t = c + \epsilon_t + \sum_{i=1}^{p} \phi_i z_t - i \tag{5.28}$$

Here ϵ_t is (Gaussian) noise.

To determine the parameter p in an AR model, one defines the *sample autocovariance function* $\hat{\gamma}(h)$ and *sample autocorrelation function (ACF)* $\hat{p}(h)$:

$$\hat{\gamma}(h) = \frac{1}{n} \sum_{n-|h|}^{t=1} (x_{t+h} - \bar{x})(x_t + \bar{x}) \hat{p}(h) = \frac{\hat{\gamma}(h)}{\hat{\gamma}(0)} \tag{5.29}$$

where $\bar{x} = \sum_{i=1}^{n} x_i / n$ is the sample mean and h is an integer between $-n$ and n. Note that $\hat{\gamma}(h)$ measures the covariance of Z_t with its shifted form, X_{t+h} and hence the name autocovariance. It is a part of the stationarity assumption that this quantity is independent of t. It can be shown that if our time series is such that $\{Z_t\}$ are identical and independently distributed (IID) noise (i.e., each Z_t being a Gaussian distribution and independent of the others), with mean zero and variance 1, then about 95% of the $°\hat{p}(h)$, for different h will lie in the bound $\pm 1.96 / \sqrt{n}$.

For forecasting, one usually assumes that $\{Z_t\}$ is *(weakly) stationary*, that is, the mean of its values, called *trend* is independent of t. There are different ways to define the mean.

$$m_t = \frac{1}{2k+1} \sum_{i+t-k}^{t+k} z_i \tag{5.30}$$

The parameter k has to be chosen carefully so as to smooth out the fluctuations in the values of $\{Z_t\}$ but not the actual trend of the time series. The mean can also be computed using *exponential smoothing*. It is defined recursively by $m_1 = Z_1$ and $m_t = \alpha Z_t + (1-\alpha)m_{t-1}$, where $\alpha \in [0,1]$ is a parameter. Unlike (5.30), the exponential smoothing depends only on past values and hence can be used for forecasting as well.

Once we compute m_t, we can take $Z_t - m_t$, which is likely to be stationary. The idea is to forecast $Z_t - m_t$ and then add the extrapolated trend back. More generally one has the *classical decomposition* of a time series:

$$Z_t = m_t + Y_t + s_t \tag{5.31}$$

where m_t is the trend, Y_t is stationary with mean zero and s_t is the *seasonal component*. It satisfies $s_{t+d} = s_t$ for a constant d called the period. Note that d can often be deduced from the source of the data (such as yearly sales) or by plotting the data. To estimate s_t one takes the average of $Z_{t+id} - m_{t+id}$ for all the i for which the latter is defined. If s_t' denotes this average then we set $s_t = s_t' - \sum_{i=1}^{d} s_i' / d$ so that $s_1 + s_2 + \cdots + s_d = 0$.

An example of a classical decomposition of a time series, for sunspot activity data, is depicted in Figure 5.12. The top plot depicts the average monthly number of Sunspots from 1749 to 1983 [139]. The second plot depicts the overall trend of this time series over the years. It is obtained by taking the moving average of the time series to smooth out small-scale fluctuations. Seasonality is depicted in the third plot, which clearly shows the 11-year Sunspot cycle. The last plot is the remaining noise (uncertainty) in the data. Decomposition is done using the Statsmodels library for Python.

Note that in the AR model, noise is associated only with time t. However in general, we can have independent noise for each time t. Considering separate noise for different time lags results in the more general autoregressive moving average (ARMA) model:

$$Z_t = c + \sum_{l=1}^{q} \theta_i \in_{t-i} + \sum_{i=1}^{p} \phi_i Z_{t-i} \tag{5.32}$$

where each $\in_{t-i}$ is Gaussian noise with mean zero and the same variance σ^2. The first sum in (5.32) is called the *moving average* part, and it expresses the current value in terms of past errors. This reflects the fact that error are caused by uncertainties in the model and environment that can be recurrent in time.

There are various enhancements of the ARMA model such as autoregressive integrated moving average (ARIMA) for non-stationary time series and Seasonal ARIMA (SARIMA) [134]. In 2018, Facebook released an enhancement of the ARIMA method, called Prophet, that takes into account the effect of holidays as well. It is also designed so as to be scalable for business forecasting [136].

Recent years have seen increased interest in probabilistic methods of time series forecasting [137–139]. These methods differ from the traditional methods in the following:

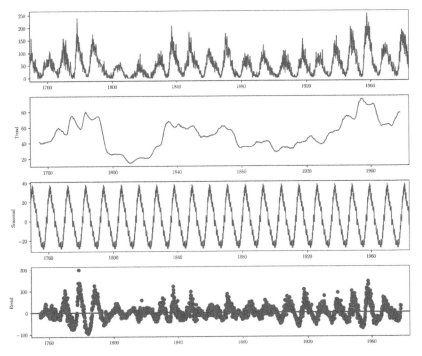

Figure 5.12 Historical Sunspot Activity Time Series and Its Classical Decomposition.

- They do not assume a linear relation between the current value and the past values and errors.
- They give probabilistic forecasts for future values instead of point estimates. This enables one to prepare for different possible outcomes instead of focusing on the most probable one.
- They model and forecast ensembles of time series. Examples of such ensembles include sales of different items for a retailer or the availability of drivers for ride hailing apps such as Uber or Gojek. This significantly improves time performance of the algorithm and its scalability. It can also improve prediction accuracy.
- They do not assume that the distribution of noise (error) is normal, an assumption that often does not hold in reality [138]. They instead estimate the distribution of data using state-of-the-art statistical methods.
- They typically use recurrent neural networks for distribution estimation and forecasting.

Bibliography

1 Davis, J., et al. (2015). Smart manufacturing. *Annu. Rev. Chem. Biomol. Eng.* 6: 141–160. 10.1146/annurev-chembioeng-061114-123255.

2 Wuest, T., Weimer, D., Irgens, C., and Thoben, K.D. (2016). Machine learning in manufacturing: Advantages, challenges, and applications. *Prod. Manuf. Res.* 4 (1): 23–45. 10.1080/21693277.2016.1192517.

3 Sony, M., Antony, J., and Douglas, J.A. (2020). Essential ingredients for the implementation of Quality 4.0. *TQM J.* 32 (4): 779–793. 10.1108/tqm-12-2019-0275.

4 Zonnenshain, A., and Kenett, R.S. (2020). Quality 4.0—The challenging future of quality engineering. *Qual. Eng.* 32 (4): 614–626.

5 Deep machine learning: GE and BP will connect thousands of subsea oil wells to the industrial Internet. https://www.ge.com/news/reports/deep-machine-learning-ge-and-bp-will-connect-2.

6 Winick, E. (2018). Uptake is putting the world's data to work. https://www.technologyreview.com/2018/09/12/140306/uptake-is-putting-the-worlds-data-to-work.

7 Pham, D.T., and Afify, A.A. (2005). Machine-learning techniques and their applications in manufacturing. *Proc. Inst. Mech. Eng. Part B J. Eng. Manuf.* 219 (5): 395–412. 10.1243/095440505X32274.

8 Harding, J.A., Shahbaz, M., Srinivas, S. and Kusiak, A. (2006). Data mining in manufacturing: A review. *J. Manuf. Sci. Eng. Trans. ASME* 128 (4): 969–976. 10.1115/1.2194554.

9 Susto, G.A., Schirru, A., Pampuri, S., McLoone, S., and Beghi, A. (2015). Machine learning for predictive maintenance: A multiple classifier approach. *IEEE Trans. Ind. Informatics* 11 (3): 812–820. 10.1109/TII.2014.2349359.

10 Filipič, B., and Junkar, M. (2000). Using inductive machine learning to support decision making in machining processes. *Comput. Ind.* 43 (1): 31–41. 10.1016/S0166-3615(00)00056-7.

11 Köksal, G., Batmaz, I., and Testik, M.C. (2011). A review of data mining applications for quality improvement in manufacturing industry. *Expert Syst. Appl.* 38 (10): 13448–13467. 10.1016/j.eswa.2011.04.063.

12 Poole, D.L., and Mackworth, A.K. (2010). *Artificial Intelligence: Foundations of Computational Agents.* Cambridge University Press.

13 Venkatasubramanian, V. (2019). The promise of artificial intelligence in chemical engineering: Is it here, finally? *AIChE J.* 65 (2): 466–478.

14 Domingos, P. (2015). *The Master Algorithm: How the Quest for the Ultimate Learning Machine Will Remake Our World.* Basic Books.

15 Goodfellow, I., Bengio, Y., and Courville, A. (2016). *Deep Learning Book* Vol. 521 (7553): 800. MIT Press.

16 Floreano, D., and Mattiussi, C. (2008). Bio-Inspired Artificial Intelligence. MIT Press.

17 Gelman, A., Carlin, J.B., Stern, H.S., and Rubin, D.B. (1995). *Bayesian Data Analysis.* Chapman and Hall/CRC.

18 Baral, C., and Gelfond, M. (1994). Logic programming and knowledge representation. *J. Log. Program.* 19: 73–148.

19 Mikolov, T., Chen, K., Corrado, G., and Dean, J. (2013). Efficient estimation of word representations in vector space. arXiv Prepr. arXiv1301.3781.

20 Vapnik, V., Golowich, S.E., and Smola, A.et al (1997). Support vector method for function approximation, regression estimation, and signal processing. In: *Proceedings of Adv. Neural Inf. Process. Syst.* 281–287.

21 Weiss, K., Khoshgoftaar, T.M., and Wang, D.D. (2016). *A Survey of Transfer Learning*, Vols. 1, 3. Springer International Publishing.

22 Weiss, K., Khoshgoftaar, T.M., and Wang, D. (2016). A survey of transfer learning. *J. Big Data* 3 (1): 1–40.

23 Urbanowicz, R., and Browne, W. (2015). Introducing rule-based machine learning: a practical guide. In: *Proceedings of the Companion Publication of the 2015 Annual Conference on Genetic and Evolutionary Computation*, 263–292.

24 Apte, C., Weiss, S., and Grout, G. (1993). Predicting defects in disk drive manufacturing. A case study in high-dimensional classification. In: *Proc. Conf. Artif. Intell. Appl* 212–218. doi:10.1109/caia.1993.366608.

25 Hackeling, G. (2017). *Mastering Machine Learning with Scikit-learn*. Packt Publishing Ltd.

26 Provost, F., and Fawcett, T. (2013). *Data Science for Business: What You Need to Know about Data Mining and Data-analytic Thinking*. O'Reilly Media, Inc.

27 Zhang, H. (2005). Exploring conditions for the optimality of naive Bayes. *Int. J. Pattern Recognit. Artif. Intell.* 19 (2): 183–198.

28 Shakhnarovich, G., Darrell, T., and Indyk, P. (2005). *Nearest-Neighbor Methods in Learning and Vision Theory and Practice*. MIT Press.

29 Tee, T.X. and Khoo, H.K. (August 2020). Facial recognition using enhanced facial features k-nearest neighbor (k-NN) for attendance system. In: *ACM International Conference Proceeding Series*, 14–18. doi: 10.1145/3417473.3417475.

30 Peng, X., Cai, Y., Li, Q., and Wang, K. (2017). Control rod position reconstruction based on K-Nearest neighbor method. *Ann. Nucl. Energy* 102: 231–235.

31 Sarwar, B., Karypis, G., Konstan, J., and Riedl, J. (2000). Application of dimensionality reduction in recommender system-a case study. https://apps.dtic.mil/sti/citations/ADA439541.

32 Izenman, A.J. (2013). Linear discriminant analysis. In: *Modern Multivariate Statistical Techniques (ed. A. J. Izenman)*, 237–280. Springer.

33 Chelali, F.Z., Djeradi, A., and Djeradi, R. (2009). Linear discriminant analysis for face recognition. In: *2009 International Conference on Multimedia Computing and Systems*, 1–10.

34 Altman, E.I. (1968). Financial ratios, discriminant analysis and the prediction of corporate Bankruptcy. *J. Finance* 23 (4): 589. 10.2307/2978933.

35 Jurkovic, Z., Cukor, G., Brezocnik, M., and Brajkovic, T. (2018). A comparison of machine learning methods for cutting parameters prediction in high speed turning process. *J. Intell. Manuf.* 29 (8): 1683–1693. 10.1007/s10845-016-1206-1.

36 Crawford, B., Khayyam, H., Milani, A.S., and Jazar, R.N. (2020). *Big Data Modeling Approaches for Engineering Applications*. Springer.

37 Rejani, Y., and Selvi, S.T. (2009). Early detection of breast cancer using SVM classifier technique. *International Journal on Computer Science and Engineering* 1: 127–130. arXiv Prepr. arXiv0912.2314.

38 Burbidge, R., Trotter, M., Buxton, B., and Holden, S. (2001). Drug design by machine learning: support vector machines for pharmaceutical data analysis. *Comput. & Chem.* 26 (1): 5–14.

39 Salahshoor, K., Kordestani, M., and Khoshro, M.S. (2010). Fault detection and diagnosis of an industrial steam turbine using fusion of SVM (support vector machine) and ANFIS (adaptive neuro-fuzzy inference system) classifiers. *Energy* 35 (12): 5472–5482. doi: 10.1016/j.energy.2010.06.001.

40 Çaydac, U., and Ekici, S. (2012). Support vector machines models for surface roughness prediction in CNC turning of AISI 304 austenitic stainless steel. *J. Intell. Manuf.* 23 (3): 639–650.

41 Widodo, A., and Yang, B.S. (2007). Support vector machine in machine condition monitoring and fault diagnosis. *Mech. Syst. Signal Process.* 21 (6): 2560–2574. doi: 10.1016/j.ymssp.2006.12.007.

42 Sun, J., Rahman, M., Wong, Y.S., and Hong, G.S. (2004). Multiclassification of tool wear with support vector machine by manufacturing loss consideration. *Int. J. Mach. Tools Manuf.* 44 (11): 1179–1187. doi: 10.1016/j.ijmachtools.2004.04.003.

43 Ribeiro, B. (2005). Support vector machines for quality monitoring in a plastic injection molding process. *IEEE Trans. Syst. Man, Cybern. Part C Applications Rev.* 35 (3): 401–410. doi: 10.1109/TSMCC.2004.843228.

44 Tipping, M.E. (Jun 2001). Sparse Bayesian learning and the relevance vector machine. *J. Mach. Learn. Res.* 1: 211–244.

45 Li, Y., et al. (2019). Data-driven health estimation and lifetime prediction of lithium-ion batteries: a review. *Renew. Sustain. Energy Rev.* 113: 109254.

46 Kotsiantis, S.B. (2014). Bagging and boosting variants for handling classifications problems: a survey. *Knowl. Eng. Rev.* 29 (1): 78–100. doi: 10.1017/S0269888913000313.

47 Nielsen, D. (2016). Tree boosting with xgboost-why does xgboost win" every" machine learning competition? Master's thesis, NTNU.

48 Gruber, M. (2017). *Improving Efficiency by Shrinkage: The James–stein and Ridge Regression Estimators.* Routledge.

49 Awad, M., and Khanna, R. (2015). Support vector regression. In: *Efficient Learning Machines: Theories, Concepts, and Applications for Engineers and System Designers,* 67–80. Apress.

50 Wang, J., Li, L., Niu, D., and Tan, Z. (2012). An annual load forecasting model based on support vector regression with differential evolution algorithm. *Appl. Energy* 94: 65–70. doi: 10.1016/j.apenergy.2012.01.010.

51 Hu, W., Yan, L., Liu, K., and Wang, H. (February 2016). A short-term traffic flow forecasting method based on the hybrid PSO-SVR. *Neural Process. Lett.* 43 (1): 155–172. doi: 10.1007/s11063-015-9409-6.

52 Golkarnarenji, G., et al. (2019). Multi-objective optimization of manufacturing process in carbon fiber industry using artificial intelligence techniques. *IEEE Access* 7: 67576–67588. doi: 10.1109/ACCESS.2019.2914697.

53 KI Williams, C. (2006). *Gaussian Processes for Machine Learning.* Taylor & Francis Group.

54 Liu, K., Hu, X., Wei, Z., Li, Y., and Jiang, Y. (2019). Modified Gaussian process regression models for cyclic capacity prediction of lithium-ion batteries. *IEEE Trans. Transp. Electrif.* 5 (4): 1225–1236.

55 Liu, K., Li, Y., Hu, X., Lucu, M., and Widanage, W.D. (2019). Gaussian process regression with automatic relevance determination kernel for calendar aging prediction of lithium-ion batteries. *IEEE Trans. Ind. Informatics* 16 (6): 3767–3777.

56 Wahba, G. (1990). *Spline Models for Observational Data. SIAM.*

57 Bookstein, F.L. (1989). Principal warps: Thin-plate splines and the decomposition of deformations. *IEEE Transactions on Pattern Analysis and Machine Intelligence* 11 (6): 567–585.

58 Khayyam, H., et al. (2017). Predictive modelling and optimization of carbon fiber mechanical properties through high temperature furnace. *Appl. Therm. Eng.* 125: 1539–1554.

59 Rand, W.M. (1971). Objective criteria for the evaluation of clustering methods. *J. Am. Stat. Assoc.* 66 (336): 846–850.

60 Sousa, L. and Gama, J. (2014). The application of hierarchical clustering algorithms for recognition using biometrics of the hand.

61 Khan, K.S., Yin, Y., and Jamalipour, A. (2019). On the application of agglomerative hierarchical clustering for cache-assisted D2D networks. 2019 16th IEEE Annual Consumer Communications Networking Conference (CCNC), 1–6.

62 Metzker, M.L., Mindell, D.P., Liu, X.-M., Ptak, R.G., Gibbs, R.A., and Hillis, D.M. (2002). Molecular evidence of HIV-1 transmission in a criminal case. *Proc. Natl. Acad. Sci.* 99 (22): 14292–14297.

63 Khan, K., Rehman, S.U., Aziz, K., Fong, S., and Sarasvady, S. (2014). DBSCAN: Past, present and future. The fifth international conference on the applications of digital information and web technologies (ICADIWT 2014), 232–238.

64 Tracking down the Villains: Outlier Detection at Netflix. http://techblog.netflix.com/2015/07/tracking-down-villains-outlier.html.

65 Mahapatro, G., Mishra, D., Shaw, K., Mishra, S., and Jena, T. (2012). Phylogenetic tree construction for DNA sequences using clustering methods. *Procedia Eng.* 38: 1362–1366.

66 Bolón-Canedo, V., Sánchez-Maroño, N., and Alonso-Betanzos, A. (2015). *Feature Selection for High-dimensional Data*. Springer.

67 Blum, A., Hopcroft, J., and Kannan, R. (2020). *Foundations of Data Science*. Cambridge University Press.

68 Shenai, P.M., Xu, Z., and Zhao, Y. (2012). Applications of principal component analysis (PCA) in materials science. *Princ. Compon. Anal. Appl.* 25–40.

69 Pasini, G. (2017). Principal component analysis for stock portfolio management. *Int. J. Pure Appl. Math.* 115 (1): 153–167.

70 Zheng, N. and Xue, J. (2009). *Statistical Learning and Pattern Analysis for Image and Video Processing*. Springer Science & Business Media.

71 Belkin, M. and Niyogi, P. (2003). Laplacian eigenmaps for dimensionality reduction and data representation. *Neural Comput.* 15 (6): 1373–1396.

72 Lurie, J. (1999). Review of spectral graph theory. *ACM SIGACT News* 30 (2): 14–16.

73 Perry, T., Zha, H., Frias, P., Zeng, D., and Braunstein, M. (2012). Supervised Laplacian eigenmaps with applications in clinical diagnostics for pediatric cardiology. arXiv Prepr. arXiv1207.7035.

74 Schölkopf, B., Smola, A., and Müller, K.-R. (1997). Kernel principal component analysis. International conference on artificial neural networks, 583–588.

75 Hoffmann, H. (2007). Kernel PCA for novelty detection. *Pattern Recognit.* 40 (3): 863–874.

76 Zomorodian, A.J. (2005). *Topology for Computing*, Vol. 16. Cambridge University Press.

77 Carlsson, G., Topology and data. *Bulleting of the American Mathematical Society* 46: 255–308.

78 Hatcher, A. (2005). *Algebraic Topology*. Cambridge University Press.

79 Cohen-Steiner, D., Edelsbrunner, H., and Harer, J. (2007). Stability of persistence diagrams. *Discret. \& Comput. Geom.* 37 (1): 103–120.

80 The GUDHI Project, GUDHI User and Reference Manual. 2021.

81 Lum, P.Y. et al. (2013). Extracting insights from the shape of complex data using topology. *Sci. Rep.* 3 (1): 1–8.

82 De Silva, V. and Ghrist, R. (2007). Coverage in sensor networks via persistent homology. *Algebr. \& Geom. Topol.* 7 (1): 339–358.

83 Horak, D., Maletić, S., and Rajković, M. (2009). Persistent homology of complex networks. *J. Stat. Mech. Theory Exp.* 03: P03034.

84 Petri, G. et al. (2014). Homological scaffolds of brain functional networks. *J. R. Soc. Interface* 11 (101): 20140873.

85 Carstens, C.J. and Horadam, K.J. (2013). Persistent homology of collaboration networks. *Math. Probl. Eng.* 2013.

86 Duponchel, L. (2018). When remote sensing meets topological data analysis. *J. Spectr. Imaging* 7: 1–10. 10.1255/jsi.2018.a1.

87 Chan, J.M., Carlsson, G., and Rabadan, R. (2013). Topology of viral evolution. *Proc. Natl. Acad. Sci.* 110 (46): 18566–18571.

88 Swarup, S. and Rezazadegan, R. (2019). Generating an agent taxonomy using topological data analysis. Proceedings of the 18th International Conference on Autonomous Agents and MultiAgent Systems, 2204–2205.

89 Zhu, X. (2005). Semi-Supervised Learning Literature Survey.

90 Settles, B. (2010). *Active Learning Literature Survey. University of Wisconsin.* Madison.

91 Hassoun, M.H., and others. (1995). *Fundamentals of Artificial Neural Networks.* MIT press.

92 Kidger, P. and Lyons, T. (2020). Universal approximation with deep narrow networks. Conference on Learning Theory, 2306–2327.

93 Albawi, S., Mohammed, T.A., and Al-Zawi, S. (2017). Understanding of a convolutional neural network. 2017 International Conference on Engineering and Technology (ICET), 1–6, doi: 10.1109/ICEngTechnol.2017.8308186.

94 Yamashita, R., Nishio, M., Do, R.K.G., and Togashi, K. (2018). Convolutional neural networks: an overview and application in radiology. *Insights Imaging* 9 (4): 611–629.

95 Simard, P.Y., Steinkraus, D., Platt, J.C., and others. (2003). Best practices for convolutional neural networks applied to visual document analysis. *Icdar* 3. 2003.

96 Medsker, L.R. and Jain, L.C. (2001). Recurrent neural networks. *Des. Appl.* 5.

97 Graves, A., Mohamed, A., and Hinton, G. (2013). Speech recognition with deep recurrent neural networks. 2013 IEEE International Conference on Acoustics, Speech and Signal Processing, 6645–6649.

98 Staudemeyer, R.C. and Morris, E.R. (2019). Understanding LSTM – a tutorial into long short-term memory recurrent neural networks. arXiv, 1–42.

99 Graves, A., Liwicki, M., Fernández, S., Bertolami, R., Bunke, H., and Schmidhuber, J. (2009). A novel connectionist system for unconstrained handwriting recognition. *IEEE Trans. Pattern Anal. Mach. Intell.* 31 (5): 855–868. doi: 10.1109/ TPAMI.2008.137.

100 Apple's Machines Can Learn Too.

101 Wu, Y. et al. (2016). Google's neural machine translation system: bridging the gap between human and machine translation. arXiv Prepr. arXiv1609.08144.

102 Vaswani, A. et al. (2017). Attention is all you need. arXiv Prepr. arXiv1706.03762.

103 Wolf, T. et al. (October 2020). Transformers: State-of-the-art natural language processing. In: *Proceedings of the 2020 Conference on Empirical Methods in Natural Language Processing: System Demonstrations*, 38–45.

104 Brown, T.B. et al. (2020). Language models are few-shot learners. arXiv Prepr. arXiv2005.14165.

105 Silva, T.C. and Zhao, L. (2016). *Machine Learning in Complex Networks*, Vol. 1. Springer.

106 Scarselli, F., Gori, M., Tsoi, A.C., Hagenbuchner, M., and Monfardini, G. (2008). The graph neural network model. *IEEE Trans. Neural Networks* 20 (1): 61–80.

107 Zhou, J. et al. (2018). Graph neural networks: a review of methods and applications. arXiv Prepr. arXiv1812.08434.

108 Rahimi, A., Cohn, T., and Baldwin, T. (2018). Semi-supervised user geolocation via graph convolutional networks. arXiv Prepr. arXiv1804.08049.

109 Hamilton, W.L., Ying, R., and Leskovec, J. (2017). Inductive representation learning on large graphs. arXiv Prepr. arXiv1706.02216.

110 Ying, R., He, R., Chen, K., Eksombatchai, P., Hamilton, W.L., and Leskovec, J. (2018). Graph convolutional neural networks for web-scale recommender systems. In: *Proceedings of the 24th ACM SIGKDD International Conference on Knowledge Discovery & Data Mining*, 974–983.

111 Bojchevski, A., Shchur, O., Zügner, D., and Günnemann, S. (2018). Netgan: Generating graphs via random walks. In: *International Conference on Machine Learning*, 610–619.

112 De Cao, N. and Kipf, T. (2018). MolGAN: an implicit generative model for small molecular graphs. arXiv Prepr. arXiv1805.11973.

113 Goodfellow, I.J., et al. (2014). Generative adversarial networks. arXiv Prepr. arXiv1406.2661.

114 De Oliveira, L., Paganini, M., and Nachman, B. (2017). Learning particle physics by example: location-aware generative adversarial networks for physics synthesis. *Comput. Softw. Big Sci.* 1 (1): 1–24.

115 Tschannen, M., Bachem, O., and Lucic, M. (2018). Recent advances in autoencoder-based representation learning. arXiv Prepr. arXiv1812.05069.

116 Kohonen, T. (2012). *Self-organizing Maps, 30*. Springer Science & Business Media.

117 Luke, S. (2013). Essentials of Metaheuristics, Second. Lulu.

118 Müller, M. (2007). Dynamic time warping. In: *Information Retrieval for Music and Motion*, 69–84.

119 Mathews, D., Sabina, J., Zuker, M., and Turner, D.H. (1999). Expanded sequence dependence of thermodynamic parameters improves prediction of RNA secondary structure. *J. Math. Biol.* 288: 911–940.

120 Floreano, D. and Mattiussi, C. (2008). *Bio-inspired Artificial Intelligence: Theories, Methods, and Technologies*. MIT Press.

121 Venkatasubramanian, V., Chan, K., and Caruthers, J.M. (1994). Computer-aided molecular design using genetic algorithms. *Comput. & Chem. Eng.* 18 (9): 833–844.

122 Hornby, G., Globus, A., Linden, D., and Lohn, J. (2006). Automated antenna design with evolutionary algorithms. in Space 2006, 7242.

123 Khayyam, H., Jamali, A., Assimi, H., and Jazar, R.N. (2020). Genetic programming approaches in design and optimization of mechanical engineering applications. In: *Nonlinear Approaches in Engineering Applications (ed. R. N. Jazar and L. Dai)*, 367–402. Springer.

124 Sutton, R.S. and Barto, A.G. (2018). *Reinforcement Learning: An Introduction*. MIT press.

125 François-Lavet, V., Henderson, P., Islam, R., Bellemare, M.G., and Pineau, J. (2018).

126 Mnih, V., et al. (2013). Playing Atari with deep reinforcement learning. arXiv Prepr. arXiv1312.5602.

127 Silver, D. et al. (2016). Mastering the game of Go with deep neural networks and tree search. *Nature* 529 (7587): 484–489.

128 Pan, X., You, Y., Wang, Z., and Lu, C. (2017). Virtual to real reinforcement learning for autonomous driving arXiv Prepr. arXiv1704.03952.

129 François-Lavet, V. (2017). *Contributions to Deep Reinforcement Learning and Its Applications in Smartgrids*. Université de Liège.

130 Levine, S., Finn, C., Darrell, T., and Abbeel, P. (2016). End-to-end training of deep visuomotor policies. *J. Mach. Learn. Res.* 17 (1): 1334–1373.

131 Pinto, L., Andrychowicz, M., Welinder, P., Zaremba, W., and Abbeel, P. *Asymmetric Actor Critic for Image-Based Robot Learning*.

132 Gandhi, D., Pinto, L., and Gupta, A. (2017). Learning to fly by crashing. In: *2017 IEEE/RSJ International Conference on Intelligent Robots and Systems (IROS)*, 3948–3955.

133 Bisgaard, S. and Kulahci, M. (2011). *Time Series Analysis and Forecasting by Example*. John Wiley & Sons.

134 Brockwell, P.J. and Davis, R.A. (2002). *Introduction to Time Series and Forecasting* 2e.

135 Guo, X., Sun, L., Li, G., and Wang, S. (2008). A hybrid wavelet analysis and support vector machines in forecasting development of manufacturing. *Expert Syst. Appl.* 35 (1–2): 415–422.

136 Taylor, S.J., and Letham, B. (2018). Forecasting at scale. *Am. Stat.* 72 (1): 37–45.

137 Alexandrov, A. et al. (2019). GluonTS: probabilistic time series models in python. *arXiv* 1–24.

138 Salinas, D., Flunkert, V., Gasthaus, J., and Januschowski, T. (2020). DeepAR: Probabilistic forecasting with autoregressive recurrent networks. *Int. J. Forecast.* 36 (3): 1181–1191.

139 Gasthaus, J., et al. (2020). Probabilistic forecasting with spline quantile function RNNs. In: *AISTATS 2019-22nd Int. Conf. Artif. Intell. Stat.*, 89.

Part II

Industry 4.0 Technologies

6

Multi-Vector Internet of Energy (IoE)

A Key Enabler for the Integration of Conventional and Renewable Power Generation

Sara Ershadi-nasab[a,1], Esmat Kishani Farahani[b,1], and Mahdi Sharifzadeh[a,]*

[a] *Sharif Energy, Water and Environment Institute (SEWEI), Sharif University of Technology, Tehran, Iran*
[b] *Institute of Electrical Engineering and Information Technology, Iranian Research Organization for Science and Technology*
[1] *Sara Ershadi-nasab and Esmat Kishani Farahani contributed equally to this work as first authors.*
[*] *Corresponding author. Sharif Energy, Water and Environment Institute (SEWEI), Sharif University of Technology, Tehran, Iran*

6.1 Challenges of the Conventional Power Grid and Need for Energy Internet

Over the past years, fossil fuel-based power grid has been used to provide the required power for industry and consumer usage. This centralized power generation system has some challenges that limit its usage. One of its main challenges is long distances between the power generators and consumers, which causes energy loss and challenges in transmitting the power. Another challenge is its production of large amount of CO_2 emissions, the leading cause in total greenhouse gas emissions since 1970 [1–3]. Furthermore, the increasing demand for energy in industrial and daily life [4] has brought about a critical energy crisis. Distributed renewable energy resources could provide the required energy demand and do not have the negative effects of fossil fuels in contributing to CO_2 emissions and climate change. Therefore, integrating renewable energy resources (RERs) with the conventional power system and improving the efficiency of energy usage are inevitable to overcoming energy limitations [5].

Wind and solar power are the two most common renewable resources that can be employed efficiently alongside conventional power generation technologies [4]. RERs are capable of providing energy demand. One of the most important features of renewable resources is that they are spatially distributed and have intermittent generation patterns. Therefore, their generation is subject to intermittency, and their exploiting within the conventional electricity grid may result in the imbalance of power supply and demand [6]. Therefore, RERs need to be interconnected with other sources of power generation as well as to storage systems. These interconnections change the structure of the power system from a radial single source with a centralized control system to a multi-source, gridding structure with distributed control systems. The existence of massively distributed generators in a power system renders the traditional control methods insufficient to satisfy both of the power demands and the stability of the whole system. The costs of renewable power generation from wind and solar energy are declining fast and have led to increases in a tendency toward their commercial utilization [3]. Furthermore, recent advancements in fast-charging

Industry 4.0 Vision for the Supply of Energy and Materials: Enabling Technologies and Emerging Applications, First Edition. Edited by Mahdi Sharifzadeh.
© 2022 John Wiley & Sons, Inc. Published 2022 by John Wiley & Sons, Inc.

batteries have opened up new possibilities in using renewable energies for vehicles in the transportation system [7].

The traditional power grid has a demand-driven and centralized architecture, with unidirectional flow pattern in which power is generated in a large-scale power plant and then transferred through transmission lines and distribution lines to reach the consumers at the end nodes. In this structure, consumers are passive and do not participate in generating power. However, renewable energy sources (RESs) such as solar, wind, hydro, tidal, and geothermal energies are inherently variable, uncertain, and distributed; hence including them into traditional power systems makes them inherently unstable [8]. To address this problem, smart grids were introduced [9,10].

A smart grid is a power grid that utilizes bidirectional multipath flow of energy and information to deliver, manage, and integrate RESs and includes communication, advanced control, and information technologies [5]. The communication architecture of the smart grid is composed of a home area network, neighborhood area network, and wide area network [11]. The home area network—sometimes called a premise area network or building area network—is responsible for providing connectivity of distributed RERs, advanced metering infrastructures (AMIs), and smart meters inside the home. It also provides the required interface for charging plug-in electric vehicles. The neighborhood area network is composed of several connected home area networks and AMIs in the locality. The wide area network is responsible for connecting neighborhood area networks with the proper electric power service for control and management of all the requests and demands in the network [11].

However, smart grids have some limitations in large scale in providing energy security, integrating the conventional and distributed power resources, and efficient use of power resources. Smart grids provide local access only to distributed and scalable energy sources; they cannot support other forms of energy to improve energy efficiency and have difficulty in providing reliable and secure energy. So Internet of Energy (IoE) has been proposed as the next generation of smart grid (smart grid 2.0 [9]) to overcome its limitations and advance it [5]. By using IoT solutions, IoE is an innovative approach to ensure the availability of energy anywhere at any time. In the IoE system, both centralized and highly distributed energy resources are existed in the network. All consumers in the network can closely interact and decide about the kind of energy they use.

6.2 Internet of Energy

6.2.1 Concept and Definition

The emergence of distributed energy resources such as solar and wind has opened up new opportunities for end users to change their role in power systems from consumer to both producer and consumer, a role called *prosumer*. Prosumers actively participate in the power market by producing electricity and selling its surplus. This changes the energy generation pattern from limited-number, high-production capacity, stationary, and centralized power generators to large-number, low-production capacity, variable, and highly distributed generators. These highly distributed energy generators (prosumers) need high connectivity and data exchange to provide energy reliably and securely that can be achieved only by the integration of advanced measuring infrastructure, low latency, secure and reliable communication and network technologies, real-time data collection, analysis, and decision-making, which are realized through the IoE concept, also known as energy Internet or energy Internet of things.

IoE is inspired by the information Internet. As the information Internet changed central computing infrastructure to distributed computing technologies, IoE changes centralized traditional power system to interconnected renewable energy systems, electric loads, storage devices, electric vehicles, distributed generators, and traditional power plants using Internet-based information and communications technology [2,12], as shown in Figure 6.1.

Analogous to the information Internet in which plug-and-play interface, information router, and open-source operating system are the key features, plug-and-play interface, energy router, and open platform are the key features of IoE. In other words, IoE unifies the IoT with smart grids [13–15] to enable the access to distributed renewable energies in a large area; it is a multi-vector energy system that integrates different forms of energy including electricity, gas, heat, and hydrogen to improve energy utilization efficiency and system resilience. IoE encompasses a variety of technologies, including IoT, advanced communication, big-data analysis, cloud computing, blockchain, and vehicle to grid and creates new market opportunities for real-time energy pricing. In an IoE system, data is collected using sensors and communication technologies from different edge devices within the system; it is transferred through the Internet and organized and analyzed to predict demand and supply and thus optimize and manage distributed energy systems [16].

6.2.2 IoE Structure

An IoE system consists of three subsystems—energy, information, and network—which are connected through energy routers, as shown in Figure 6.2 [2]. The energy subsystem is devoted to energy transformation and scheduling; the information

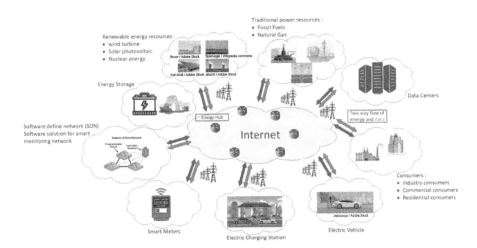

Figure 6.1 General View of IoE. Based on K. Wang, et al., 2017.

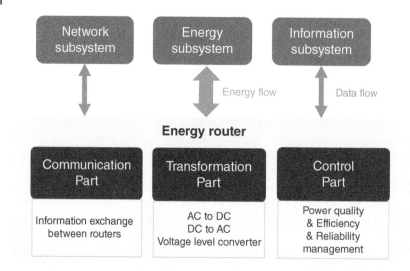

Figure 6.2 IoE Subsystems and Energy Router Parts. Based on K. Wang, et al., 2017.

subsystem consists of smart sensing and computing parts [17] and provides an open platform for data collection, processing, and management; and the network subsystem connects all the IoE system devices through wired or wireless technologies to make real-time monitoring, controlling, management, and pricing possible [2]. The energy router is the key component of the IoE and interacts with three main IoE subsystems through two-way communication links for both energy and information flows [2].

6.2.2.1 Energy Router

The concept of energy routers, inspired by Internet routers, was first introduced by Xu et al. [18] for dynamic scheduling of energy flow and real-time communications between power devices [2]. An energy router is the core of an IoE system and provides two-way flows of information and energy between the IoE components to facilitate the integration of RESs and storage devices into the traditional power grid and to manage energy optimally. An energy router, as shown in Figure 6.2, consists of three main parts: communication, transformation, and control. The communication part is responsible for the exchange of information between routers; the transformation part, which is usually a solid-state transformer (SST), connects DC/AC sources and loads at different voltage levels; and the control part optimizes energy demand and supply to improve system efficiency, reliability, and power quality. Three main features of energy routers are (1) plug-and-play interface for providing a seamless connection for end users; (2) bidirectional energy conversion among different kinds of terminals (AC, DC) with different voltage levels; and (3) optimal energy management in the local grid [2].

Several energy routers with different architectures have been proposed in the literature. In future renewable electric energy delivery and management (FREEDM) systems, an energy router is proposed that includes an SST device, distributed grid intelligent software, and communication ports [19]. Researchers at the Swiss Federal Institute of Technology developed an energy hub for energy conversion and storage. The federation of the digital power grid in Japan proposed a digital grid router that allocates IP addresses to different power devices to connect an existing power grid to an IoE and coordinate power management in certain areas [2], [20]. Shanghai researchers proposed an energy router for power dispatching and managing numerous decentralized RESs in low-voltage networks . Energy routers can be classified into SST-based, power line communication (PLC)-based, and multi-port converter (MPC)-based energy routers [21]. The PLC-based energy routers use modulation and multiplexing techniques such as time division multiple access to transfer both energy and information flows through the same transmission line. They reduce the cost and device volume [21]. MPC-based energy routers can realize energy balance among multiple RESs and loads. They have a high degree of reuse and integration compared with the SST-based energy routers. In MPC-based energy routers, each DC port can be expanded into an independent DC bus system. In MPC proposed in [22] an independent control center is employed to realize the energy exchange among all kinds of subsystems [23]. The subsystems of the aforementioned MPC-based energy router contain distributed energy sources, storage devices, and loads that can maintain energy balance under the control of three-port bidirectional converters [23].

6.2.2.2 Energy Hub

In a traditional power grid, only one form of energy (electricity) is considered, while IoE is a multi-vector energy system and considers the interactions of several energy forms together [24]. Various energy forms have many interdependencies, and a holistic expansion approach is needed for considering multi-vector energies in IoE [25].

As shown in Figure 6.3, an energy hub (EH) as the core of multi-energy systems is a place where different forms of energy can be converted, conditioned, or stored [26]. An

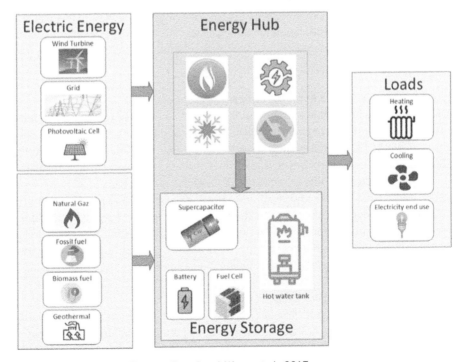

Figure 6.3 Energy Hub Concept. Based on J. Wang, et al., 2017.

EH is defined in the vision of future energy networks to integrate and interconnect the energy systems [15]. The operations of electricity and natural gas networks and their interactions and possible security issues are studied in [27–29]. Geidl et al. [30] proposed an EH model that optimizes the power dispatch economically. In [31,32], the role of EH is considered as an interface between the consumer and energy producers with various energy forms. An EH model is presented in [24] for exchanging electricity, natural gas, and heat resources. Electricity and natural gas supplies are inputs of EH, and in the output ports, electricity, and heating loads are supplied by multiple energy carriers. EH is considered as a unit that supports energy conversion, energy storage, and input–output functions for multiple energy carriers.

There are multiple energy hardware devices inside the EH for energy conversion, including distribution transformers, combined heat and power, micro-turbines, heat pumps, energy storage systems, gas furnaces, air conditioning systems, and thermal energy systems. For supplying a load using the EH, there are several options based on the availability and cost of energy carriers. For example, in a combined power and gas system [33], with electricity and natural gas as inputs of the EH and electricity, cold, and thermal as outputs of the EH, EH can convert input gas to electricity through micro-turbines to supply output electric loads at peak hours to optimize the multi-energy system. Furthermore, it can provide thermal energy to the building during these hours. On the other hand, at the hours during which the price of electricity is low, appliances can receive electricity directly from the power system [26]. Therefore, EH can optimize the energy system and increase its flexibility and economic performance [26].

6.2.2.3 The IoE Network

The IoE network connects all devices within the IoE using a combination of advanced wired and wireless communication technologies such as a software-defined network (SDN) [34], cognitive radio [34], cellular communications [35], PLC, WiMAX [36], and Zigbee [37].

6.2.2.3.1 Software-Defined Networks: Concept and Features

SDNs are proposed to enable optimal monitoring and management of a network and are widely used in many network applications such as data centers [38,39], wireless sensor networks [40], wireless networks [41], wide area networks [42], underwater sensor networks, and enterprises. SDNs separate the data plane from the control plane, which increases the flexibility of the network for global management.

SDNs add several features and capabilities to the network such as programmability, protocol independency, traffic identification at any flow, packet-level determination, service quality control, throughput identification and delay handling, faster and easier network management, configuring a large list of vendors with different protocols, and resiliency against failures, link failures, and malicious attacks.

6.2.2.3.2 SDN-Based Solutions for Smart Grid Problems

Smart grids in the current state face lots of limitations and challenges such as difficulty in large-scale integrating of centralized power generators and highly distributed energy resources; difficulty in control, management operation, and dispatching of power; security issues; and energy trading problems to work in the proper and expected state [43]. SDN as a software solution is proposed to overcome some of these limitations in the following ways.

There are lots of different devices in the smart grid, such as smart meters, energy routers, different sensors, and energy devices. These devices should establish a connection between each other to enable data communication and energy exchange. Different companies and vendors have their own policy for establishing a connection between their product devices. In the large-scale smart grid network, it is not guaranteed that all devices in the network come from the same manufacturer, and therefore different communication protocols should be synchronized to work properly. Usually, an operator is needed to interfere and perform the initial setting for correct communication between different devices. To avoid manual interference, software solutions are proposed to provide the seamless connections among different energy devices from different vendors and manufacturers, each with its own communication protocol. The smart grid heavily relies on communication systems and uses a heterogeneous combination of different communication protocols and standards. Since software solutions do not rely on protocols and standards, they can overcome the protocol dependency limitations of smart grids.

The final goal of a smart grid is to provide reliable electric power by incorporating renewable energy resources to the power grid and automating its control and management operation, making it possible to monitor the network in real time and satisfy all demand and requests. Software solutions such as SDN help the smart grid to provide the capability for such real-time monitoring of the network in terms of information flow rate or packet-level traffic control to manage the network in an optimal way.

The communication system in a smart grid should be resilient to attacks and failures and remain able to adapt to the new condition and restore the network operational state without losing important data. For this reason, the system and devices in the grid need to be easily programmable. But since current smart grids consist of many vendor-specific devices, each of them uses its own protocols for communication, compatibility to the new condition and overall control of the grid are not easily possible.

Network granularity is another issue, despite the fact that two different vendors may design switches that monitor the network in two different protocols—at the flow level or the packet level. Synchronizing diverse applications in very different network conditions with different protocols and standards leads to several problems in managing and controlling the power and information flows. Smart meter data should be safe, secure, and protected against attacks to the network, and the smart grid data and information privacy should be guaranteed.

Current variants of smart grids need manual configuration settings, while network management in advanced smart grids should be performed without manual interference. Currently, parameters in the local area network should not be set by the administrator or network engineers in various network conditions and attacks because this increases risk. In addition, traditional Internet switches and routers need to be configured by the network operators. Configuration policies for using these devices are not flexible enough to adapt easily to the Internet. Therefore, capabilities for automatic management and control of the global network using optimization methods should be provisioned in new systems. An SDN with programmable solutions manage the network in a smart way [44].

6.2.2.3.3 SDN Architecture for IoE

SDN is a software solution for enabling programmability of the network, which decouples the data plane from the control plane [11]. SDN architecture is composed of a data plane, control plane, and application plane [11]. The data plane, also known as the forwarding plane, is responsible for data and statics gathering, processing, local information monitoring, analyzing, and data forwarding. The control plane is responsible for programming and managing the forwarding plane (data plane). It defines network operation and routing using information provided by the data plane. It comprises one or more software controllers that communicate with the forwarding network elements through standardized, or southbound, interfaces [45]. The application plane has two components, namely, a management center and an intelligent decision-making center. Applications such as generation management, data management, and communication resource management are performed in the management center, while intelligent decisions are made with the help of the intelligent decision-making center by using learning algorithms [11].

Software-defined energy Internet [46] architecture is proposed to enhance the flexibility and efficiency of the power network. As can be seen in Figure 6.4, there are three different planes in software-defined energy Internet architecture: data, control, and energy. Furthermore, the IoE devices are classified into control, network, and power grid devices. The hierarchical energy control architecture proposed in [46] enables the programmability of the energy flow for point-to-point energy delivery. The data plane is responsible for the provision of energy-related data services. Data in the data plane is generated and transmitted in point-to-point fashion. The control plane has a global view of the energy Internet and dynamically optimizes the energy router configuration. In the control plane, there are software-defined data and energy controllers responsible for data flow control and energy flow control, respectively [46]. The energy plane is responsible for physical energy flow control.

Network function virtualization (NFV) is a software solution that enables the programmability of the network. Through software virtualization techniques, NFV performs network functions in dedicated hardware in data centers. For improving the flexibility and simplicity of networks and service delivery over them, a method for

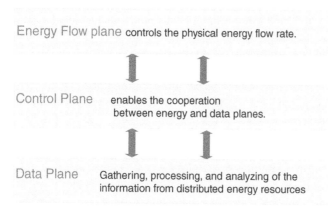

Energy Flow plane controls the physical energy flow rate.

Control Plane enables the cooperation
between energy and data planes.

Data Plane Gathering, processing, and analyzing of the
information from distributed energy resources

Figure 6.4 SDN Architecture for IoE.

integrating NFV and SDN has been proposed in [47]. Furthermore, a method for integrating SDN with NFV specifically for the IoE ecosystem by leveraging the advantages of edge computing has been proposed in [48].

6.2.2.4 Information Technology for IoE

6.2.2.4.1 *Features of IoE Data*

In an IoE system, a large volume of data is generated by devices periodically or by events that come from different sources with a variety of types. Big data is defined as high-volume data with high frequency, variety, and variability [20]. According to this, IoE data can be called big data because it requires more advanced capture, management, and analysis methods than the traditional tools and signal processing methods [49].

Various architectures are proposed for big data. They usually have layer-like structures including a data collection, storage, and management layer; a data analytics layer; and an application layer (Figure 6.5). The data collection, storage, and management layer is the basis of the whole data architecture. In this layer, all types of data (structured, semi-structured, or unstructured) are collected and then stored as SQL or NoSQL in a distributed file system. Then computation power and data usage are managed by a global resource manager. Different data management tools are also employed to handle data with different formats. In a nutshell, this layer provides suitable data for further utilization by other layers [20]. The data analytics layer performs various operations or processes on the stored data to provide results for applications. The application layer provides object-oriented services [20].

An important issue in IoE systems is how to integrate the big data stack's software and IoE, considering the existing energy and communication infrastructure, in order to create an efficient and affordable solution [20]. To stablish the big data stack, knowing the characteristics of data generated in the IoE system is essential. Generally, the data in IoE has three features [20]. First, data volume is very large and is produced with fast speed. In an IoE structure, each device can keep producing its status log such

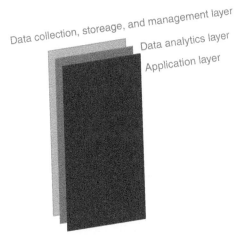

Figure 6.5 Big Data Layer-Like Structure.

as voltage, current, active power, and reactive power with quite enough frequency. Therefore, collecting, processing, analyzing, and answering via a big data stack with very high throughput is a challenge. Second, a significant amount of generated data is monitoring data that are not critical for IoE operation unless a situation changes. Generated data by the devices in a regular situation are used to confirm forecast and management results. But if a sudden situation happens, dense data should be analyzed to give real-time responses. Third, the main portion of generated data are internal to the IoE system, and a small portion of it comes from external sources (e.g., weather forecasts, holiday pattern announcements).

6.2.2.4.2 Architectural Requirements of IoE Big Data

IoE aims to facilitate access to renewable energy and to promote reliability, safety, and power quality. To accomplish this, the following requirements are needed for its big data architecture [20]:

- Real-time processing and monitoring of massive streaming data
- Performing process and analysis in an enormous number of nodes located in large-scale parallel structures
- Low-latency real-time response, decision-making, feedback, and support
- Scalable, stable, and robust in data analysis and prediction
- Secure data, communication, and decision-making
- Protection of user privacy

Most of the traditional analytical methods can be used in big data stacks, with only a low degree of complexity. To have great analytical capabilities, machine learning is used widely for big data analysis and is separated into two categories: supervised and unsupervised learning [20].

6.2.2.4.3 IoE Big Data Platform

There are different tools for building a big data platform. Each provides various services with different features. For choosing among them, scale and speed should be considered. For trade-offs between speed and scale in IoE big data stack, real-time and offline analysis should be gathered. Real-time analysis computes the shortest time to access, process, and respond to a data changing situation [50]. Two types of structures are usually used to implement this analysis: (1) a common relational database in a parallel processing cluster (nodes/computers); and (2) integration of in-memory analytics platforms with a distributed file system (DFS) [50]. In in-memory computing technology, both data processing and storing are performed in the same place (array of memory devices), unlike the conventional von Neumann architecture, that processing operations are performed in arithmetic logic unit and data are stored in slow disk drivers, respectively. Using in-memory computing removes the bottleneck of data transfer between memory and logic circuits and leads to high throughput, which improves speed and performance [51–55]. Having a file system as a part of an operating system helps performs file management activities such as organization, naming, storing, retrieval, and sharing of a single node. DFS has emerged as a response to the growth of distributed systems and computing to facilitate access to the files on different nodes. In addition to the file system functionalities, DFS allows for transparent remote-file sharing, use of diskless workstations, user mobility between different nodes, and availability by keeping multiples copies of files that increases fault tolerance. They bring scalability, fault tolerance, and high concurrency with them. The first type of structure does not meet the growing demand for speed and scale. The second one has a good performance and high speed; however, only its simple operations are implemented, and its more complex operations need to be developed. Real-time analysis that can meet the requirements of big data is not yet fully achieved, and the existing tools such as Spark and Storm have implemented the big data real-time analysis partially [20].

Offline analysis uses larger and more complex data to do more comprehensive data processing and analysis.

The big data workflow stages are acquisition, storage, management, analysis, and response. To improve data acquisition efficiency and decrease data conversion cost in the first stage, several acquisition tools based on DFS have been developed, such as Kafka, Chukwa, Scribe, and TimeTunnel. The second stage, data storage, can be implemented by a Hadoop distributed file system (HDFS) [56] tool. HDFS is the most well-known DFS; it is highly fault-tolerant and designed for deployment on low-cost hardware [56]. It has a master–slave structure and consists of multiple nodes. The incoming file is broken into several blocks with standard sizes and is allocated to each node by the master node. The master node also manages namespace of the file system and regulates the file access from clients. To ensure reliability, each block is replicated several times and stores in several nodes. The next stage is resource management, in which the cluster utilization is scheduled and optimized. Yet Another Resource Negotiator (YARN) is a famous resource manager that distributes the resources (e.g., CPU, memory, network) of the cluster among all running applications on the file system. It assigns a master manager to each application to negotiate with resource manager and run the application. Furthermore, it assigns each node a node manager to report resource usage regularly. During the analytics stage, a parallel programing framework is used to process the massive data efficiently. Hadoop MapReduce is an open-source analytics tool with a parallel and distributed algorithm on a cluster. It has a processing model with two sequential stages: map and reduce. The map stage performs filtering and sorting operations, and the reduce stage performs summary operations. The reduce stage realizes specific services and is usually object-oriented. Various or a combination of tools such as Apache Hive can be used to realize this section.

Open-source tools for the big data platform are summarized in Table 6.1.

Table 6.1 Open-source Tools for Big Data Workflows.

Big Data Workflow	Open-Source Tool
Acquisition	Kafka, Chukwa, Scribe, TimeTunnel
Storage	Hadoop Distributed File System (HDFS) [56]
Resource management	Yet Another Resource Negotiator (YARN)
Analytics	MapReduce, Spark
Application	Hive

6.2.2.4.4 IoE Big Data Challenges

There are several main challenges for using big data in the IoE.

Data privacy and security: In practice, power companies are reluctant to let a large amount of stream data or recodes be taken outside their walls. Furthermore, both power companies and individuals are conservative about their privacy and security. Hence, finding an efficient way to obtain data with protection of privacy and security is a challenge for researchers and the industrial sector. Blockchain technology has the potential to tackle this challenge. Data authorization, including the data abstract, authorized personnel, access location, and access time, can be published on a chain. In this way, data usage can be monitored easily by the public or the owner.

Data throughput: A large amount of data is generated within the IoE system that can change rapidly depending on the situations. Therefore, fast processing and analyzing of the records is required to provide suggestions. Furthermore, a big data stack needs the ability to collect, store, manage, process, and analyze as much as data it can to have better performance. Hence, increasing data throughput remains as a continuing challenge of IoE big data. One possible solution to tackle this challenge is using fog computing that pushes data processing toward the IoE sensors.

Data storage: Huge amounts of data are generated in the IoE system with fast speed that requires data storages with high speed and high capacity. In-memory database technology can be a potential solution to increase the speed of data storage. In this technology data is stored in memory instead of disks or flashes that improves the speed of reading and writing. However, in-memory databases technology currently has some challenges: it cannot maintain the data for a long time; it cannot communicate with other databases easily; it should be loaded from/to disk images before/after using, and data should fit in-memory processing.

Data stream processing: The IoE system requires real-time processing of all valuable data for realizing consumption pattern, adjusting real-time power, scheduling, management, and so forth. In addition to the real-time data streams, historic data should be extracted from the file system and interact with the incoming data. Hence, timeliness of big data real-time processing is a challenge. To tackle this, several processing frameworks for streaming processing have been developed, such as Flink and Beam.

Data opening: The records of energy data and data streams are valuable for research and future usage. However, they should be stored as anonymous data for reuse. Data anonymization is a type of data sanitization in which personally identifiable information is removed from data sets to preserve privacy. Data opening is the procedure of extracting data from anonymous data records, and it requires financial support. Furthermore, more accurate data requires more investment as well as more precise analysis and management, which pose a challenge for data opening. Another challenge of data opening is its management including levels of access rights. Blockchain is a possible solution for data opening challenges [20].

6.2.2.4.5 IoE Data Uncertainties Management

High penetration of RESs in IoE, which is associated with uncertainty, leads to many problems such as frequency fluctuations, voltage instability, and power imbalance [8].

Therefore, new distributed energy management methods are required that are able to accommodate the stochastic and intermittent nature of renewable energy [57].

To manage data uncertainties in energy management problems, two main methodologies are used broadly: robust optimization and stochastic optimization [58]. The former uses moderate data to give a model of uncertainties that is distribution-free [59]. It considers the worst-case operation scenarios and mitigates the negative effects of uncertainties [57]. The latter is applied in the cases in which the uncertainty in the data has a well-known probability distribution. Considering the worst-case scenarios in the robust optimization method leads to high protection but degrades the performance, as the worst case often represents a rare scenario that seldom happens [57]. On the other hand, precise estimation in the stochastic optimization method is very challenging in practical applications due to practical constraints and complexity induced by details of operation [57].

Considering large volumes of data collected from devices and equipment in IoE, forecasting approaches based on big data analytics and machine-learning can be considered in order to provide short-term estimations of uncertainties [57]. For wind power forecasting, the prediction model can be extracted through data-centric persistence, linear, and nonlinear methods [57]. The persistence method is used as a benchmark method to know if a predicted model provides better results than any trivial reference model (persistence model). It assumes a high correlation between wind speeds in short prediction horizons [60]. Linear methods such as autoregressive moving average and autoregressive integrated moving average have better performance than persistence methods for short-term prediction [61]. Nonlinear methods such as artificial neural networks [43] and support vector machines [62,63] outperform linear methods when the system behavior is inherently nonlinear [57]. Artificial neural networks are simple to implement, accurate for prediction, and fast in self-learning. However, wind data are high-dimensional time series, and developing a closed-form description of them with accurate parameters is hard to achieve, which may degrade optimal performance or decrease the speed of learning convergence [57]. Deep learning methods for short-term/long-term wind power prediction are proposed in which [57, 64, 65] are able to manage high-dimensional nonlinear and high-volume data [66].

6.2.3 Applying Blockchain in IoE: Capabilities and Challenges

6.2.3.1 Blockchain Definition

Blockchain technology was introduced in 2008 to enable online payment on the Internet without any central authority and through a peer-to-peer (P2P) system. It is a combination of a digital data structure and a decentralized consensus mechanism to maintain the consistency and precision of the data stored in a ledger where the ledger is distributed through a P2P system [67]. A chain of blocks can be created, maintained, and stored by multiple entities in the network, and each entity is able to verify that the chain order and data are not tampered with [5]. A blockchain network is a P2P network of independent nodes that are randomly connected. Nodes are controlled by users, and there is no central control; every node uses the same blockchain algorithm to perform operations. Each user of the network is able to generate any number of addresses. Each address uniquely identifies a user and is derived from a public key. In a blockchain network, all nodes are equally privileged to send and receive messages to and from each other. There are two types of blockchains: permissioned (private) blockchain and permissionless (public) blockchain. The former has limited admission, so only known nodes can participate in consensus, and there is no need to proof of work (PoW) consensus protocols due to no risk of a Sybil attack. This kind of blockchain is called a Byzantine fault tolerance blockchain. The latter is unlimited, everyone can participate in consensus, and this type of blockchain needs PoW consensus; it's called PoW blockchain [67].

6.2.3.2 Opportunities for the Application of Blockchains in IoE

Blockchains with the features of immutability, irreversibility, consensus, security, decentralization, and automation are very well posed to address IoE applications [16]. Blockchain has the potential to be used for automated data exchange, demand response management, and complex energy transactions [16]. Opportunities for using blockchains in energy systems and their level of feasibility and impact are discussed in Wu et al. [68]. According to Wu, their feasibility is moderate in P2P microgrids, smart meters, and carbon credit trading and is high in renewable energy certificates and IoT utilities management. Furthermore, the impact of blockchain on these cases increases from low to high, respectively. Kang et al. [69] reported the contribution of the most popular startups and use cases deployed around blockchain in different parts of energy systems. According to it, around 40% of theses startups are active in the field of decentralized generation, 20% of them are in the fields of grid management and electric vehicle charging, and 10% of them are in the fields of IoT and metering.

6.2.3.3 Challenges of the Application of Blockchains in IoE

There are challenges in applying blockchain in IoE systems [68], including blockchain bottlenecks, physical nature of energy networks, high concentration of energy industry, standardization, and the unbalance between supply and demand ratio of professional talents in the blockchain field.

The challenges associated with blockchain bottlenecks are as follows:

- By increasing the number of access blockchain nodes, the response speed decreases due to the decreasing efficiency of consensus algorithms.
- Large-capacity storage systems are needed so that all participating actors be able to store transaction records and related information synchronously for support.
- Significant amount of energy is consumed due to the participation of all blockchain nodes in accounting.

The challenges with the physical nature of energy networks originate from the fact that energy conversion and distributions associated with loss. Therefore, applying blockchains to energy system needs to consider the physical laws of the system. Blockchain technology is based on value transmission, and value transmission in IoE happens around the energy transmission. So, mapping a digital world to a physical world is a challenge. Nonetheless, currently, the application of blockchain technology in the energy field is relatively immature, and with a high risk of early adoption [68]. In addition, due to the energy industry being dominated by multiple major companies, more than 50% of computing resources will be controlled by a special group, which will pose security and reliability challenges for blockchains applied in the IoE.

6.2.4 Energy Trading in the IoE

In the emerging IoE, centralized energy transactions in traditional power systems will be replaced by distributed energy transactions. Therefore, energy transactions need decentralized management, too [15]; otherwise, by a single point failure, the whole system would collapse, and the system will not be secure. Therefore, decentralization is one of the requirements of an IoE system. Blockchain as an emerging technology has the potential to address such requirement [5].

At present, the most common application of blockchains is in the energy market to realize P2P energy trading [70]. In this way, trading models based on blockchain have received a lot of attention [69,71–73]. However, most of these models suffer from privacy disclosure. To address this challenge, some research has been conducted. Hiding the data on the blockchain and the connections between users [74,75], using cryptographic primitives, and combining blockchain with traditional privacy protection plans [76] are some of the proposed schemes to protect privacy. Gai et al. [77] investigated the privacy problems in blockchain-based neighboring energy trading systems. These trading systems are not protected against linking attacks that link the open information recorded in blocks with the information from other data sets.

Many data-mining algorithms can launch this type of attack. They use an account mapping mechanism to entirely change the distributions and trends. They propose a consortium blockchain system as a basic platform and implement a privacy protected module of smart contract (a computing protocol for enabling transaction services and corresponding functionalities) on it. The module creates mapping accounts for energy vendors by which the trading distribution will be screened. In this way, attackers cannot touch the data directly [74]. Guan et al. [77] proposed a privacy-preserving blockchain energy trading scheme that can control the access through transaction arbitration in the ciphertext form finely. It divides all the nodes of the blockchain into three different groups: trading, arbitration, and accounting. The trading nodes participate in the transactions by their own pseudonyms and can sell and buy electricity. The accounting node is highly credible and packages the transaction information to make a block and adds it to the blockchain. The arbitration node is responsible for making a judgment based on the meter reading in the situations that a dispute occurs over the amount of transmitted electricity between the buyer and the seller.

Certification authority is a trusted third-party center in the system that is responsible for key generation, distributing, and management. For selling electricity, first a transaction initiator is sent out. It drafts the information and formulates the access strategy. Then the encrypted transaction request is broadcast in the blockchain network that can be seen by the users who meet the access policy. The buyers based on their need can choose to buy electricity and send their transaction applications. The transaction initiator selects the transaction party from the applicants and negotiates to reach an agreement. Then the two parties jointly draft the transaction to form a contract and send it to the accounting node for consensus confirmation. The accounting node packs and records all transaction information. After completing the transaction, if there is a dispute between two parties they can apply it to the arbitration node.

Generating the result of the arbitration, the original ciphertext will be updated to form a new transaction record. A two-level ciphertext is suggested. Only participants who satisfy the access policy can view the detailed information of the transaction. At the transaction initiation, the first level of transaction information is encrypted according to the determined access policy. Only users who meet the requirements can view it. After completing the transaction, only the participants in the transaction can obtain the detailed information. If there is a dispute, the system will update the key after arbitration to prevent the arbitration node from using the same key to obtain new transaction information [77].

Tian et al. [78] modeled the IoE system as consisting of wide-area IoE, regional IoE, energy routers, energy chain, and users, where the energy router is responsible for connecting the wide-area IoE and the regional area IoE and performs energy storage and conversion. All energy routers make a blockchain called an energy chain, and users are electric vehicles, power plants, and family users. Tian et al. proposed an IoE identity authentication scheme based on blockchain and cryptographic accumulator. The scheme consisted of three stages: system initialization, energy router authentication, and user authentication. This scheme improves authentication efficiency [78].

6.2.5 Transactive IoE

6.2.5.1 Transactive IoE: The Concept and Enablers

The aim of transactive IoE is to optimally use distributed energy resources to meet both the economical and operational objectives of energy producers and end consumers [79,80]. It manages the flow of energy in the IoE system and coordinates the operation of intelligent devices to minimize carbon footprint, maintain optimal balance between the supply side and demand side in real time, and maximize interactivity of producers and consumers. In other words, IoE transaction provides functionalities of both market and control. Two key enablers of transactive IoE are digitalization and decentralization, which can be realized by IoT and blockchain technologies, respectively.

6.2.5.2 IoT Architecture for Transactive IoE

IoT is the basic architecture for the digitalization of IoE that enables the integration of different kinds of hardware, software, platforms, networks, services, and applications [79]. IoT architecture for transactive IoE consists of several layers that are as following from bottom to top (Figure 6.6) [79]:

- Physical things layer: Consists of all energy devices of the supply side and demand side. Energy transactions happen in this layer. Data from the devices of this layer are collected and sent to the upper layer. In return, this layer also receives control signals for distributed automation after its data was processed, analyzed, and managed by the upper layers.
- Perception and actuation layer: Consists of sensors and actuators. The sensors are responsible for data reception of both the supply side and demand side, and the actuators are responsible for precise control.
- Connectivity and network layer: Different networks including low power networks (LoRa, narrowband IoT, and Sigfox), mobile networks (2G, 3G, 4G, and 5G), wireless networks (IEEE802.11 And IEEE 802.15), and wired networks (RS-232/485, PLC, IEEE 802.3) are integrated and communicate in this layer.
- Middleware layer: Acts as the intermediary between IoT devices and cross-domain applications. Designing this layer is very challenging because it very demanding regarding the architectural features, function management, programming and

implementation. There are functional and nonfunctional requirements. Functional requirements are those concerning with data processing, storage, and management. Nonfunctional requirements are needed for security, privacy, scalability, reliability, availability, and real time. The middleware layer provides data services for the upper layers to carry out their functional goals.

- Service-based layer: Consists of applications or energy-aware services from the supply and demand sides, both of which can communicate at the cloud level. Furthermore, the interoperability of cross-domain applications is possible. Energy efficiency improvement, renewables penetration increase, flexible balancing of the supply side and demand side, energy management optimization, and automatic demand response can take place in this layer.

Serval research institutions and commercial companies proposed standard reference IoT architecture models: Cisco (IoT); IBM Watson (IoT); Huawei (EC-IoT); and Microsoft Azure (IoT reference).

FIWARE is one of the popular open-source platforms based on IoT. It is a standard architecture that connects, combines, and organizes IoT devices to achieve a smart digital energy system [81]. IoT architecture aims to provide cross-device, network, data, service, and domain integration in the power system.

Different numbers of layers for IoT architecture have been proposed in the literature: three-layer [82,83], five-layer, and fog- and edge-based [84, 85]. These variations IoT provide opportunities to speed up the digitalization of IoE using innovative ICT solutions, such as big data, cloud technology, and artificial intelligence (AI).

In IoE, consumers turn into prosumers who can trade energy with the grid as supplier or consumer; with each other in the same community as business to business, vehicle to building, and vehicle to vehicle; and with other energy communities in a wide range of time scale and space scale [79]. Although IoT has played an important role in connecting distributed prosumers, however, it suffers from security and technical issues such as network security, data privacy, software faults, and counterfeit hardware, which need to be addressed to realize the IoE concept [86].

6.2.5.3 Transactive IoE Infrastructure

Transactive IoE requires integrated infrastructures including AMI, smart router, energy router, microgrids, and P2P energy trading platforms. AMI is the key component to create the link between suppliers and consumers: smart routers guarantee power quality and distributed energy automation, and energy routers are responsible for plug-and-play interface and energy flow adjustments. Microgrids are controllable regulators that coordinate energy cells, and P2P energy trading platforms are responsible for enabling online trading markets [79].

6.2.5.3.1 Ami

AMI is a system on the demand side that is responsible for measuring, processing, analyzing, communicating, and storing energy consumption data. It supports two-way power flow monitoring and bidirectional data flow between the consumer and the utility and also enables energy-aware programs. AMI consists of three core components: smart meter, communication network, and meter data management system (MDMS) [79].

The smart meter is the key component of AMI. It stores total energy consumption by a predefined period from 5 to 60 minutes and reports them in real time. It is also able

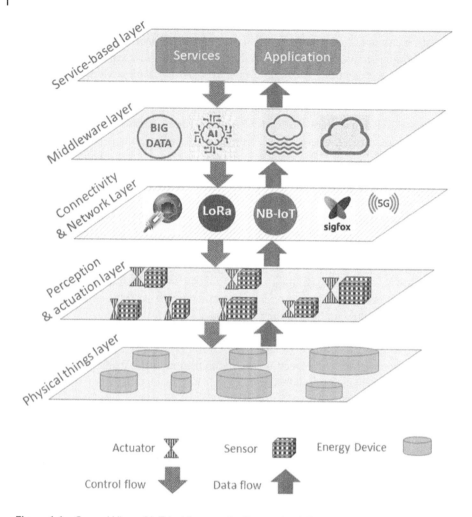

Figure 6.6 General View of IoT Architecture for Transactive IoE. Based on Y. Wu, et al., 2021.

to measure power quality and current and voltage phase angles [87]. It can communicate with other meters, show real-time pricing information, and support remote switch control [79].

Communication network provides bidirectional energy data flows between the consumer and utility. Through this network, data from different meters are collected and transmitted to a central server; furthermore, control signals are sent to the meters. It contains different communication technologies such as PLC, fiber optic cable, broadband over power lines, digital subscriber lines, and wireless.

MDMS imports smart meters' data from the data concentrator and, after processing, stores it. The stored data can be used by the application systems for billing, geographic information, distribution management, and energy management [79].

6.2.5.3.2 *Smart Inverter*

The high penetration of distributed energy resources potentially destabilizes the power system due to power quality issues such as frequency fluctuation, voltage fluctuation, and synchronization. Controlling this system is very difficult and complicated because of fundamental, switching, and Nyquist frequency dynamics [88]. Therefore, two kinds of high-impact control strategies should be applied: high-level and low-level. A microgrid is responsible for high-level control, and a smart inverter is responsible for low-level control. Low-level control handles transiences, synchronization with the main grid, frequency stability, voltage stability, and inter-area oscillation damping. Smart inverters provide bidirectional communication between producers and applications and can communicate with IoT devices. They have the ability to respond quickly, self-maintain, diagnose faults, and decrease autonomic peak power [89].

6.2.5.3.3 Energy Router Differences in Different Energy Domains

Energy routers are responsible for physical energy control and energy routing in the IoE [18]. To deploy energy management from a local grid to the whole energy domains (generation, transmission, distribution, consumption), energy-associated data services can be embedded into the energy routers. Furthermore, energy routers in each domain have different components and functionalities [79], as discussed in the following.

In the generation domain, some SST-based, energy hub–based, and multi-mode energy routers have building blocks like converters, inverters, power-to-x techniques, combined heat and power/CHPP, transformers, controllers, and software [18,21,33,79,90]. These routers support high penetration of RESs, multi-vector energy generation, balance maintenance between the supply side and the demand side (including local demand and global demand), and adjustment of energy generation dynamically.

The transmission domain contains SST-based, MPC-based, and PLC-based energy routers composed of transformers, static synchronous compensators (which quickly provide or absorb the reactive current to regulate the voltage at the point of connection to a power grid), unified power flow controllers (which quickly balance the reactive power on high-voltage transmission networks by injecting current into a transmission line through a series transformer), software, and AC bus. These routers are responsible for flexible control of active and reactive power, power quality control, dynamic energy dispatching, high-voltage power routing in a wide area network, multiple energy conversion and coupling, and providing a hierarchical multi-vector energy system [18,21,90].

For the distribution domain, SST-based, MPC-based, PLC-based, and energy hub–based energy routers [18, 21, 90] consisting of transformers, DC/DC converters, AC/AC converters, AC/DC converters, AC bus, DC bus, software, and power-to-x techniques are proposed. These routers are responsible for ensuring the balance between demand side and supply side in a local area, demand change real-time tracking and resource allocation dynamically, keeping energy distribution balance among residential, commercial and industrial energy sectors, providing multi-carrier energy conversions, and energy resource coordination in local area [21, 79, 90].

Energy routers in the consumption domain are SST-based, energy hub-based, and multi-mode. The main components of these routers are smart meters, converters, smart prosumers, smart switches, and power flow controllers. They provide dynamic pricing, energy-aware decision-making, demand response, selling renewables to the main grid, optimal management energy, and bidirectional power flow services [18, 21, 33, 91].

The control section of an energy router contains an intelligent control module that handles the needs of each end user, who can customize their requirements such as cost, reliability, and efficiency. Furthermore, energy solutions can be personalized for customers. Energy router receives information (e.g., load, storage status, price) from monitoring devices and data centers, shows the related information based on the determined requirements of the user, and then generates corresponding control strategies based on the user's personalized energy solution (which can contain metrics such as energy cost, reliability, and energy storage status). Energy consumption in an area can be forecasted by monitoring energy consumption behavior and be used to make optimal control strategies [20].

6.2.5.3.4 Microgrid

A microgrid is a distributed level energy system that provides energy for local loads. It consists of various small power generating sources (micro sources) that make it very efficient and highly flexible. Its key components are distributed generators, loads, and storage systems. It can work in two modes of connected and island. In the former case, the microgrid is connected to the grid through a point of common coupling that allows it to sell or buy its excess or shortage power to or from the grid; in the latter case, it is isolated from the grid and acts as an independent energy system. Microgrid size is limited to megavolt amperes and encourages the use of renewable energy.

Due to the increasing number of IoT energy devices and an increasing growth in renewable energies, it is necessary to find effective ways to coordinate and automate distributed energy resources to achieve self-sufficiency in local or regional areas. IoT-equipped microgrids can play this role by providing interaction among the grid, distributed energy resources, and customers. The control structure of microgrids is hierarchical and includes five levels. First, the control level is devoted to the internal control of generation processes. This level is responsible for handling the uncertainties of RERs by controlling DC voltage, maximum power point tracking (i.e., obtaining the maximum power from sources with variable power in any environmental condition), and smart inverter interface. The second level is responsible for solving primary stability problems caused by the integration of renewables in the grid. The third level is responsible for providing grid reliability and sustainability by controlling frequency, voltage, and grid synchronization. The fourth layer is responsible for power flow management through energy-aware services. The fifth level performs policy control to achieve high-level business planning [79, 91]. In a nutshell, the first three levels are dedicated to power quality control, and the last two levels perform power management. Therefore, due to the ability of microgrids in the coordination of distributed energy systems, they can be seen as controllable regulators to coordinate RERs, the power grid, and producers or customers. Microgrids can provide services for the integration of renewable energies, ensuring the stability and sustainability of the grid, and optimization and management of energy, based on demand response [79, 92].

6.2.5.3.5 P2P Energy Trading Platforms

P2P energy trading is one of IoE requirement that its implementation needs effective platforms to provide secure decentralized services. Ethereum, Hyperledger Fabric, R3 Corda, and Quorum are some blockchain platforms that can be integrated with IoT architecture. Ethereum is an open-source decentralized permissionless blockchain with smart contract functionality. It uses PoW consensus algorithms, and its cryptocurrency is Ether. Its throughput is larger than 2000 transaction per second (tps) and is mostly used in business-to-consumer operations. It is in a high state of development in GitHub. Hyperledger Fabric is an open-source decentralized permissioned blockchain with smart contract functionality. It uses pluggable framework consensus algorithms and is highly modular. Its throughput is around 200 tps and is mostly used in business-to-business operations. R3 Corda is an open-source decentralized permissioned blockchain with smart contract functionality. It uses pluggable framework consensus algorithms and is highly modular. Its throughput is around 170 tps and is mostly used in financial services. Quorum is an open-source decentralized permissioned blockchain with smart contract functionality. It uses majority voting consensus algorithms. Its throughput is around 100 tps and is mostly used in financial services. It is in a high state of development in GitHub. These platforms have different features, so the right platform should be chosen based on its application and solution criteria [79].

6.2.6 Integrated Demand Response as the Next Generation of Demand Response

Demand response (DR) refers to all the intentional actions of end users to modify electricity consumption patterns by changing the timing, level of instantaneous use, or total electricity consumption [93]. Customers' responses can be different: They may decrease their electricity usage in peak hours when the price is high without changing their consumption pattern in other hours, leading to temporary discomfort; they may shift their operations from peak hours to off-peak hours, which may not be suitable for industrial customers; or they may establish their own distributed generation [94]. DR programs can be categorized into two main groups: price-based programs (PBP) and incentive-based programs (IBP) [94, 95].

PBP consider dynamic pricing for the electricity tariffs, and their goal is to flatten the demand curve by offering high prices in peak hours and lower prices in off-peak hours. Their price can be subdivided into real time, multi-step, time of use, and critical peak.

IBP are further subdivided into classical and market-based programs. Classical programs include direct load control programs and interruptible/curtailable load programs. In direct load control programs, utilities can shut down the customers' equipment remotely on short notice. In interruptible/curtailable programs, utilities ask the participants to decrease their load to predefined values or else potentially face penalties according to the programs' conditions. The customers who participate in these programs receive payments as a bill credit or discount rate. The market-based programs include demand bidding, emergency DR programs, capacity market, and the ancillary services market [94]. Participating customers in these programs receive money proportionate to the amount of load reduction during peak hours [94].

Integrating different forms of energy such as electricity, natural gas, and thermal energy is necessary in multi-vector energy systems [26] to increase energy efficiency [96], [97]. Integrated demand response (IDR) is one of the main approaches to promote the consumption of RESs and to improve energy efficiency in these systems [98]. As DR stimulates the demand-side resources to interact with renewable generation in the power system through price- and incentive-based programs [99, 100], the goal of IDR is to fully exploit the DR capabilities of all users in multi-vector energy systems to make it reliable and economic. In single-energy systems, DR programs cannot be applied to must-run loads, whereas IDR programs consider them by converting various forms of energy to the one desired in peak hours in response to the real price. IDR programs allow users to shift their consumption from peak hours to valley hours and to change the source of energy they use [26]. The energy hub has a key role in multi-energy systems and is responsible for converting different forms of energy [26].

6.3 Summary and Conclusion

In this chapter, first the challenges of the traditional power system and the limitations of the smart grids are discussed. Then IoE as the next generation of smart grids is introduced. The growing demand for electricity, lack of enough fossil fuels, environmental concerns about CO_2 emissions, and the emergence of renewable energy resources make integration of the conventional and renewable power generation inevitable. The IoE structure and its key components are explained. The promising technologies for realizing the IoE system such as big data analysis, blockchain, energy routers, and SDN are introduced. The current challenges of each technology for implementing the IoE system are discussed and any applicable tools, platforms, and implementation models introduced. Transactive IoE is explained as the provider of market as well as control functionalities in the IoE system. Finally, integrated demand response is introduced in the IoE system as a replacement for demand response in the traditional power system.

References

1 Firouzi, F., Chakrabarty, K., and Nassif, S. (2020). *Intelligent Internet of Things: From Device to Fog and Cloud*. Springer Nature.

2 Wang, K., Hu, X., Li, H., Li, P., Zeng, D., and Guo, S. (2017). A survey on energy Internet communications for sustainability. *IEEE Trans. Sustain. Comput* 2 (3): 231–254.

3 Zhang, Y. (2020). Distributed renewable energy in China: Current state and future outlook. In: *Annual Report on China's Response to Climate Change (2017)* (ed. W. Wang and Y. Liu),129–144. Springer.

4 IEA World energy outlook 2020. vol. 2020.

5 Mollah, M.B., et al. (2020). Blockchain for future smart grid: A comprehensive survey. *IEEE Internet Things J* 8 (1): 18–43.

6 Owusu, P.A., and Asumadu-Sarkodie, S. (2016). A review of renewable energy sources, sustainability issues and climate change mitigation. *Cogent Eng* 3 (1): 1167990.

7 Wang, W., and Liu, Y. (2020). *Annual Report on China's Response to Climate Change (2017)*. Springer.

8 Clement-Nyns, K., Haesen, E., and Driesen, J. (2009). The impact of charging plug-in hybrid electric vehicles on a residential distribution grid. *IEEE Trans. Power Syst.* 25 (1): 371–380.

9 Glinkowski, M., Hou, J., and Rackliffe, G. (2011). Advances in wind energy technologies in the context of smart grid. *Proc. IEEE* 99 (6): 1083–1097.

10 Kong, F., Dong, C., Liu, X., and Zeng, H. (2014). Quantity versus quality: Optimal harvesting wind power for the smart grid. *Proc. IEEE* 102 (11): 1762–1776.

11 Rehmani, M.H., Davy, A., Jennings, B., and Assi, C. (2019). Software defined networks-based smart grid communication: A comprehensive survey. *IEEE Commun. Surv. Tutorials* 21 (3): 2637–2670.

12 Joseph, A., and Balachandra, P. (2020). Energy Internet, the future electricity system: Overview, concept, model structure, and mechanism. *Energies* 13 (16): 4242.

13 Kabalci, Y., Kabalci, E., Padmanaban, S., Holm-Nielsen, J.B., and Blaabjerg, F. (2019). Internet of things applications as energy Internet in smart grids and smart environments. *Electronics* 8 (9): 972.

14 Mahmud, K., Khan, B., Ravishankar, J., Ahmadi, A., and Siano, P. (2020). An Internet of energy framework with distributed energy resources, prosumers and small-scale virtual power plants: An overview. *Renew. Sustain. Energy Rev.* 127: 109840.

15 Saleem, Y., Crespi, N., Rehmani, M.H., and Copeland, R. (2019). Internet of things-aided smart grid: Technologies, architectures, applications, prototypes, and future research directions. *IEEE Access* 7: 62962–63003.

16 Miglani, A., Kumar, N., Chamola, V., and Zeadally, S. (2020). Blockchain for Internet of Energy management: Review, solutions, and challenges. *Comput. Commun.* 151: 395–418.

17 Kabalci, E. (2019). *From Smart Grid to Internet of Energy*. Elsevier.

18 Xu, Y., Zhang, J., Wang, W., Juneja, A., and Bhattacharya, S. (2011). Energy router: Architectures and functionalities toward energy internet. In: *2011 IEEE International Conference on Smart Grid Communications (SmartGridComm)*, 31–36.

19 Huang, A.Q., Crow, M.L., Heydt, G.T., Zheng, J.P., and Dale, S.J. (2010). The future renewable electric energy delivery and management (FREEDM) system: The energy Internet. *Proc. IEEE* 99 (1): 133–148.

20 Zobaa, A.F., and Cao, J. (2020). *Energy Internet*. Springer.

21 Guo, H., Wang, F., Luo, J., and Zhang, L. (2016). Review of energy routers applied for the energy internet integrating renewable energy. In: *2016 IEEE 8th International Power Electronics and Motion Control Conference (IPEMC-ECCE Asia)*, 1997–2003.

22 Tao, H., Kotsopoulos, A., Duarte, J.L., and Hendrix, M.A.M. (2006). Family of multiport bidirectional DC–DC converters. *IEE Proceedings-Electric Power Appl* 153 (3): 451–458.

23 Wu, H., Zhang, J., and Xing, Y. (2014). A family of multiport buck–boost converters based on DC-link-inductors (DLIs). *IEEE Trans. Power Electron* 30 (2): 735–746.

24 Zhang, X., Shahidehpour, M., Alabdulwahab, A., and Abusorrah, A. (2015). Optimal expansion planning of energy hub with multiple energy infrastructures. *IEEE Trans. Smart Grid* 6 (5): 2302–2311.

25 Reynolds, J., Ahmad, M.W., and Rezgui, Y. (2018). Holistic modelling techniques for the operational optimisation of multi-vector energy systems. *Energy Build* 169: 397–416.

26 Wang, J., Zhong, H., Ma, Z., Xia, Q., and Kang, C. (2017). Review and prospect of integrated demand response in the multi-energy system. *Appl. Energy* 202: 772–782.

27 Li, T., Eremia, M., and Shahidehpour, M. (2008). Interdependency of natural gas network and power system security. *IEEE Trans. Power Syst.* 23 (4): 1817–1824.

28 Liu, C., Shahidehpour, M., Fu, Y., and Li, Z. (2009). Security-constrained unit commitment with natural gas transmission constraints. *IEEE Trans. Power Syst.* 24 (3): 1523–1536.

29 Shahidehpour, M., Fu, Y., and Wiedman, T. (2005). Impact of natural gas infrastructure on electric power systems. *Proc. IEEE* 93 (5): 1042–1056.

30 Geidl, M. and Andersson, G. (2007). Optimal power flow of multiple energy carriers. *IEEE Trans. Power Syst.* 22 (1): 145–155.

31 Kienzle, F., Favre-Perrod, P., Arnold, M., and Andersson, G. (2008). Multi-energy delivery infrastructures for the future. In: *2008 First international conference on infrastructure systems and services: Building networks for a brighter future (INFRA)*, 1–5.

32 Favre-Perrod, P., Kienzle, F., and Andersson, G. (2010). Modeling and design of future multi-energy generation and transmission systems. *Eur. Trans. Electr. Power* 20 (8): 994–1008.

33 Geidl, M., Koeppel, G., Favre-Perrod, P., Klockl, B., Andersson, G., and Frohlich, K. (2006). Energy hubs for the future. *IEEE Power Energy Mag.* 5 (1): 24–30.

34 Aydeger, A., Akkaya, K., and Uluagac, A.S. (2015). SDN-based resilience for smart grid communications. In: *2015 IEEE Conference on Network Function Virtualization and Software Defined Network (NFV-SDN)*, 31–33.

35 Xu, Y., and Fischione, C. (2012). Real-time scheduling in LTE for smart grids. In: *2012 5th International Symposium on Communications, Control and Signal Processing*, 1–6.

36 Aalamifar, F., and Lampe, L. (2016). Optimized WiMAX profile configuration for smart grid communications. *IEEE Trans. Smart Grid* 8 (6): 2723–2732.

37 Zhang, Q., Sun, Y., and Cui, Z. (2010). Application and analysis of ZigBee technology for smart grid. In: *2010 International Conference on Computer and Information Application*, 171–174.

38 Aujla, G.S., and Kumar, N. (2018). SDN-based energy management scheme for sustainability of data centers: An analysis on renewable energy sources and electric vehicles participation. *J. Parallel Distrib. Comput.* 117: 228–245.

39 Cui, Y., Xiao, S., Liao, C., Stojmenovic, I., and Li, M. (2013). Data centers as software defined networks: Traffic redundancy elimination with wireless cards at routers. *IEEE J. Sel. Areas Commun.* 31 (12): 2658–2672.

40 Kobo, H.I., Abu-Mahfouz, A.M., and Hancke, G.P. (2017). A survey on software-defined wireless sensor networks: Challenges and design requirements. *IEEE Access* 5: 1872–1899.

41 Haque, I.T., and Abu-Ghazaleh, N. (2016). Wireless software defined networking: A survey and taxonomy. *IEEE Commun. Surv. Tutorials* 18 (4): 2713–2737.

42 Ahmed, R., and Boutaba, R. (2014). Design considerations for managing wide area software defined networks. *IEEE Commun. Mag.* 52 (7): 116–123.

43 Kim, J., Filali, F., and Ko, Y.-B. (2015). Trends and potentials of the smart grid infrastructure: From ICT sub-system to SDN-enabled smart grid architecture. *Appl. Sci.* 5 (4): 706–727.

44 Nunes, B.A.A., Mendonca, M., Nguyen, X.-N., Obraczka, K., and Turletti, T. (2014). A survey of software-defined networking: Past, present, and future of programmable networks. *IEEE Commun. Surv. Tutorials* 16 (3): 1617–1634.

45 Braun, W., and Menth, M. (2014). Software-defined networking using openflow: Protocols, applications and architectural design choices. *Futur. Internet* 6 (2): 302–336.

46 Zhong, W., Yu, R., Xie, S., Zhang, Y., and Tsang, D.H.K. (2016). Software defined networking for flexible and green energy internet. *IEEE Commun. Mag.* 54 (12): 68–75.

47 Matias, J., Garay, J., Toledo, N., Unzilla, J., and Jacob, E. (2015). Toward an SDN-enabled NFV architecture. *IEEE Commun. Mag.* 53 (4): 187–193.

48 Garg, S., Kaur, K., Kaddoum, G., and Guo, S. (2021). SDN-NFV-Aided edge-cloud interplay for 5G-Envisioned energy internet ecosystem. *IEEE Netw.* 35 (1): 356–364.

49 Hossain, E., Khan, I., Un-Noor, F., Sikander, S.S., and Sunny, M.S.H. (2019). Application of big data and machine learning in smart grid, and associated security concerns: A review. *IEEE Access* 7: 13960–13988.

50 Laplante, P.A. (2004). *Real-Time Systems Design and Analysis.* Wiley Online Library.

51 Ielmini, D., and Wong, H.-S.P. (2018). In-memory computing with resistive switching devices. *Nat. Electron.* 1 (6): 333–343.

52 Mutlu, O., Ghose, S., Gómez-Luna, J., and Ausavarungnirun, R. (2019). Processing data where it makes sense: Enabling in-memory computation. *Microprocess. Microsyst.* 67: 28–41.

53 Sebastian, A., Gallo, M.L., Khaddam-Aljameh, R., and Eleftheriou, E. (2020). Memory devices and applications for in-memory computing. *Nat. Nanotechnol.* 15 (7): 529–544.

54 Wong, H.-S.P., and Salahuddin, S. (2015). Memory leads the way to better computing. *Nat. Nanotechnol* 10 (3): 191–194.

55 Wulf, W.A., and McKee, S.A. (1995). Hitting the memory wall: Implications of the obvious. *ACM SIGARCH Comput. Archit. News* 23 (1): 20–24.

56 Borthakur, D. (2021). HDFS architecture guide. https://hadoop.apache.org/docs/r1.2.1/hdfs_design.html#Introduction.

57 Zhou, Z. et al. (2017). Game-theoretical energy management for energy Internet with big data-based renewable power forecasting. *IEEE Access* 5: 5731–5746.

58 Wei, W., Liu, F., Mei, S., and Hou, Y. (2014). Robust energy and reserve dispatch under variable renewable generation. *IEEE Trans. Smart Grid* 6 (1): 369–380.

59 Valencia, F., Collado, J., Sáez, D., and Marín, L.G. (2015). Robust energy management system for a microgrid based on a fuzzy prediction interval model. *IEEE Trans. Smart Grid* 7 (3): 1486–1494.

60 Khalid, M. and Savkin, A.V. (2012). A method for short-term wind power prediction with multiple observation points. *IEEE Trans. Power Syst.* 27 (2): 579–586.

61 Zhang, J. and Wang, C. (2013). Application of ARMA model in ultra-short term prediction of wind power, in *2013 International Conference on Computer Sciences and Applications* 361–364.

62 Gu, B. and Sheng, V.S. (2016). A robust regularization path algorithm for v-support vector classification. *IEEE Trans. Neural Networks Learn. Syst.* 28 (5): 1241–1248.

63 Gu, B., Sheng, V.S., Tay, K.Y., Romano, W., and Li, S. (2014). Incremental support vector learning for ordinal regression. *IEEE Trans. Neural Networks Learn. Syst.* 26 (7): 1403–1416.

64 Thirukovalluru, R., Dixit, S., Sevakula, R.K., Verma, N.K., and Salour, A. (2016). Generating feature sets for fault diagnosis using denoising stacked auto-encoder. in *2016 IEEE International Conference on Prognostics and Health Management (ICPHM)* 1–7.

65 Zhang, C.-Y., Chen, C.L.P., Gan, M., and Chen, L. (2015). Predictive deep Boltzmann machine for multiperiod wind speed forecasting. *IEEE Trans. Sustain. Energy* 6 (4): 1416–1425.

66 Hosseini-Asl, E., Zurada, J.M., and Nasraoui, O. (2015). Deep learning of part-based representation of data using sparse autoencoders with nonnegativity constraints. *IEEE Trans. Neural Networks Learn. Syst.* 27 (12): 2486–2498.

67 Bolfing, A. (2020). *Cryptographic Primitives in Blockchain Technology: A Mathematical Introduction.* Oxford University Press.

68 Wu, J. and Tran, N.K. (2018). Application of blockchain technology in sustainable energy systems: An overview. *Sustainability* 10 (9): 3067.

69 Kang, J., Yu, R., Huang, X., Maharjan, S., Zhang, Y., and Hossain, E. (2017). Enabling localized peer-to-peer electricity trading among plug-in hybrid electric vehicles using consortium blockchains. *IEEE Trans. Ind. Informatics* 13 (6): 3154–3164.

70 Foti, M. and Vavalis, M. (2021). What blockchain can do for power grids? *Blockchain Res. Appl.* 2 (1): 100008.

71 Wang, N. et al. (2019). When energy trading meets blockchain in electrical power system: The state of the art. *Appl. Sci.* 9 (8): 1561.

72 Mengelkamp, E., Notheisen, B., Beer, C., Dauer, D., and Weinhardt, C. (2018). A blockchain-based smart grid: Towards sustainable local energy markets. *Comput. Sci. Dev.* 33 (1): 207–214.

73 Li, M., Hu, D., Lal, C., Conti, M., and Zhang, Z. (2020). Blockchain-enabled secure energy trading with verifiable fairness in industrial Internet of things. *IEEE Trans. Ind. Informatics* 16 (10): 6564–6574.

74 Gai, K., Wu, Y., Zhu, L., Qiu, M., and Shen, M. (2019). Privacy-preserving energy trading using consortium blockchain in smart grid. *IEEE Trans. Ind. Informatics* 15 (6): 3548–3558.

75 Kvaternik, K. et al. (Sep. 2017). Privacy-Preserving Platform for Transactive Energy Systems. [Online]. Available: http://arxiv.org/abs/1709.09597.

76 Abidin, A., Aly, A., Cleemput, S., and Mustafa, M.A. (Jan. 2018). Secure and Privacy-Friendly Local Electricity Trading and Billing in Smart Grid. [Online]. Available: http://arxiv.org/abs/1801.08354.

77 Guan, Z., Lu, X., Yang, W., Wu, L., Wang, N., and Zhang, Z. (2021). Achieving efficient and Privacy-preserving energy trading based on blockchain and ABE in smart grid. *J. Parallel Distrib. Comput.* 147: 34–45.

78 Tian, X., Chen, X., and Li, S. (2021). An identity authentication scheme of energy Internet based on blockchain. *Int. J. Netw. Secur.* 23 (2): 261–269.

79 Wu, Y., Wu, Y., Guerrero, J.M., and Vasquez, J.C. (2021). Digitalization and decentralization driving transactive energy Internet: Key technologies and infrastructures. *Int. J. Electr. Power Energy Syst.* 126: 106593.

80 (2015). GridWise Transactive Energy Framework Version 1.0 The GridWise Architecture. [Online]. Available: https://www.gridwiseac.org/pdfs/te_framework_report_pnnl-22946.pdf.

81 Sotiriadis, S., Stravoskoufos, K., and Petrakis, E.G.M. (2017). Future Internet systems design and implementation: Cloud and IoT services based on IoT-A and FIWARE. In: *Designing, Developing, and Facilitating Smart Cities* (V. Angelakis, E. Tragos, H.C. Pöhls, A. Kapovits, and A. Bassi Eds.), 193–207. Springer.

82 Mashal, I., Alsaryrah, O., Chung, T.-Y., Yang, C.-Z., Kuo, W.-H., and Agrawal, D.P. (2015). Choices for interaction with things on Internet and underlying issues. *Ad Hoc Networks* 28: 68–90.

83 Qin, Y., Sheng, Q.Z., Falkner, N.J.G., Dustdar, S., Wang, H., and Vasilakos, A.V. (2016). When things matter: A survey on data-centric Internet of things. *J. Netw. Comput. Appl.* 64: 137–153.

84 Omoniwa, B., Hussain, R., Javed, M.A., Bouk, S.H., and Malik, S.A. (2018). Fog/edge computing-based IoT (FECIoT): Architecture, applications, and research issues. *IEEE Internet Things J.* 6 (3): 4118–4149.

85 Moghaddam, M.H.Y. and Leon-Garcia, A. (2018). A fog-based Internet of energy architecture for transactive energy management systems. *IEEE Internet Things J.* 5 (2): 1055–1069.

86 Cui, P., Guin, U., Skjellum, A., and Umphress, D. (2019). Blockchain in IoT: Current trends, challenges, and future roadmap. *J. Hardw. Syst. Secur.* 3 (4): 338–364.

87 Depuru, S.S.S.R., Wang, L., Devabhaktuni, V., and Gudi, N. (2011). Smart meters for power grid—Challenges, issues, advantages and status in *2011 IEEE/PES Power Systems Conference and Exposition* 1–7.

88 Wang, X. and Blaabjerg, F. (2018). Harmonic stability in power electronic-based power systems: Concept, modeling, and analysis. *IEEE Trans. Smart Grid* 10 (3): 2858–2870.

89 SPW (2014). What Is A smart solar inverter? https://www.solarpowerworldonline.com/2014/01/smart-solar-inverter.

90 Hussain, S.M.S., Nadeem, F., Aftab, M.A., Ali, I., and Ustun, T.S. (2019). The emerging energy Internet: Architecture, benefits, challenges, and future prospects. *Electronics* 8 (9): 1037.

91 Wu, Y. et al. (2019). Two thermally stable and AIE active 1, 8-naphthalimide derivatives with red efficient thermally activated delayed fluorescence. *Dye. Pigment.* 169: 81–88.

92 Wu, Y., Wu, Y., Guerrero, J.M., Vasquez, J.C., Palacios-Garcia, E.J., and Li, J. (2020). Convergence and interoperability for the energy Internet: From ubiquitous connection to distributed automation. *IEEE Ind. Electron. Mag.* 14 (4): 91–105.

93 Jones, M. (Ed.) (2003). *The Power to Choose: Demand Response in Liberalised Electricity Markets.* OECD.

94 Qdr, Q. (2006). Benefits of demand response in electricity markets and recommendations for achieving them *US Dept. Energy, Washington, DC, USA, Tech. Rep.*

95 Albadi, M.H. and El-Saadany, E.F. (2007). Demand response in electricity markets: An overview. In *EEE Power Engineering Society General Meeting* 1–5.

96 Shabanpour-Haghighi, A. and Seifi, A.R. (2015). Multi-objective operation management of a multi-carrier energy system. *Energy* 88: 430–442.

97 Krause, T., Andersson, G., Fröhlich, K., and Vaccaro, A. (2010). Multiple-energy carriers: Modeling of production, delivery, and consumption. *Proc. IEEE* 99 (1): 15–27.

98 Wang, D. et al. (2019). Integrated demand response in district electricity-heating network considering double auction retail energy market based on demand-side energy stations. *Appl. Energy* 248: 656–678.

99 Biegel, B., Andersen, P., Stoustrup, J., Hansen, L.H., and Birke, A. (2016). Sustainable reserve power from demand response and fluctuating production—Two Danish demonstrations. *Proc. IEEE* 104 (4): 780–788.

100 Zhong, H., Xie, L., and Xia, Q. (2012). Coupon incentive-based demand response: Theory and case study. *IEEE Trans. Power Syst.* 28 (2): 1266–1276.

7

The Economic Implication of Integrating Investment Planning and Generation Scheduling

The Application of Big Data Analytics and Machine Learning

Yue Ma[a] and Mahdi Sharifzadeh[b],*

[a] Electronic and Electrical Engineering Department, University College London, Torrington Place, London, UK
[b] Sharif Energy, Water and Environment Institute (SEWEI), Sharif University of Technology, Tehran, Iran
* Corresponding author. Sharif Energy, Water and Environment Institute (SEWEI), Sharif University of Technology, Tehran, Iran

7.1 Introduction

The increasing demand of electricity and the limited conventional resource such as coal and natural gas have encouraged intensive research into achieving a sustainable development, including the mixed use of renewable resources such as wind, solar, geothermal, biomass, and hydro-power. A recent and comprehensive study by NREL [1] suggests that 80% of total US electricity demand could be supplied by renewable energy by 2050. The incorporation of renewable power (such as wind and solar power) into the existing electricity grid not only can meet the growing electricity demand of human over the next several decades but also can relieve the global problem of climate change. The high penetration of wind and solar photovoltaics (PV) can significantly reduce greenhouse gas emissions [2].

However, integrating a large amount of wind and solar energy into the power system poses operational challenges due to the characteristics of wind and solar PV generation. One of the biggest features of wind and solar resource is their uncertainty depending on the natural factors. This adds to significant uncertainties already imposed from the demand side due to the stochastic behavior of consumers. Balancing the deficit of power supply and demand requires a combination of electricity storage and standby power generation from conventional generators that are often driven by fossil fuels. These not only incur economic losses but also result in reduced conversion efficiency and additional greenhouse gas emissions. Therefore, the aim of the present study is to find a method for enhancing the predictability of electricity grid from both renewable supplies and demand sides. We will further quantify these in terms of the improvement it can make in terms of economic saving.

Various methods for dealing with the uncertainty in electricity grid have been extensively studied in the literature. Table 7.1 summarizes representative research in the field. The approaches can be divided into two categories. The first category applies optimization in scheduling issues to deal with the uncertainty. The second category applies machine-learning algorithms in forecasting of renewable generation and

Industry 4.0 Vision for the Supply of Energy and Materials: Enabling Technologies and Emerging Applications, First Edition. Edited by Mahdi Sharifzadeh.
© 2022 John Wiley & Sons, Inc. Published 2022 by John Wiley & Sons, Inc.

Table 7.1 Research in the Field of Predictive Models for Renewable Power Generation.

Authors	Solar Uncertainty	Wind Uncertainty	Demand Uncertainty	Method
Jin et al. [3]	No	Yes	No	Stochastic optimization
Wang et al. [4]	No	Yes	No	Deterministic and stochastic optimization
Tuohy et al. [5]	No	Yes	Yes	Stochastic optimization
Constantinescu et al. [6]	No	Yes	No	Stochastic optimization
Aghaei et al. [7]	No	Yes	Yes	Stochastic optimization
Pereira et al. [7]	No	Yes	No	Stochastic optimization
Sharafi et al. [8]	No	No	No	Multi-objective optimization
Wang et al. [9]	No	Yes	Yes	Multi-objective optimization
Baker et al. [10]	No	Yes	No	Two-stage stochastic optimization
Wang et al. [11]	No	Yes	No	Kernel density forecast and stochastic optimization
Voyant et al. [12]	Yes	No	No	Machine-learning methods
Bae et al. [13]	Yes	No	No	Support vector machine
Gutierrez-Corea et al. [14]	Yes	No	No	Artificial neural network
Ramasamy et al. [15]	No	Yes	No	Artificial neural network
Doucoure et al. [16]	No	Yes	No	Artificial neural network
Okumus et al. [17]	No	Yes	No	Machine-learning methods and optimization
Jiang et al. [18]	No	No	Yes	Support vector machine and optimization
Quan et al. [19]	No	No	Yes	A neural network–based method and optimization
Sikinioti-Lock [20]	Yes	Yes	Yes	Artificial neural network, support vector machine, and Gaussian process regression

demand to deal with the uncertainty. The details of these approaches will be briefly discussed.

In the first category, a unit commitment problem is formulated as an optimization problem with a least-cost objective function [3]. Wang et al. [4] compared the deterministic and stochastic approaches and concluded that stochastic optimization can be applied in a scheduling issue to consider uncertainties in the renewable generation. Typically, stochastic unit commitment (UC) models were constructed in power transmission scheduling with penetration of wind power [4, 19, 20]. Aghaei et al. [6] accounted for the forecast errors of wind power and load in a stochastic programming approach by treating them as random variables with known probability distribution functions (PDF). Pereira et al. [7] proposed a mixed integer optimization model to analyze different electricity scenarios. The impact of different wind power scenarios was addressed, demonstrating the importance of the tool. Sharafi et al. [8] proposed a multi-objective optimization model for the optimal design of hybrid renewable energy systems applying various generators and storage devices. A particle swarm optimization simulation-based approach was applied to solve the optimization problem. Wang et al. [9] proposed a multi-objective algorithm combining probability distributions and fuzzy set theory to deal with the uncertainties. Baker et al. [10] introduced two-stage stochastic optimization. The first-stage variables included the energy storage, the initial level of stored energy, and the maximum charge or discharge rate of storage. These were common to all scenarios. The second-stage variables included the current level of energy storage, power output, power charged into the storage, and power discharged from the storage. These were scenario-specific and inherited the value of first-stage variables. In terms of the inclusion of uncertainty, the distribution of errors was incorporated into chance constraints for the upper and lower storage bounds. The criteria for generation up-ramping and down-ramping capacity was also added to the constraints.

In the second category of research studies, machine-learning algorithms are used to enhance the predictability of renewable power. Wang et al. [11] proposed a unit commitment model considering the sub-hourly variability of wind power. A time-adaptive quantile-copula estimator-based kernel density forecast was used. Conditional kernel density estimation provides a tool to build models. Its output is a PDF of the forecasted wind power. The time-adaptive estimator allows for updating of the density function [21]. Voyant et al. [12] reviewed machine-learning method used to forecast solar radiation, including support vector regression (SVR), artificial neural network (ANN), k-nearest neighbor, regression tree, boosting, bagging, and random forests. It was concluded that SVR and ANN were applied more widely and performed better than others. Bae et al. [13] proposed a method using a support vector machine to predict solar irradiance one hour ahead. The proposed scheme was compared with the nonlinear autoregressive and ANN schemes in terms of three performance evaluation metrics, which were the root mean square error, mean relative error, and R^2 (coefficient of determination). Gutierrez-Corea et al. [14] used ANN for predicting solar irradiance. They applied multi-layer perceptron (MLP)-type ANN architecture with both endogenous variables and exogenous variables (atmospheric pressure, cloudiness, temperature, or the number of the day of the year) as inputs. Time intervals of estimation were 1 hour, and the sliding window method was adopted as an ANN input. Besides, the selection of ANN architecture for different prediction hours was studied. It was observed that for a time frame of 15 minutes to 1 hour, the output (solar irradiance) had a strong linear relationship with its input. For a time frame of 2 or more hours, the prediction was no longer a linear relationship problem. Ramasamy et al. [15] proposed an ANN model to forecast the wind speeds. The inputs were temperature, air pressure, solar radiation, and altitude, and the output was wind speed. The correlation coefficient between the predicted and measured statistics was found to be 4.55% and 0.98, respectively. Doucoure et al. [16] developed a prediction method for wind speeds using wavelet decomposition and artificial neural networks. Analysis of predictability using the Hurst coefficient eliminated unnecessary components of input data to reduce the computation complexity of the algorithm. Prediction errors for different time horizons (1–5 h) were calculated. Okumus et al. [17] concluded that different machine-learning methods perform better depending on the prediction horizons. For 1 h ahead forecast, the adaptive neuro-fuzzy inference system was combined with the feedforward artificial neural network, whereas for 3 h ahead forecast, a combined method applying the mutual information, wavelet transform, evolutionary particle swarm optimization, and ANFIS was the best.

Demand side load forecasting has also been the subject of intensive research. In [18], short-term load forecasting scheme utilizing SVR was proposed. A hybrid optimization algorithm was applied to get the best parameters. A designed grid traverse algorithm and particle swarm optimization were used in two steps. Performance of minutes-ahead and hours-ahead forecasting was listed for each season according to the mean absolute percentage error (MAPE). Quan et al. [22] implemented a neural network–based method combined with particle swarm optimization to predict short-term load. Sikinioti-Lock [23] applied predictive methods including ANN, SVR, and Gaussian process regression for both renewable power generation and demand. She also looked into the response of the model when the prediction time horizon increases.

In brief, most of these studies use only one method (optimization programming or machine-learning algorithm) to include the uncertainty of wind, solar power, and electricity demand. In the current study, a method that combines optimization programming with machine learning is employed to investigate the economic profit of reducing the contribution of stand-by units by utilizing the demand and renewable power forecasting models. In layman's terms, this work aims to give an accurate answer to the following question: How much economic value can be generated, if forecasting methods are applied in planning power dispatching? The chapter is organized as follows: The case study problem is introduced, and then the methodology is presented, which includes the forecasting model and mathematical formulation of cost function and constraints of the optimization problem. Next, the optimization results are presented and carefully explained. The chapter concludes with a proposal of future works.

7.2 Case Study

The present research was built on the results of a recent study by Sharifzadeh et al. [24], who discussed retrofitting the existing electricity grid in the United Kingdom to benefit from 50% renewable power generation by 2030. The historic meteorological, as well as wind and solar power data of Britain, was collected from a website called Ninja Renewables [25, 26]. To enhance the accuracy of the prediction, the data was divided into three geographical zones, shown in Table 7.2. The neural network prediction model is fitted on the data sets, corresponding to each zone. More details can be found in the Appendix.

Table 7.2 The Three Zones for Wind and Solar Historical Data.

Station Name	Coordinates	Zone
Cairngorn	57.1164, −3.64165	Wind 1
Aviemore	57.2065, −3.82693	Solar 1
Bramham	53.8687, −1.31722	Wind 2, Solar 2
Wisley	51.3103, −0.47478	Wind 3, Solar 3

7.3 Methodology

The research methodology includes two parts. In the first part, machine learning was applied for developing predictive models. In the second part, the results of machine learning in terms of the prediction errors were incorporated into an optimization program concerned with the investment planning and operational scheduling of an electricity grid with a high penetration of wind and solar energy.

7.3.1 Machine-Learning Algorithms for Predictive Models

As mentioned earlier, the neural network prediction model is fitted to the data sets, corresponding to the zones presented in Table 7.2. The architecture of the neural network applied to forecast wind power is shown in Figure 7.1.

Table 7.3 shows the input and structure of each model. The structure of the ANN for the solar and demand forecasting are shown in Figures A7.1 and A7.2 in the Appendix.

The hourly variable indicates the change of time within 24 h. The seasonal variable indicates the change of time within a year. The wind power, wind speed, and temperature are the values at a specified location. To train the model, the Levenberg-Marquardt training algorithm was applied to train the neural networks. Compared with other two options—the Bayesian regularization and the scaled conjugate gradient—the Levenberg-Marquardt is most commonly used because of its computational efficiency [23]. The performance of predictions was measured by the mean square error.

7.3.2 Simultaneous Optimization Programming of Generation Expansion and Electricity Dispatch

The optimization model is presented in this section. The model incorporates the power dispatch, generation expansion planning, and prediction models. Moreover, it is modified to include the prediction errors as follows.

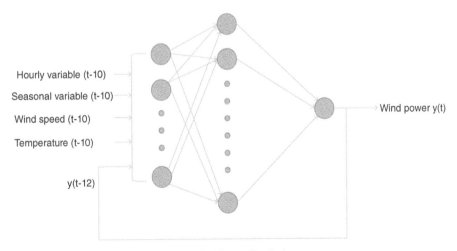

Figure 7.1 NARX Neural Network for Wind Power Prediction.

Table 7.3 The Parameters for Wind Power, Solar Power, and Electricity Demand Forecasting.

Prediction	Input Vectors	Neural Network Architecture	Training Algorithm	Data Division
Wind power	Hourly variable Seasonal variable Wind power Temperature Wind speed	10 input delays 12 feedback delays 10 neurons (a hidden layer)	Levenberg-Marquardt	70% training 15% validation 15% testing
Solar power	Hourly variable Seasonal variable Solar power Temperature Direct irradiance Diffuse irradiance	12 input delays 12 feedback delays 14 neurons (a hidden layer)	Levenberg-Marquardt	
Electricity demand	Hourly variable Seasonal variable Electricity demand	22 input delays 24 feedback delays 1 neuron (a hidden layer)	Levenberg-Marquardt	

7.3.2.1 Objective Function

The total objective function [24] aims to minimize the costs of electricity grid considering all the prediction error scenarios.

$$
Total\ Objective = \sum_{(sc=1)}^{SC} prob_{sc} \times obj_{sc} \tag{7.1}
$$

where sc is the prediction error scenario, and $prob_{sc}$ is the probability of the scenario sc.

Equation (7.2) shows the objective function formulated as a total hourly cost (£/h) and includes four terms. The first term $cc_{sc,g}$ refers to the capital costs and is the multiplication of the required investment for a generation technology (wind, solar, natural gas, nuclear). It is calculated as the product of the costs of a unit power (£/MW) and the unit's capacity (MW) divided by the expected life span (hours) of that technology. The second term $fc_{sc,g}$ is the fixed costs such as the employee salaries and is calculated as the multiplication of the fixed cost of a power generation unit (£/MW) and the unit's generation capacity (MW) divided by the total hours of a year (8760 h). The third term $oc_{sc,g}$ is the hourly operating costs and is calculated by the multiplication of the operational cost of a power generation unit (£/MWh) and the generated power (MW). The fourth term $vc_{sc,g}$ is the variable costs such as hourly fuel costs, which is the multiplication of the variable cost of a power generation unit (£/MWh) and the generated power (MW).

$$
obj_{sc} = \sum_{t=1}^{T}\sum_{g=1}^{G} cc_{sc,g}(t) + fc_{sc,g}(t) + oc_{sc,g}(t) + vc_{sc,g}(t) \tag{7.2}
$$

where t is a time period, g is the generation technology, $cc_{sc,g}$ are the capacity costs, $fc_{sc,g}$ are the fixed costs, $oc_{sc,g}$ are the operating costs, and $vc_{sc,g}$ are the variable costs.

7.3.2.2 Optimization Decision Variables

The decision variables can be divided into the categories of design and operational variables. The key difference is that once the design variables are decided, they will be fixed, all through the life cycle of the infrastructure. By comparison, the operational variable is available to be adjusted according to the electricity demand and availability of renewable resources. The design decisions are often integer variables that are concerned with investment in renewable technologies. For instance, IWC_{no} is a design variable that decides the installed wind capacity at each node of the grid. ISC_{no} decides the installed solar capacity at each node. $IT_{no,j}$ decides the installed capacity from natural gas power plants at each node. IPC_{no} indicates installed electricity storage capacity.

Operational variables are continuous variables, which change hourly. For instance, $PT_{sc,no,j}$ represents the hourly output power of a natural gas power plant, and $PNU_{sc,no,m}$ represents the hourly output power of a nuclear power plant. There is also a continuous variable $FL_{sc,l}$ regarding the power dispatch, which indicates not only the value but also the direction of the transmitted flow in a line. More precisely, if the power flows out of the node, then its sign is considered negative; otherwise, its sign is supposed to be positive. Operational variables regarding the output power of wind and solar ($PW_{sc,no}$ and $PS_{sc,no}$) are decided by their availability. The wind and solar power are utilized at 100% because of the very low operational capacity.

7.3.2.3 Constraints of Electricity Balance

Equation (7.3) is the electricity balance constraint. The left side of the equation includes eight terms. The first two terms are the output power of conventional natural gas and nuclear power plants. The third and fourth terms represent the pumped electricity from the storage device and the electricity stored considering a conversion efficiency n_p. The fifth and sixth terms are the renewable wind and solar power output. The seventh and eighth terms are the power leaving the node through the transmission lines considering the associated loss. Factor 0.5 is to avoid counting the transmission loss twice, because this term is common among connected nodes. The right side is the demand.

$$\sum_{j=1}^{JN} PNG_{sc,no,j}(t) + \sum_{m=1}^{MN} PNU_{sc,no,m}(t) + PP_{sc,no}(t) - \eta_p PSt_{sc,no}(t) +$$
$$PS_{sc,no}(t) + \sum_{l=1}^{LN} \left(FL_{sc,l}(t) - 0.5 LO_{sc,l}(t) \right) = D_{sc,no}(t) \tag{7.3}$$

Where sc is the prediction error scenario, no is the node, t is any certain hour, j is the natural gas generator, m is the nuclear generator, η_p is the conversion efficiency of storage, l is the transmission line, and FL is the power entering the node. Additional constraints, including the constraints on the rate of changes in the output power of the generators, storage charge and discharge rates, and constraints on transmission line capacity, can be found in the Appendix.

7.3.3 Incorporating the Prediction Errors into the Programming of the Stochastic Scenarios

The prediction errors were included in the optimization program by the constructing stochastic scenarios. Without the loss of generality, and to manage computational complexity in the stochastic optimization programs, three levels of forecasting error (low, medium, and high) were considered in predictions of wind power, solar power, and demand, respectively. In this case, 27 scenarios are constructed, and they are shown in Figure 7.2. The probabilities of these scenarios are calculated in terms of the error distribution of each scenario, using machine-learning algorithm results, as discussed earlier. Furthermore, two deterministic scenarios—perfect and inefficient—were programmed to compare with the stochastic optimization. In a perfect scenario, the perfect knowledge of wind and solar power and demand are available. Here, the average historical values for wind power, solar power, and electricity demand were considered. This is an ideal scenario, with minimal need for standby power and energy storage. The inefficient scenario was formulated as the worst situation where there is a low generation of wind power, solar power, and high demand and no prediction model is applied. As a result, large standby capacities and storage devices are needed to ensure the security of the power system, which is inefficient.

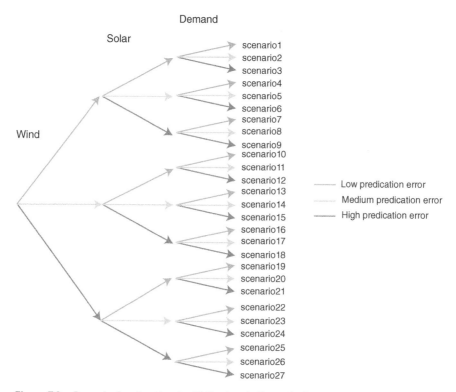

Figure 7.2 Scenario Construction for 27 Stochastic Scenario Cases.

7.4 Results

In the following, the results of the (1) machine-learning algorithm and the (2) optimization programming are presented. The performance of predictive models and the error distribution will be discussed in the first part. In the second part, the cost of electricity grid considering the prediction errors will be compared with the result of deterministic perfect and inefficient scenarios.

7.4.1 Predictive Modeling Results

The performance of prediction was measured by mean square error, and it is shown in Table 7.4. It illustrates that the predictions of electricity demand and solar power perform better than the prediction of wind power in this study, which means that the prediction error for wind power is larger. The model performance for wind power prediction can be seen in the response of the neural network in Figure 7.3. (The performance for solar power and demand prediction can be found in Figure A7.3 and Figure A7.4). The blue line shows the value of the target series, while the red line

Table 7.4   The Performance of Wind Power, Solar Power, and Electricity Demand Forecasting.

Prediction	Mean Square Error
Wind power	6.97E-04
Solar power	3.98E-05
Electricity demand	3.29E-05

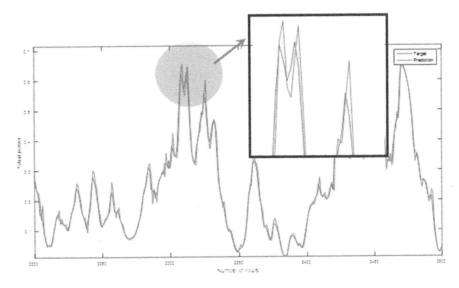

Figure 7.3   The Response of Neural Network for Predicting Wind Power for a Number of Time Horizons.

Table 7.5 The Probabilities of Three Levels of Prediction Error in Solar Power, Wind Power, and Demand Forecasting.

Type of electricity	Level of prediction errors	Error range of the level	Probabilities
Solar	Low	$-0.005 \leq e \leq 0.005$	76.1%
	Medium	$0.005 \leq e \leq 0.01 \, or \; -0.01 \leq e \leq -0.005$	12.5%
	High	$0.01 \leq e \, or \, e \leq -0.01$	11.4%
Wind	Low	$-0.025 \leq e \leq 0.025$	78.3%
	Medium	$0.025 \leq e \leq 0.05 \, or -0.05 \leq e \leq -0.025$	15.2%
	High	$0.05 \leq e \, or \, e \leq -0.05$	6.5%
Demand	Low	$-0.005 \leq e \leq 0.005$	66.2%
	Medium	$0.005 \leq e \leq 0.01 \, or -0.01 \leq e \leq -0.005$	25.9%
	High	$0.01 \leq e \, or \, e \leq -0.01$	7.9%

shows the value of the predicted series. The figures illustrate overall good responses for forecasting solar power, wind power, and electricity demand. More illustrative figures are presented in the Appendix.

7.4.1.1 Prediction Errors in Machine Learning

The distribution of prediction errors of three solar zones is shown in Figure 7.4, the distribution of prediction errors of three wind zones is shown in Figure 7.5, and the error distribution of demand prediction is shown in Figure 7.6. It should also be mentioned that the prediction errors are scaled in the range of [–1, 1]. The normalization operation can be reverted by multiplying the prediction values by the installed capacity of the generators. As mentioned earlier, the prediction error of wind forecasting is larger than that of solar power and demand forecasting. It can also be seen in the error distributions.

Furthermore, the probabilities of three levels (low, medium, and high) are discretized using the error distribution functions. The corresponding error ranges for three levels can be seen in Table 7.5. It can be found that the probability of low prediction error is much higher than the probabilities of medium and high prediction error. This indicates the high accuracy of machine-learning predictions for wind power, solar power, and electricity demand.

7.4.2 Generation Expansion and Electricity Dispatch Results

The results of stochastic optimization including predictive models were compared with the deterministic optimization (the perfect scenario and the inefficient scenario). The perfect scenario illustrates the average value of electricity demand as well as the generated wind and solar power for a whole year. The predictions are perfect without any errors. The inefficient scenario illustrates the situation when the electricity demand is high and the generation of wind and solar power is low. The balance of supply and demand was achieved by applying more natural gas power, nuclear power,

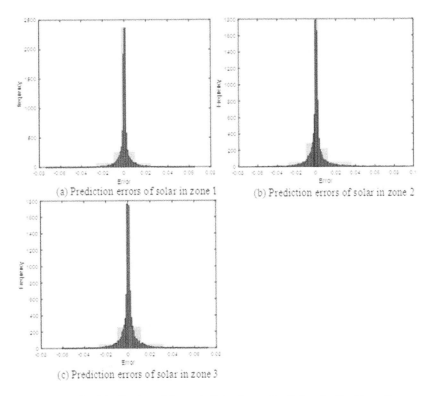

Figure 7.4 The Distribution of Errors in Solar Power Prediction for the Three Zones.

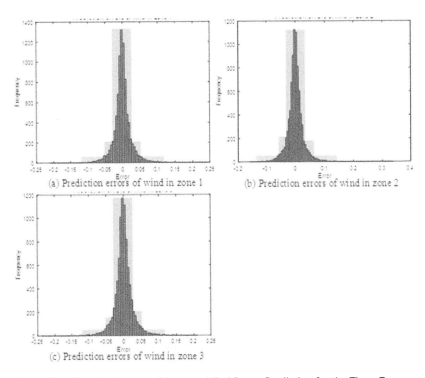

Figure 7.5 The Distribution of Errors in Wind Power Prediction for the Three Zones.

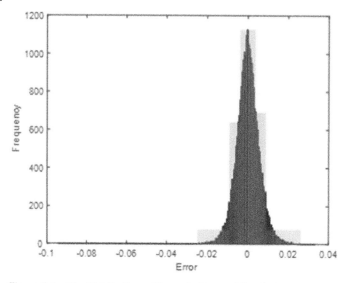

Figure 7.6 The Distribution of Errors in Demand Prediction.

and storage devices, which is inefficient. The considered scenarios for the behavior of electricity demand, solar power generation, and wind power generation in deterministic optimization are shown in Figure 7.7 The solar irradiation in southern nodes is higher than that in northern nodes. On the contrary, the wind speed in northern nodes is slightly higher than southern nodes.

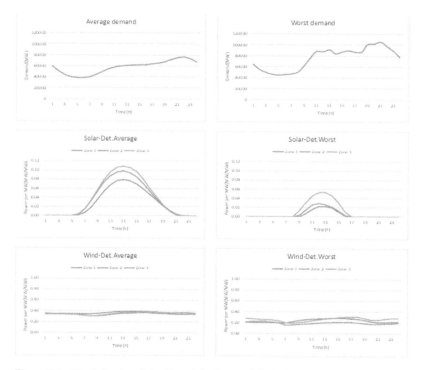

Figure 7.7 The Behavior of the Electricity Demand, Solar Power Generation, and Wind Power Generation in Deterministic Optimization for the Three Zones.

Figure 7.8 The Behavior of the Wind Power Generation in Stochastic Optimization with Three Levels of Prediction Errors for the Three Zones.

The uncertain behavior of wind power generation with three levels of prediction errors for the three zones is shown in Figure 7.8. The behavior of solar power and electricity demand can be seen in Figure A7.5 and Figure A7.6 in the Appendix. The demand, solar power generation, and wind power generation with high prediction errors present large fluctuation, whereas those with low prediction errors present smooth curves. Those with medium prediction errors stand between these extremes.

7.4.2.1 Economic Performance

The objective values reflecting the hourly capital and operational costs are illustrated in Table 7.6. The first column presents the cost of the perfect scenario. The last column presents the cost of the inefficient scenario. The value of the scenario with smart control can be seen in the columns between them. The main part of the costs is the capital investment for the installation of renewable power generation, natural gas power generation, and storage devices. It can be observed that the inefficient scenario has the highest investment in capital costs. This is because the scenario of high demand and low renewable power generation needs more construction of standby capacity to ensure the operation of the power system. The scenario with smart control considering the prediction errors also requires a higher investment than the perfect scenario. More

Table 7.6 The Objective Values for Various Deterministic and Stochastic Scenarios.

	The Perfect Scenario	The Scenario with Smart Control [a]	The Inefficient Scenario
Objective value(£/h)	1.5076E+09	1.5114E+09	1.7009E+09

[a] Wind prediction errors variation (3 levels), solar prediction errors variation (3 levels), and demand prediction errors variation (3 levels).

standby capacity and storage devices are built to compensate for the prediction errors. The economic loss for the scenario with predictive models is 3.3585E+10 £ per year (0.25%), while that for the inefficient scenario is 1.6939E+12 £ per year (12.83%). The economic saving is up to 1.6603E+12 £ per year (12.57%). This indicates that using machine-learning prediction for renewable power generations and electricity demand can greatly reduce the investment in the system of more than 50% power generation from wind and solar energies.

Figure 7.9 illustrates the economic savings in the geographic distribution of newly installed wind and solar power generation, natural gas power generation, and

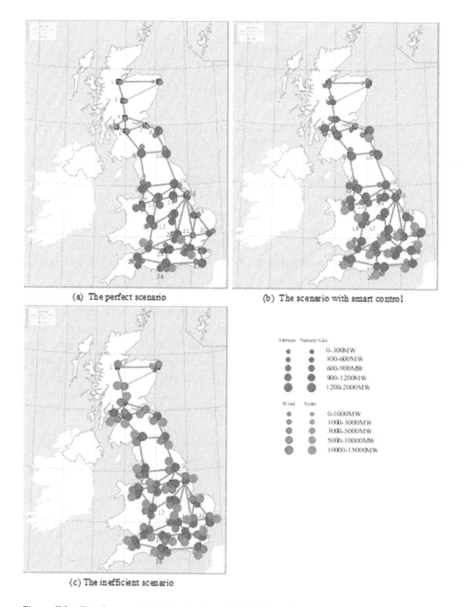

(a) The perfect scenario

(b) The scenario with smart control

(c) The inefficient scenario

Figure 7.9 The Geographical Distribution of Newly Installed Wind and Solar Power Generation, Natural Gas Power Generation, and Storage Devices Sites.

storage sites. The UK electricity grid used in this study can be found in Figure A7.5.The geographic distribution of installed generations of the smart controlled scenario is more similar to the distribution of the perfect scenario: the construction of solar power generators mainly lies in the southern nodes, which can make the most of the solar power while the wind power generators are built less in the southern nodes compared with the case of inefficient scenario, which can also increase the utilization of wind power.

In addition, the operational scheduling for the perfect scenario, the inefficient scenario, and the scenario with predictions for wind power are presented in Figure 7.10. It can be observed that the use of natural gas power and nuclear power in the scenario with prediction errors for wind power is much less than that in the inefficient scenario. This means that the application of machine-learning prediction can reduce the use of natural gas power and thus greenhouse gas emissions. Moreover, the solar power is utilized more in the scenarios with prediction errors than the inefficient scenario, which implicates that the efficiency of using solar power is enhanced.

It can be concluded that the application of machine-learning prediction can optimize the geographic distribution of power generations with wind and solar energies. Besides, the operational scheduling is also optimized.

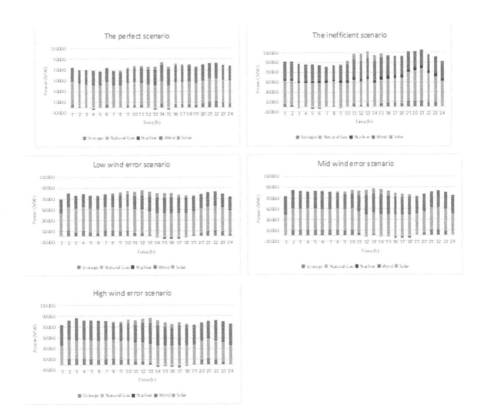

Figure 7.10 The Operational Scheduling for the Deterministic Scenarios and Stochastic Scenarios.

7.5 Conclusions

Renewable resources such as wind and solar power can be applied for electricity generation to reduce the associated greenhouse gas emissions. However, renewable power generation from wind and solar energy inevitably suffer from intermittency and uncertainty. In addition, the stochastic behavior of consumers also leads to significant uncertainty on the demand side. The uncertainty can be compensated by the construction of natural gas power generators and storage facilities. However, this will also lead to the increase of economic expenditure and greenhouse gas emissions. In this study, machine-learning algorithms were applied to enhance the accuracy of wind power, solar power, and demand prediction for economic saving. The prediction errors are divided into three levels (low, medium, and high), and the probabilities of each level were calculated according to the error distributions from predictive models. Then the probabilities of low, medium, and high prediction errors can be used in the stochastic optimization to obtain the best solution of investment in the construction of wind and solar power generators as well as natural gas power generators and storage facilities. The optimization results demonstrate that the application of machine-learning predictions in the stochastic optimization enables the reduction of capital cost of standby capacity and more efficient use of wind and solar power. It was also illustrated that the effect of uncertainty on the power system can be greatly reduced with highly accurate prediction methods. To improve the accuracy and applicability of the proposed approach, the effect of microgrids can be considered in the economic analysis. Moreover, the effect of the gas and electricity prices volatility, seasonality, and emission production constraints can be investigated in future studies.

7.6 Nomenclature

sc	prediction error scenario	$oc_{sc,g}$	operating costs
no	node	$vc_{sc,g}$	variable costs
$prob_{sc}$	probability of the scenario	IWC_{no}	installed wind capacity at each node
t	time period	ISC_{no}	installed solar capacity at each node
g	generation technology	$IT_{no,j}$	new installed capacity from natural gas power plants at each node
j	natural gas generator	IPC_{no}	installed electricity storage capacity
m	nuclear generator	$PT_{sc,no,j}$	hourly output power of a natural gas power plant
η_p	conversion efficiency of storage	$PNU_{sc,no,m}$	hourly output power of a nuclear power plant
l	transmission line	$FL_{sc,l}$	transmitted flow
$cc_{sc,g}$	capacity costs	$PW_{sc,no}$	output power of wind
$fc_{sc,g}$	fixed costs	$PS_{sc,no}$	output power of solar

References

1 National Renewable Energy Laboratory (NREL). (2012). Renewable Electricity Futures Study. *U.S. Dep. Energy* 1: 280.

2 Mai, T., Mulcahy, D., Hand, M.M., and Baldwin, S.F. (2014). Envisioning a renewable electricity future for the United States. *Energy* 65: 374–386.

3 Jin, S., Botterud, A., and Ryan, S.M. (2014). Temporal versus stochastic granularity in thermal generation capacity planning with wind power. *IEEE Trans. Power Syst.* 29 (5): 2033–2041.

4 Wang, J., et al. (2011). Wind power forecasting uncertainty and unit commitment. *Appl. Energy* 88 (11): 4014–4023.

5 Osório, G.J., Lujano-Rojas, J.M., Matias, J.C.O., and Catalão, J.P.S. (2015). A fast method for the unit scheduling problem with significant renewable power generation. *Energy Convers. Manag.* 94: 178–189.

6 Aghaei, J., Niknam, T., Azizipanah-Abarghooee, R., and Arroyo, J.M. (2013). Scenario-based dynamic economic emission dispatch considering load and wind power uncertainties. *Int. J. Electr. Power Energy Syst.* 47 (1): 351–367.

7 Pereira, S., Ferreira, P., and Vaz, A.I.F. (2017). Generation expansion planning with high share of renewables of variable output. *Appl. Energy* 190: 1275–1288.

8 Sharafi, M., and ELMekkawy, T.Y. (2014). Multi-objective optimal design of hybrid renewable energy systems using PSO-simulation based approach. *Renew. Energy* 68: 67–79.

9 Wang, B., Wang, S., Zhou, X., and Watada, J. (2016). Multi-objective unit commitment with wind penetration and emission concerns under stochastic and fuzzy uncertainties. *Energy* 111: 18–31.

10 Baker, K., Hug, G., and Li, X. (2017). Energy storage sizing taking into account forecast uncertainties and receding horizon operation. *IEEE Trans. Sustain. Energy* 8 (1): 331–340.

11 Wang, J., Wang, J., Liu, C., and Ruiz, J.P. (2013). Stochastic unit commitment with sub-hourly dispatch constraints. *Appl. Energy* 105: 418–422.

12 Voyant, C., et al. (2017). Machine learning methods for solar radiation forecasting: A review. *Renew. Energy* 105: 569–582.

13 Bae, K.Y., Jang, H.S., and Sung, D.K. (2016). Hourly solar irradiance prediction based on support vector machine and its error analysis. *IEEE Trans. Power Syst.* 32 (2): 1–1.

14 Gutierrez-Corea, F.V., Manso-Callejo, M.A., Moreno-Regidor, M.P., and Manrique-Sancho, M.T. (2016). Forecasting short-term solar irradiance based on artificial neural networks and data from neighboring meteorological stations. *Sol. Energy* 134: 119–131.

15 Ramasamy, P., Chandel, S.S., and Yadav, A.K. (2015). Wind speed prediction in the mountainous region of India using an artificial neural network model. *Renew. Energy* 80 (March 2014): 338–347.

16 Doucoure, B., Agbossou, K., and Cardenas, A. (2016). Time series prediction using artificial wavelet neural network and multi-resolution analysis: Application to wind speed data. *Renew. Energy* 92: 202–211.

17 Okumus, I., and Dinler, A. (2016). Current status of wind energy forecasting and a hybrid method for hourly predictions. *Energy Convers. Manag.* 123: 362–371.

18 Jiang, H., Zhang, Y., Muljadi, E., Zhang, J., and Gao, W. (2016). A short-term and high-resolution distribution system load forecasting approach using support vector regression with hybrid parameters optimization. *IEEE Transactions on Smart Grid.* 1–1.

19 Tuohy, A., Meibom, P., Denny, E., and O'Malley, M. (2009). Unit commitment for systems with significant wind penetration. *IEEE Trans. Power Syst.* 24 (2): 592–601.

20 Constantinescu, E.M., Zavala, V.M., Rocklin, M., Lee, S., and Anitescu, M. (2011). A computational framework for uncertainty quantification and stochastic optimization in unit commitment with wind power generation. *IEEE Trans. Power Syst.* 26 (1): 431–441.

21 Connolly, D., Lund, H., and Mathiesen, B.V. (2016). Smart energy Europe: The technical and economic impact of one potential 100% renewable energy scenario for the European Union. *Renew. Sustain. Energy Rev.* 60: 1634–1653.

22 Quan, H., Srinivasan, D., and Khosravi, A. (2014). Short-term load and wind power forecasting using neural network-based prediction intervals. *IEEE Trans. Neural Networks Learn. Syst.* 25 (2): 303–315.

23 Sikinioti-lock, A. (2016). *Big-Data for Integrated Renewable Energy Systems.* Imperial College London.

24 Sharifzadeh, M., Lubiano-Walochik, H., and Shah, N. (2017). Integrated renewable electricity generation considering uncertainties: The UK roadmap to 50% power generation from wind and solar energies. *Renew. Sustain. Energy Rev.* 72: 385–398.

25 Pfenninger, S., and Staffell, I. (2016). Long-term patterns of European PV output using 30 years of validated hourly reanalysis and satellite data. *Energy 114: 1251–1265.*

26 Staffell, I., and Pfenninger, S. (2016). Using bias-corrected reanalysis to simulate current and future wind power output. *Energy* 114: 1224–1239.

Appendix

In this section, details about the study will be presented, including two parts. The first part includes the constraints of generators and storage, constraints of transmission and prediction models. The second part includes the details of the existing electricity grid.

Nomenclatures

sc	prediction error scenario	$oc_{sc,g}$	operating costs
no	node	$vc_{sc,g}$	variable costs
$prob_{sc}$	probability of the scenario	IWC_{no}	installed wind capacity at each node
t	time period	ISC_{no}	installed solar capacity at each node
g	the generation technology	$IT_{no,j}$	new installed capacity from natural gas power plants at each node
j	natural gas generator	IPC_{no}	installed electricity storage capacity
m	nuclear generator	$PT_{sc,no,j}$	hourly output power of a natural gas power plant
η_p	conversion efficiency of storage	$PNU_{sc,no,m}$	hourly output power of a nuclear power plant
l	transmission line	$FL_{sc,l}$	transmitted flow
$cc_{sc,g}$	capacity costs	$PW_{sc,no}$	output power of wind
$fc_{sc,g}$	fixed costs	$PS_{sc,no}$	output power of solar
$PT_{no,j\min}$	minimum power of existing natural gas generator	$SP_{sc,no}$	output power of the solar panel unit
$PT_{no,j\max}$	maximum power of existing natural gas generator	ISC_{no}	number of installed solar capacity
$IT_{no,j}$	new installed capacity	IPC_{no}	installed storage capacity
$x_{sc,no,j}$	operational status of the generator	EN_{no}	new installed capacity
$RUT_{no,j}$	upward rates of changing power of the generator	EO_{no}	old storage capacity
$RDT_{no,j}$	downward rates of changing power of the generator	$PP_{sc,no}$	pumped power out of the storage
$P_{sc,no,j}$	generated power	$PSt_{sc,no}$	stored power into the storage
$a_{no,j}, b_{no,j}$	parameters of conversion efficiency	$ES_{sc,no}$	stored power

Nomenclatures

$y_{no,m}$	operational status of the nuclear generator	$ES_{no,max}$	maximum stored power
$PN_{no,m_{max}}$	maximum power generated by a nuclear power plant	$\theta_{sc,NF}, \theta_{sc,NT}$	angles of nodes on the transmission line
$PW_{sc,no}(t)$	wind power generation	$f_{l,max}$	maximum flow on the transmission line
$WP_{sc,no}$	output power of the wind turbine unit	$LO_{sc,l}$	loss in the transmission
IWC_{no}	number of installed wind capacity		

7.1.1 Constraints of Generators and Storage

The constraints of generators include the models of natural gas power generation, nuclear power plant, wind power generation, and solar power generation.

The model of the natural gas power plant includes three constraints: (A7.1) limits the amount of power that can be generated by a plant, and (A7.2) and (A7.3) limit the upward and downward rates of changing output power, respectively. Moreover, an additional constraint (A7.4) represents the amount of the fuel required when the power plant is not operating on a nominal point.

$$PT_{no,jmin}x_{sc,no,j} \leq PT_{sc,no,j}(t) \leq (PT_{no,jmax} + IT_{no,j})x_{sc,no,j}$$
$$PT_{no,jmin}x_{sc,no,j} \leq PT_{sc,no,j}(t) \leq (PT_{no,jmax} + IT_{no,j})x_{sc,no,j}$$

$$x_{sc,no,j} \in [0,1] \tag{A7.1}$$

where $PT_{no,jmin}$ is the minimum power of existing natural gas generator, $PT_{no,jmax}$ is the maximum power of existing natural gas generator, $IT_{no,j}$ is the new installed capacity, $x_{sc,no,j}$ is the operational status of the generator.

$$PT_{sc,no,j}(t) - PT_{sc,no,j}(t-1) \leq RUT_{no,j} \tag{A7.2}$$

where $RUT_{no,j}$ is the upward rates of changing power of the generator.

$$PT_{sc,no,j}(t-1) - PT_{sc,no,j}(t) \leq RDT_{no,j} \tag{A7.3}$$

where $RDT_{no,j}$ are the downward rates of changing power of the generator.

$$fuel_{sc,no,j}(t) = a_{no,j}P_{sc,no,j}(t) + b_{no,j} \tag{A7.4}$$

where $P_{sc,no,j}$ is the generated power, $a_{no,j}$ and $b_{no,j}$ are the parameters of conversion efficiency.

The generation of the nuclear power is modeled by (A7.5). The binary variable $y_{no,m}$ decides whether the power generated is at the largest capacity or zero. $PN_{no,mmax}$ is the maximum power generated by a nuclear power plant

$$PNU_{sc,no,m}(t) = PNU_{no,mmax} y_{no,m}$$

$$y_{no,m} \in [0,1] \tag{A7.5}$$

The wind power generation $PW_{sc,no}(t)PW_{sc,no}(t)$ can be calculated by multiplying the output of each wind turbine unit by the number of units installed (wind capacity), as shown in (A7.6). The number of the installed wind capacity is an integer variable in the design of the system.

$$PW_{sc,no}(t) = WP_{sc,no}(t) \times IWC_{no} \tag{A7.6}$$

where $WP_{sc,no}$ is the output power of the wind turbine unit, IWC_{no} is the number of installed wind capacity.

Similarly, the solar power generation is shown in (7A7.).

$$PS_{sc,no}(t) = SP_{sc,no}(t) \times ISC_{no} \tag{A7.7}$$

where $SP_{sc,no}SP_{sc,no}$ is the output power of the solar panel unit, $ISC_{no}ISC_{no}$ is the number of installed solar capacity.

The model of the storage is based on pumped hydroelectric systems. Constraints (A7.8) and (A7.9) indicate the limits of pumped power out of the storage and the stored power into the storage. The installed storage capacity IPC_{no} includes the new installed capacity EN_{no} and old storage capacity EO_{no}.

$$PP_{sc,no}(t) \le IPC_{no} \tag{A7.8}$$

where $PP_{sc,no}$ is the pumped power out of the storage, IPC_{no} is the installed storage capacity.

$$PSt_{sc,no}(t) \le IPC_{no} \tag{A7.9}$$

where $PSt_{sc,no}$ is the stored power into the storage.

$$PC_{no} = EN_{no} + EO_{no} \tag{A7.10}$$

where EN_{no} is the newly installed storage, EO_{no} is the old storage.

Besides, (A7.11) represents a balance for any period and η_p represents an efficiency coefficient considering the conversion losses. The maximum power that can be stored is restricted by the capacity of the reservoir (A7.12).

$$ES_{sc,no}(t) = ES_{sc,no}(t-1) + \eta_p PSt_{sc,no}(t) - \eta_p PSt_{sc,no}(t) \tag{A7.11}$$

where $ES_{sc,no}$ is the stored power.

$$ES_{sc,no}(t) \le ES_{no,max} \tag{A7.12}$$

where $ES_{no,max}$ is the maximum stored power.

Furthermore, there is a balance for the stored electricity through all time periods and over all scenarios and nodes.

$$\sum_{sc=1}^{SC}\sum_{t=1}^{T}\sum_{no=1}^{NO}(\eta_p PSt_{sc,no}(t)+PP_{sc,no}(t))=0 \tag{A7.13}$$

7.1.2 Constraints of *Transmission*

There are also constraints limiting the transmitted flow in a line. The flow is related to the angle between the nodes connected to that line. To consider the losses $LO_{sc,l}$ in the transmission, they are obtained by multiplying the conductance of the line and the cosine of the angle between nodes and limited by the line capacity.

$$FL_{sc,l}(t)=B_{sc,l}(\theta_{sc,NF}(t)-\theta_{sc,NT}(t)) \tag{A7.14}$$

where $\theta_{sc,NF}, \theta_{sc,NT}\theta_{sc,NF}, \theta_{sc,NT}$ are the angles of nodes on the transmission line.

$$-f_{l,max}\leq FL_{sc,l}(t)\leq f_{l,max} \tag{A7.15}$$

where $f_{l,max}$ is the maximum flow on the transmission line.

$$LO_{sc,l}(t)=2G_l(1-cos(\theta_{sc,NF}(t)-\theta_{sc,NT}(t)) \tag{A7.16}$$

where $LO_{sc,l}$ is the loss in the transmission.

$$0\leq LO_{sc,l}(t)\leq f_{l,max}(l) \tag{A7.17}$$

7.1.3 Prediction Models for Solar Power and Electricity Demand

The structure of the ANN for the solar and demand are shown in Figures A7.1 and A7.2.

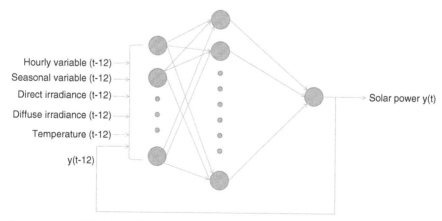

Hourly variable (t-12)
Seasonal variable (t-12)
Direct irradiance (t-12)
Diffuse irradiance (t-12)
Temperature (t-12)

y(t-12)

Solar power y(t)

Figure A.71 NARX Neural Network for Solar Power Prediction.

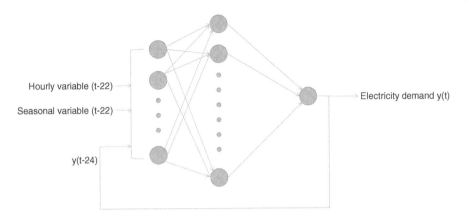

Figure A7.2   NARX Neural Network for Electricity Demand Prediction.

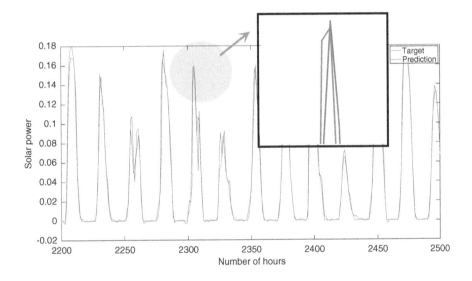

Figure A7.3   The Response of Neural Network for Predicting Solar Power for a Number of Time Horizons.

The model performance for solar power and demand prediction can be seen in Figure A7.3 and Figure A7.4.

The uncertain behavior of solar power and electricity demand with three levels of prediction errors for the three zones is shown in Figure A7.5 and Figure A7.6.

7.2   Generation Sites and Transmission Lines

The UK electricity grid used in this study is shown in Figure A7.7. There are 29 generation sites and 99 transmission lines. Table A7.1 includes the parameters (line capacity, susceptance and admittance) of the electricity network [1].

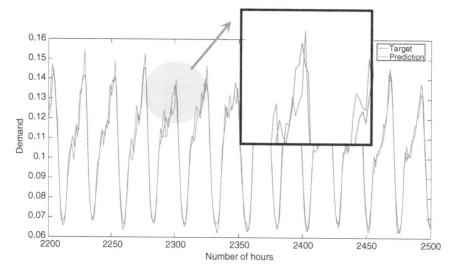

Figure A7.4 The Response of Neural Network for Predicting Electricity Demand for a Number of Time Horizons.

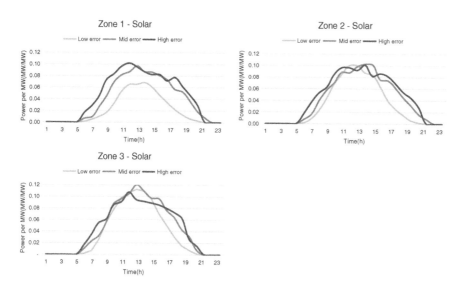

Figure A7.5 The Behavior of the Solar Power Generation in Stochastic Optimization with Three Levels of Prediction Errors for the Three Zones.

7.2.2 The Existing Natural Gas Generation Sites

There are 33 existing natural gas generation sites in the United Kingdom, and the type of these power plants are natural gas combined cycle (NGCC). The minimum output power for each power plant is half of the power plant capacity [2]. The maximum power for each plant and the parameters for calculating conversion efficiency are

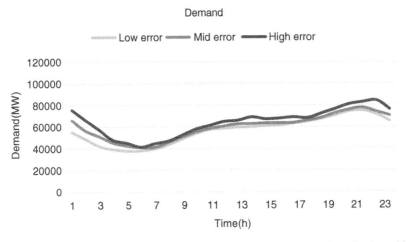

Figure A7.6 The Behavior of the Electricity Demand in Stochastic Optimization with Three Levels of Prediction Errors for the Three Zones.

Nodes

1. Beauly
2. Peterhead
3. Errochty
4. Denny
5. Neilston
6. Strathaven
7. Torness
8. Eccles
9. Harker
10. Stella West
11. Penwortham
12. Deeside
13. Daines
14. Th. Marsh
15. Thornton
16. Keadby
17. Ratcliffe
18. Feckenham
19. Walpole
20. Bramford
21. Pelham
22. Sundon
23. Melksham
24. Bramley
25. London
26. Kemsley
27. Sellindge
28. Lovedean
29. SW. Peninsula

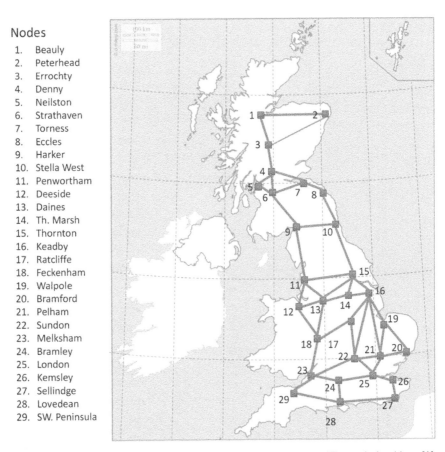

Figure A7.7 The Geographical Distribution of Generation Sites and Transmission Lines [1] / With permission of Elsevier.

Table A7.1 The Parameters of the Transmission Lines Based on [1].

Line	Node from	Node to	Capacity (MW)	B	G
l1	1	2	525		3050.00
l2	1	2	525	5000.00	3050.00
l3	1	3	132	666.67	31.11
l4	1	3	132	666.67	31.11
l5	2	4	760	1538.46	9.47
l6	2	4	760	1538.46	9.47
l7	3	2	652	1298.70	506.66
l8	3	4	648	2439.02	178.47
l9	3	4	648	2439.02	178.47
l10	4	5	1000	4166.67	173.61
l11	4	5	1000	4166.67	173.61
l12	4	6	1120	4347.83	245.75
l13	4	6	1500	4347.83	245.75
l14	4	7	1090	7407.41	1157.75
l15	4	7	1090	7407.41	1152.26
l16	5	6	1390	9514.75	769.51
l17	5	6	1390	6199.63	580.37
l18	6	9	2100	11737.09	1074.52
l19	6	9	2100	11737.09	1074.52
l20	7	6	950	500.00	7.50
l21	7	6	950	500.00	7.50
l22	7	8	2500	58823.52	235294.11
l23	7	8	2180	58823.52	235294.11
l24	8	10	3070	5714.29	271.02
l25	8	10	3070	5714.29	271.02
l26	9	10	855	4076.64	584.99
l27	9	10	775	2915.45	418.19
l28	9	11	1390	6134.97	617.26
l29	9	11	1390	6134.97	617.26
l30	10	15	4020	11976.05	760.16
l31	10	15	4840	15873.02	1310.15
l32	11	12	3320	11764.71	138.41
l33	11	12	3320	11764.71	138.41
l34	11	13	2210	19230.77	1479.29
l35	11	13	2170	19230.77	1479.29
l36	11	15	2520	2380.95	39.68
l37	11	15	2520	2380.95	56.12
l38	12	13	3100	9276.44	826.10

Table A7.1 (Continued)

Line	Node from	Node to	Capacity (MW)		G
139	12	13	3100	9276.44	826.10
140	12	18	2400	11111.11	913.58
141	12	18	2400	11111.11	1197.53
142	13	14	1040	8598.45	791.09
143	13	14	1040	8326.39	568.50
144	13	15	955	4347.83	258.98
145	13	15	1240	4347.83	310.02
146	13	18	2400	14285.71	1000.00
147	13	18	2400	14285.71	1714.29
148	14	16	625	5555.56	1543.21
149	14	16	2580	6250.00	195.31
150	15	14	5000	45045.05	3855.21
151	15	14	5000	45045.05	3652.30
152	15	16	5540	58139.53	5408.33
153	15	16	2270	19230.77	1220.41
154	16	19	2780	7092.20	281.68
155	16	19	3820	7092.20	281.68
156	17	16	1890	9328.36	870.18
157	17	16	2150	9328.36	870.18
158	17	22	2100	10309.28	722.71
159	17	22	2100	10309.28	733.34
160	18	17	3460	55555.56	12962.96
161	18	17	3100	55555.56	12962.96
162	18	23	1970	10416.67	1497.40
163	18	23	1970	10416.67	1269.53
164	20	19	1590	43478.26	6616.26
165	20	19	1590	43478.26	6616.26
166	20	26	2780	4694.84	392.34
167	20	26	2780	6993.01	645.51
168	21	16	2780	5482.46	435.83
169	21	16	2780	5482.46	435.83
170	21	19	2780	16949.15	1062.91
171	21	19	3030	16949.15	1062.91
172	21	20	2780	20833.33	5208.33
173	21	20	2780	20833.33	5208.33
174	21	25	2780	10000.00	250.00
175	21	25	2780	10000.00	250.00

(Continued)

Table A7.1 (Continued)

Line	Node from	Node to	Capacity (MW)		G
176	22	16	2010	5813.95	601.68
177	22	16	2010	5813.95	601.68
178	22	21	2780	90090.09	15420.83
179	22	21	2780	16393.44	1289.98
180	22	25	3275	24390.24	2201.07
181	22	25	3275	24390.24	2022.61
182	23	22	2770	33333.33	4333.33
183	23	22	2780	33333.33	6111.11
184	23	24	4400	142857.14	46938.78
185	23	24	2780	125000.00	134375.00
186	23	29	2010	5494.51	455.86
187	23	29	2010	5494.51	455.86
188	24	25	1390	10989.01	1255.89
189	24	25	1390	10989.01	1255.89
190	24	28	2210	14285.71	1387.76
191	24	28	2210	14285.71	1387.76
192	25	26	5540	17543.86	615.57
193	25	26	6960	17543.86	615.57
194	27	26	3100	19880.72	790.49
195	27	26	3100	19880.72	790.49
196	28	27	3070	14064.70	751.70
197	28	27	3070	14064.70	751.70
198	29	28	2780	12562.81	804.90
199	29	28	2780	12562.81	804.90

shown in Table A7.2. The ramp-up and ramp-down rates are 20% of the of the plant's capacity each hour [3].

7.2.3 The Existing Nuclear Generation Sites

The parameters of nuclear power plants are shown in Table A7.3.

7.2.4 The Wind Power Generation

The wind power generation can be divided into the existing wind farms and newly built wind farms. The existing wind power generation sites are shown in Table A7.4. The wind turbine is the Vestas V90 and the specifications are illustrated in Table A7.5 [4].

Table A7.2 The Parameters of Existing Natural Gas Power Plants Based on [1].

Node	Maximum Capacity (MW)	Efficiency	$a_{n,i}$	$b_{n,i}$
2	1180	57%	0.0284	5,6212
11	810	55%	0.0294	3,9989
12	1380	58%	0.0279	6,4606
12	515	53%	0.0305	2,6385
16	1565	53%	0.0305	8,0179
16	1310	55%	0.0294	6,4674
16	1200	55%	0.0294	5,9243
16	395	58%	0.0279	1,8492
16	1748	58%	0.0279	8,1834
16	1410	60%	0.0269	6,3810
19	880	58%	0.0279	4,1198
19	240	48%	0.0337	1,3577
19	819	58%	0.0279	3,8432
19	405	56%	0.0289	2,0365
21	720	54%	0.0299	3,6204
22	401	50%	0.0323	2,1777
23	1470	55%	0.0294	7,2573
23	1232	55%	0.0294	6,0823
23	510	60%	0.0269	2,3080
23	140	44%	0.0367	0,8640
23	850	58%	0.0279	3,9794
23	2180	60%	0.0269	9,8657
25	1000	54%	0.0299	5,0284
25	408	58%	0.0279	1,9101
25	715	50%	0.0323	3,8829
26	805	55%	0.0294	3,8742
26	1290	58%	0.0279	6,0393
26	700	53%	0.0306	3,5863
26	800	57%	0.0284	3,8110
28	420	55%	0.0294	2,0735
28	824	58%	0.0279	3,9419
29	905	58%	0.0279	4,2368

Table A7.3 The Capacity of Nuclear Power Generation Sites Based on [1].

Node	Capacity (MW)
11	3450
12	2760
18	2760
20	3200
26	2300
29	3340

Table A7.4 The Parameters of Existing Wind Power Generation Sites Based on [1].

Node	Maximum Capacity (MW)
1	2435.5
2	1330
3	529
4	223
6	2071.6
7	592
8	99
10	184
11	604
12	2424
19	1466
20	1004
26	630
27	300

Table A7.5 The Specifications of the Vestas v90 Wind Turbine.

Vestas V90 Specification	Value
Cut-in speed, v_{ci}	4 m/s
Rated speed v_r	12 m/s
Cut-out speed v_{co}	25 m/s
Hub height	80

7.2.5 The Solar Power Generation

There were only new built solar power generation sites in the study, and the solar panels were chosen as the Sunpower E20–435 model [5]. The specifications are shown in Table A7.6.

Table A7.6 The Specification of SunPower e20–435
Photovoltaic Solar Panel Based on [5].

Sunpower E20–435 Specification	Value
Nominal power at 1000 W/m2 and 25ºC	435 W
Area	2,162 m2
Efficiency	20,70%

7.2.5.1. The Energy Storage Capacities

The existing energy storage sites and their capacities can be seen in Table A7.7. The newly built energy storage devices were also considered, whose maximum value was 10000 MW. The efficiency of storage and discharge of energy is 90% and the round-trip efficiency is 81% [6].

Table A7.7 The Capacity of Existing
Energy Storage Sites.

Node	Capacity (MW)
1	300
4	400
12	1644
12	360

7.2.5.2 The Capacity Factors

Table A7.8 Capacity Factors for Generation
Technologies Based on [7, 8].

Technology	Capacity Factor
Natural Gas	56.3%
Nuclear	92.2%
Wind	32.5%
Solar	28.6%
Pumped Hydro	35.9%

7.2.5.3 The Parameters of the Objective Function for Each Generation Technology

Table A7.9 The Parameters of the Economic Objective Function Based on [1].

Technology	cc_g(£/installed MW)	fc_g(£/installed MW)	oc_g(£/produced MW)	Vc_g(£/produced MW)
NGCC	600,000	22,000	0.1	2161
Nuclear	4,100,000	72,000	3	5
Wind power	1,500,000	30,100	5	-
Photovoltaic	1,000,000	22,600	-	-
Pumped storage	3,400,000	26,300	-	-

References

1 Sharifzadeh, M., Lubiano-Walochik, H., and Shah, N. (2017). Integrated renewable electricity generation considering uncertainties: The UK roadmap to 50% power generation from wind and solar energies. *Renew. Sustain. Energy Rev.* 72 (August 2016): 385–398.

2 DTI. (2004). The future value of storage in the UK with generator intermittency. Electricity Networks Strategy Group, UK, Rep. 04/1877.

3 Kumar, N., Besuner, P., Lefton, S., Agan, D., and Hilleman, D. (2012). Power plant cycling costs, NREL. USA: Rep. NREL/SR-5500-55433.

4 Vestas. (2014). Vestas V90 Technical Specifications, https://en.wind-turbine-models.com/turbines/603-vestas-v90-3.0.

5 Sunpower. (2016). E-series commercial solar panels. https://us.sunpower.com/sites/default/files/media-library/data-sheets/ds-e20-series-327-310-commercial-solar-panels-helix-compatible.pdf.

6 Energy Storage Technologies. (ESA). (2018). http://energystorage.org/why-energy-storage/technologies/.

7 US Energy Information Administration. (2013–2016). Table 6.7a Capacity Factors for Utility-Scale Generators Primarily Using Fossil Fuels. https://www.eia.gov/electricity/monthly/epm_table_grapher.php?t=epmt_6_07_a.

8

A Systematic Method for Wireless Sensor Placement

A Fault-Tolerant Communication Solution for Monitoring Water Distribution Networks

Sarah Ershadi-nasab[a], Shunyao Wang[b], and Mahdi Sharifzadeh[a,b,]*

[a] Sharif Energy, Water and Environment Institute (SEWEI), Sharif University of Technology, Tehran, Iran
[b] Department of Electrical Engineering, University College London, London UK
* Corresponding author. Sharif Energy, Water and Environment Institute (SEWEI), Sharif University of Technology, Tehran, Iran

8.1 Introduction

Water plays a vital role in sustaining life on earth. Water distribution networks are the fundamental infrastructure in all areas humans inhabit. Providing sustainable potable water and maintaining and monitoring the networks that deliver it are of critical importance. The advent of Industry 4.0 has brought innovative solutions to address challenges in such areas as manufacturing, energy, electronics, health care, and the water sector [1]. Industry 4.0 drivers such as big data, Internet of things (IoT), and cloud computing offer promising solutions that are effective in smart water management systems [2]. The most advanced communication technologies for water transfer and distribution are needed to minimize water waste in distribution systems and to optimize the utilization of freshwater resources. Urban water monitoring and municipal water and wastewater infrastructure monitoring systems should be able to immediately warn of any problems.

There are two significant problems to address in water networks: leakage and contamination. Aging infrastructure may cause water pipes and junctions to leak, and detecting and managing potential leakage can have serious economic and social impacts [3]. Leakage is the most significant problem in the distribution and transfer of water and is caused by wear and tear on pumps, distribution pipes, and junctions. In addition, pollution in the water may cause a major health crisis. Contaminants may inadvertently enter the water through industrial runoff and chemical waste; in the event of war, a malicious enemy may inject pollution into municipal water [4]. Water monitoring networks then have two important goals. The first is to evaluate the quality of potable water and to avoid contaminations in potable water [5], [6]. In case of contamination in the distribution network, it is critical to determine how to securely and rapidly detect the fault and avoid its spread to the water network in the best manner. The second goal of the monitoring system is to detect leakage in the water distribution network to decrease the water waste and potential damage to the water infrastructure and buildings [7], [8]. To address both problems, it is important to monitor the network continuously. Traditionally, measurements have been taken manually at random points in the water system to measure different water parameters, which has led to wrong samples, wrong times, and inadequate data for management.

Industry 4.0 Vision for the Supply of Energy and Materials: Enabling Technologies and Emerging Applications, First Edition. Edited by Mahdi Sharifzadeh.
© 2022 John Wiley & Sons, Inc. Published 2022 by John Wiley & Sons, Inc.

When designing water networks, some unique features should be accounted for: their massive scale, the vulnerability to vandalism and the consequences of a compromised system, fluctuation and stochastic demand, and a loopy distribution network that can lead to contaminant dilution [9]. This is accomplished using a network of intelligent sensors installed in the water network to continuously measure various parameters (e.g., turbidity, pH, dissolved oxygen, conductivity, temperature) of the water to ensure its quality and to monitor water pressure and hydraulics to ensure the pipes are not leaking. Smart sensors provide significant opportunities to detect contamination and leakage in the water pipes or junctions by measuring several important parameters of water. One solution is to install a water sensor at every node. However, it is neither technically possible nor economically feasible to place sensors in all water pipes or junctions because of economic costs, maintenance, and hard access locations in the network. Therefore, the minimum number of necessary sensors and the best location for the installation of sensors to satisfy all demand is an important measure to quantify. Optimal sensor placement solutions have been proposed in the literature [10], [11]. Nonetheless, due to the multifaceted nature of water distribution networks, such a problem has several objectives to be satisfied. Optimal sensor location has been considered extensively in various research studies in recent years [4], [10], [12], [13].

Several methods are reported in the literature for monitoring and maintaining water distribution networks [5], [10], [12]. For example, Jha et al. [14] proposed a smart monitoring system for real-time water quality and usage monitoring. Five qualitative parameters—pH, temperature, turbidity, dissolved oxygen, and conductivity—are measured to ensure that the potable water does not contain any health hazards or potential threats caused due to accidental seepage of sewage or farm release into the potable water. Al-Khashab et al. [15] designed a smart water monitoring system to be used in reconstructing the water infrastructure of war-torn Mosul, Iraq. Various drinking water parameters such as temperature, potential of pH, total dissolved solids, turbidity, and conductivity were measured by the sensors and data transmitted to the Arduino Uno [16] as the core controller module via a Wi-Fi ESP 8266 module [17]. ThingSpeak was used as a free Web service for collecting and storing sensor data in the cloud and developing IoT applications [15]. Miry et al. [18] also used ThingSpeak in their smart water quality monitoring for Baghdad. The design would monitor and analyze data gathered by different installed sensors; if the quality of water degraded below a predefined value, an online notification system would send an alert via IFTTT protocol to the authorities to notify them [18]. Miry et al. also created a system to monitor the environment and climate change; it measured the temperature, humidity, and moisture through installed sensor, and then ThingSpeak and an Arduino Uno module created a database of environmental factors for direct access through an Android application.

Rathi et al. [5] studied the optimal sensor placement in water distribution networks and proposed use of a software tool named S-PLACE for both water monitoring and contamination detection. Demetillo et al. [12] proposed a water monitoring network based on a wireless sensor network in which preinstalled sensors would use the signals from a global system for mobile communications to send measured data to the central Arduino module.

The desired methodology for optimal sensor location for contamination detection in water distribution network should have two main important goals: (1) to reduce the number of installed sensors to consider the cost of sensor placement and maintenance;

and (2) to minimize the amount of water that is distributed and consumed after contamination occurred by a malicious enemy. These two goals will help to prevent the negative consequence on public health and socioeconomic implications [9] and to protect the society from the use of unsafe water.

The problem of finding the optimal time for contamination detection and optimal sensor placement in case of water contamination from a malicious enemy is considered in a competition named as the Battle of Water challenge [4], and several solutions have been suggested [4], [6], [19]. Ung et al. [20] proposed using the binary output of sensors to decide on the source location, starting time, and duration of contamination. A criterion was defined based on the binary outputs of the installed sensors and a greedy algorithm; an optimization method based on Monte-Carlo analysis was used to find the best sensor placement.

Giudicianni et al. [21] proposed a method for optimal sensor placement that relies not on the network's hydraulic model but instead on its topology. A priory clustering algorithm was performed on the water distribution network and based on defined centrality measures; sensors were placed at the central part of each cluster. A bi-objective optimization method based on the NSGAII genetic algorithm was used to find the optimal sensor location. The objectives in the minimization were the number of sensors and the exposed population to contamination. They did not use multi-objective functions to avoid computation complexity in the optimization part.

Casillas et al. [11] proposed a method for improving the efficiency of leak detection in the water distribution network that finds the optimal sensor location by an integer optimization approach. In a low complexity scenario, the initial sensor location has been estimated by a semi-exhaustive search method based on a lazy evaluation mechanism. In more complex scenarios, a stochastic optimization process based on either genetic algorithms or particle swarm optimization is used to find the optimal sensor location.

Xie et al. [8] proposed a sensor placement optimization strategy based on the theory of compressed sensing. The measured data are transformed in sparse leak space, and based on the average mutual coherence measure the decision has been made about the possibility of a potential leak in the network. Then, the sensor placement problem is reformulated as binary {0,1} quadratic knapsack problem. For finding the optimal solution, a binary version of an artificial bee colony algorithm enhanced by genetic operators was used. The crossover and swap measures are used to solve the binary knapsack problem. A semi-supervised optimal sensor placement method was proposed in [7] that considers some unknown leak locations assumptions. The fuzzy c-means clustering algorithm was used to improve the leak localization performance.

For analyzing the quality of the water in the municipal water distribution network and to compute the water levels and pressures in different parts of the network, an integrated water monitoring network is needed [22]. The water monitoring network should be reliable in detecting malfunctions and resolving the problem. Sampling the water from the predefined base stations for quality measurement and test can be used for such purpose [23]–[25]. In [23], a water level measuring sensor was used for monitoring the level of water in a different part of the network for flood disaster prediction. Monitoring networks that are based on sampling might be prone to error due to wrong sampling time, inadequate data, or distant location. Several sensor placement techniques are proposed in the literature[7], [26].

For better monitoring over the network, sensors should have the communication ability to send and receive packet data to each other. Therefore, researchers have proposed solutions to address challenges associated with water monitoring networks using IoT solutions. Low power wide area network is one of the recently proposed protocols for implementing the IoT large infrastructure with low bandwidth, low cost, low power, and long communication distance ability [27]. Long range, SigFox, and narrowband-IoT are instances of the low power wide area network communication technologies. Researchers focused on the problem of monitoring the water level in several water tanks over the network. In [25] several sensors were used to measure the properties of the water. A raspberry PI B + core controller processed the measured values by the sensors. In [28] a method to automate and centralize monitoring the water level in dams in India was proposed. The installed sensors collect and send real-time information to the center.

Krishna et al. [29] proposed a water monitoring network using the global system for mobile communications, and a flask server. The EXO Sonde [30] is the smart water sensor that could measure the amount of conductivity, dissolved oxygen content, pH, rhodamine, total algae, and turbidity absolute pressure, ammonia, depth, salinity, suspended solids, and density of water [29]. Several water monitoring systems for measuring the pH of the water were proposed based on IEEE 802.15.4 Zigbee protocol [31]–[33] . A UHF 920 MHz transceiver, which is XBee-Pro S3B with Digi Mesh network, was proposed in [32] to avoid the complexity of the network by so many routers. An IoT-based water monitoring system was implemented in [33] using an Arduino UNO ATmega32 as the base processor and some water sensors. The gathered data by the sensors convert to the digital format using an analog to digital convertor. The collected data by the sensor is sent to the central processor using WI-FI protocol and will be processed by the webserver. Different wireless technologies and their potential to be used in the water monitoring network were studied in [34]. Liang et al. [22] proposed a method for maximizing the coverage ratio while minimizing the delay for monitoring. In this work, the flow rate of the water was considered a measurement. The pollution was modeled in a single location, and the dynamic nature of water flow was not considered in the pollution modeling.

Cassiolato et al. [26] proposed a deterministic mathematical programming approach to minimize the cost of a looped water distribution network. They considered known pipe lengths and a discrete set of available commercial diameters. In the optimization model, constraints include the mass balances in nodes, energy balances in loops, and hydraulic equations. They used generalized disjunctive programming to reformulate the discrete optimization problem to a mixed-integer non-linear programming problem. The general algebraic modeling system environment was used to solve the problem.

Ciaponi et al. [13] proposed an optimal sensor placement method that is based on clustering the water distribution network to district meter areas. The clustering method minimized the number of edge-cuts (boundary pipes) while simultaneously balancing the number of nodes of each cluster. The clustering was performed using the properties of a normalized Laplacian matrix. For separating clusters from each other, isolation valves and flow meters were installed at alternating boundary pipes.

Brentan et al. [35] considered a multi-objective optimization framework and use NSGA II to solve the optimization problem. The objectives in their problem were the optimal number of sensors and the best location for installation, the contamination detection time, and detection probability with cost. Also, the clustering algorithm was used to avoid the computational burden in finding the optimum solution in the defined multi-objective function. Using the resulted Pareto-front curves, the optimal location for the sensor placement problem was identified.

Upon careful review of the state of the art, presented in this section, it is clear that systematic methods for placing emerging sensor and communication technologies are in need. In the present research, an optimum sensor placement algorithm is proposed that makes it possible to monitor the network simultaneously and guarantees that the network preserves crucial requirements such as connectivity and observability measures. In addition, the robustness of the network against arbitrary sensor failures at multiple points is evaluated. The sensor placement problem is approached using a multi-objective genetic algorithm. Four objectives are considered: (1) the minimum number of sensors; (2) the minimum observability radius; (3) the connectivity of the network; and (4) the maximum number of sensor failures that would not affect the connectivity and observability of the network. Using Pareto fronts, the optimum locations of sensors are found versus the number of arbitrary permissible sensor failures in the network and the resulted minimum observability radius in such sensor placement scenario.

The rest of this chapter is organized as follows. The problem statement is described and connectivity constraint introduced. Then the multi-objective optimization programming method and applied genetic algorithm are explained and the check connectivity procedure presented. An explanation of the observability constraint and the algorithm for checking n-node failure connectivity is given, and then the check observability algorithm is proposed for the network that preserves its connectivity against n number of sensor failures. For reducing the computational complexity of the proposed method, the hierarchical clustering algorithm is proposed and is described next. Last, results and discussion are presented, along with conclusions.

8.2 Problem Statement

The problem statement is as follows:

Given: We have a communication device with the communication range of w (m). The communication range represents the device price implicitly.

Desired/design objective: Design a sensor network that enables observation and monitoring of a water network over a resolution of r (largest (worst-case) circle radius), which is robust against the failure of n_f. This represents various types of failures that might occur in the pipes, such as pipe breakage, bursting, and water leakage.

In the proposed method, the urban water distribution pipeline is monitored using a communication network structure to monitor the health condition of the urban water distribution. With the help of communication sensors, the whole network condition is monitored and offers capabilities for fast failure detection and resolve.

The urban water distribution pipeline is modeled as a graph G with several nodes and links. Water pipes are the links of the corresponding graph (as the sensors are located on them), while the junctions between water pipes are corresponding nodes of the graph.

This communication network aims to guarantee that using the smart sensors which are installed on pipeline distribution water, the whole network could be monitored and the connectivity and observability of the network are supported. To tackle the network failures, several decisions need to be made, with the following considerations:

1. What is the minimum number of sensors that are required to guarantee that the whole network is connected and observable?
2. In the case of several simultaneous node failures, what is the minimum radius beyond which the network remains observable?
3. Where are the optimum places of each sensor over the water distribution network, such that the observability and connectivity of the network could be guaranteed?
4. What is the trade-off between the minimum number of sensors versus the effective range of the communication technology (e.g., Wi-Fi, Bluetooth, long range)?
5. What is the suitable communication technology for wireless communication?
6. What is the minimum number of failures that make the network disconnected?

8.2.1 Connectivity Constraint

The connectivity constraint formulated in the present work is based on *graph theory*. Given the placement of sensors, a matrix is constructed such that each entity is 1 if the Euclidean distance between the corresponding nodes is fewer than the communication range. Then, the difference between the matrix and its Laplacian is computed. The connectivity of the graph is determined using the sign of the second eigenvalue of the matrix calculated from the difference between this matrix and the Laplacian matrix. If the sign of the second eigenvalue is positive, the graph (network) is connected [36].

8.3 Multi-Objective Optimization Programming

The solution strategy was to use optimization techniques such as a genetic algorithm for finding the best places of sensors, the minimum number of sensors, minimum communication technology range, and minimum observability radius while the connectivity of the network was guaranteed.

Given the water distribution pipeline as a graph G, at the first step, we discretize the possible sensor location over the edges. In this step, we consider a predefined distance D for normalization.

8.3.1 Multi-objective Genetic Algorithm

Since the problem of sensor placement in a water distribution network is related to several interactive objectives, a multi-objective optimization algorithm should be used to concurrently optimize the objective functions of the problem. The multi-objective function, Obj_{tot}, was defined as

$$Obj_{tot} = \sum_{i=1}^{N} w_i \times Obj_i \tag{8.1}$$

in which the Obj_i is the i-th objective function, w_i are the coefficient. N is the total number of objective functions. By using the multi-objective genetic algorithm, the best place of sensors could be found such that the Obj_{tot} is minimized.

As it can be seen in Figure 8.1, four objective functions should be minimized concurrently for optimizing the sensor placement problem in the water monitoring network by using the proposed communication approach.

In Figure 8.2, the block diagram of the proposed method is shown. $n_{failure}$ is the minimum number of failures that makes the system disconnected. Higher values for $n_{failure}$ are more desired since it guarantees that the network communication would remain functional even in the case of losing n_{fail} number of sensors in every possible location of the network. Obj_{total} is the total objective function. w_i are the weights of individual objective functions. Obj_i is the i-th objective functions that is being minimized or maximized, concurrently. n_{sensor} is the total number of required sensors. $r_{observability}$ is the minimum observability circle radius. r_{range} is the sensor communication range that is related to the communication technology that is used for sending and receiving packet data via one of the available ways such as Bluetooth or Wi-Fi.

8.3.2 Check Connectivity

In **Algorithm 1** the procedure for checking the connectivity of the communication network is presented, as described in the previous section.

———————————————————

Algorithm 1: *Check Connectivity* of the communication network

———————————————————

Input: (X and Y positions of selected sensors location, the radius of Wi-Fi range)
Output: s
N: Number of nodes in the network.

Compute connectivity matrix, $N \times N$, such that each entity is 1 when Euclidean distance between two nodes is less than Wi-Fi range (110 meters).

Compute $D =$ degree matrix such that for each node, it counts the number of connections to other nodes for which the requirement for Wi-Fi Euclidean distance is satisfied.

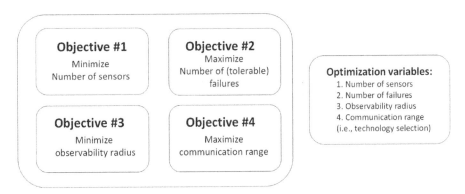

Figure 8.1 Multi-Objective Optimization Programming.

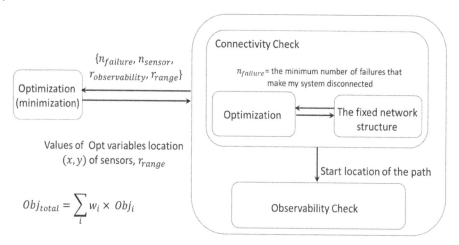

Figure 8.2 Block Diagram of the Proposed Method.

Compute Eigenvalues of the difference between matrix D and connectivity matrix.

If s is positive the connectivity constraint is satisfied,

Else
The connectivity constraint is violated.

Return the sign of the second smallest eigenvalue of the resulted matrix as s.
_____

8.3.3 Observability Constraint

For satisfying the observability constraint, it should be guaranteed that on each water distribution pipeline, at least there is one sensor that is functional and connected to the network. In the case of one or more sensor failures, the observability of the network might be limited and as a result, the observability radius will be increased. The observability radius is defined as the minimum distance, where for any larger section of the network, the state of the system in terms of the flow rate of entering and leaving water streams, and potential leakage can be accurately estimated. In Figure 8.3, one slice of the network is presented to demonstrate how in each junction (node) the summation of input flow rates is equal to the summation of output flow rates.

As can be seen in Figure 8.3, the total amount of input water stream is equal to the total amount of output water stream by considering the amount of leakage in the node. F_l is the total amount of leakage in the node, which ideally should be zero and could be estimated as

$$F_l = F_1 + F_2 + F_3, \qquad (8.2)$$

in which the values of F_1, F_2, and F_3 are the water streams that are measured by the sensors over the pipelines.

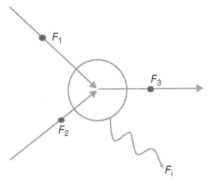

Figure 8.3 In Each Node, the Total Amount of Input Water Stream Is Equal to the Total Amount of Output Water Stream.

As can be seen in Figure 8.4, in the case of sensor failures, there might be leakages in the internal nodes of the radius $r_{observability}$ that cannot be immediately detectable. In such case, the observability of the network is reduced, and only over a larger rad ius $r_{observability}$, can leakages be guaranteed to be detectable.

8.3.4 Checking Connectivity in the Case of the Failure of n Nodes

In **Algorithm 2**, the aim is to find the minimum number of failures that make the network disconnected. For finding this number, the connectivity of the network (checked using **Algorithm 1**) is analyzed assuming that some of the sensors have been failed. A brute-force (exhaustive) search procedure was applied. The procedure starts by considering one node failure scenario and considering the possibility of all possible nodes failing. If the sensor network remains connected for all possible one-node failures, the corresponding network is guaranteed to be connected against one node failure.

For finding the corresponding $n_{failure}$ for the given sensor locations, it is supposed that in a probable scenario one of the possible combinations of i sensors from the total

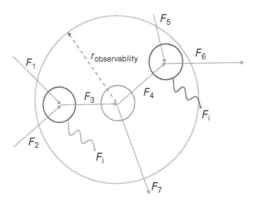

Figure 8.4 Observability Check, Although There Is Some Leakage in the Internal Nodes, It Is Guaranteed That in Radius $r_{observability}$, the Amount of Input Water Stream Is Equal to the Output Water Stream.

number of sensors, might be failed. Then the connectivity of the remained sensors is analyzed such that the requirement for the communication range (e.g., Wi-Fi distance 110 m) be satisfied based on Algorithm 1.

The genetic algorithm returns a large penalty and increases the n and checks all possible combinations of n nodes failures. If the network is disconnected in each occurrence of n node failure the genetic algorithm breaks and returns the n as the connectivity measure of the network against failures. This means that the network is guaranteed to be connected while $n-1$ number of sensors are failed. The number of failures is the minimum number of failures that make the network disconnected.

Algorithm 2: n node failure check connectivity

——————————————-

Input: *X, Y positions of selected discretized sensor location for placing sensors.*

Execute the *Check Connectivity* procedure (Algorithm 1).
For $i = 1:n$

Suppose for the given discretized sensor's locations, all possible combinations of i sensors from the total number of sensors, supposed to be failed. Compute the connectivity of the remained sensors such that the requirement for the communication range (e.g., Wi-Fi distance 110 m) is satisfied based on Algorithm 1.

Break and return i as the $n_{failure}$ in case of disconnectedness.
Return large penalty (pushing optimizer to select a larger value for n) in case the network preserves its connectivity concerning every occurrence of i number of sensors fail.

End For

——————————————

8.3.5 Check Observability Algorithm for the Network That Is Connected Against $n_{failure}$

In the proposed optimization algorithm, one arbitrary discretized sensor location is given as the input to the genetic algorithm and at first, it is checked if the whole nodes of the network are connected. At the second step, the connectivity of the network will be checked against some node failures. Then, the Check Observability algorithm checks the observability of the network while $n_{failure}$ of sensors are failed. In **Algorithm 3**, the procedure for checking the observability of the network is explained. Inputs to the Check Observability Algorithm are links of the graph, nodes, and the selected discretized sensor locations. Output is the observability resolution as $r_{observability}$. The idea for finding $r_{observability}$ is based on constructing an adjacency matrix. In the original graph, available edges between nodes are considered. Every edge is discretized and possible sensor locations are assigned. Each of these possible sensor locations might be selected or not in the sensor placement problem. The selected sensor locations are analyzed for computing the adjacency matrix. For every link, if there is not any installed sensor according to the selected sensor locations, it assigns 1 in the adjacency matrix on the location specified by the corresponding nodes of that link. Then the adjacency matrix is analyzed as a graph for finding the connected components. Figure 8.5 depicts the corresponding graph of the adjacency matrix. In this figure, it is clear that which nodes make a connected component. In fact, the edges in the graph of Figure 8.5

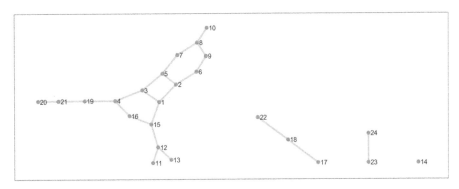

Figure 8.5 An Illustration of the Clustered Nodes and the Corresponding Graph.

highlights the edges that exist in the original graph and are connected to each other but do not contain any installed sensor in the selected sensor locations. This situation restricts the observability of the network.

Algorithm 3: Check Observability

Input: links of the graph, nodes, selected discretized sensor locations
Output: Observability resolution as the $r_{observability}$
// **Construct** new adjacency matrix with new definition
For each link between two sensors in the main graph
If there is not any selected sensor on that link
Assign 1 to the corresponding location in the new adjacency matrix
End If
End For

Compute *connected components* of the new adjacency matrix
Compute the outer circle for each connected component of the sensor locations s
$r_{observability}$ = maximum radius among the outer circle radii.

In the next step, the algorithm clusters the nodes versus the connected components that they belong to. In each of the clustered groups (connected component), the algorithm finds the outer circle that contains all of the nodes in the corresponding cluster. Figure 8.6 illustrates the connected component corresponding to the adjacency matrix. The figure shows three connected components, in which the outer circle is visualized. For guaranteeing the observability of the network, the outer circle of nodes in each cluster (connected component) is considered. The maximum circle radius will be selected as the observability radius that guarantees the observability of the network for any larger resolution. Therefore, the minimum observability radius, $r_{observability}$, is the maximum outer circle radius among all clusters (connected components).

In Figure 8.7, another adjacency graph is visualized for a better illustration of the procedure of Algorithm 3. In this example, there are more edges that do not contain any installed sensors. Some of these edges construct a connected component with the connected nodes. The aim of Algorithm 3 is to find these connected components that lack any installed sensors.

For the graph visualized in Figure 8.7, the corresponding connected components are shown in Figure 8.8. In each of the connected components, the outer circle is found and the maximum outer circle of the connected components is considered as the minimum observability radius.

In Figure 8.8 the procedure for finding the $r_{observability}$ is illustrated. Only the nodes with no sensor in the corresponding link between them in the suggested discretized sensor location are plotted. **Algorithm 3** tries to construct a new graph based on these nodes. The edge marked with red color is the corresponding graph. In this new graph, **Algorithm 3** finds the connected component and, in contrast to the connected component, performs the clustering operation on the nodes of the graph. In each of the clustered groups, **Algorithm 3** finds the outer circle that contains all of the nodes in the corresponding cluster. These outer circles are plotted in each cluster with the same color as the node color to better illustrate the method. **Algorithm 3** computes the maximum circle radius between the outer circle radii corresponding to each cluster and returns it as the minimum observability radius corresponding to the suggested discretized sensor location.

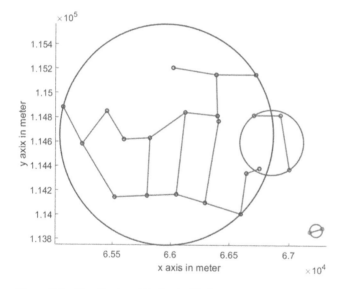

Figure 8.6 The Clustered Nodes for Finding the $r_{observability}$.

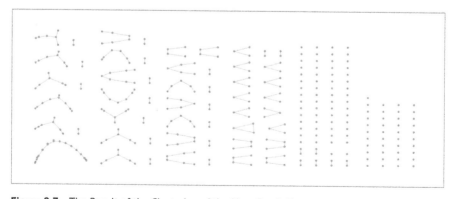

Figure 8.7 The Result of the Clustering of the New Graph G.

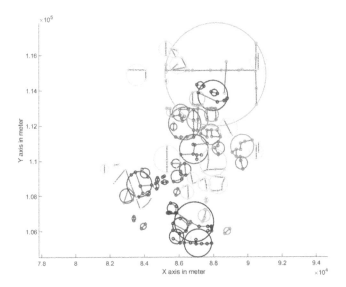

Figure 8.8 The Illustrated Procedure for Finding the $r_{observability}$. According to Algorithm 3 the Maximum Circle Radius between the Outer Circle Radiuses Corresponding to Each Cluster Is the Minimum Observability Radius Corresponding to the Suggested Discretized Sensor Location.

8.4 Hierarchical Clustering

The proposed method for finding the best sensor locations in the water distribution network is based on the hierarchical clustering method. The whole municipal water distribution network contains an enormous number of nodes and the sensor placement problem is a multi-objective one that needs to minimize several objective functions concurrently. For increasing the convergence speed and also for reducing the computational complexity the hierarchical clustering algorithm was used. Also, the hierarchical clustering algorithm guaranteed the scalability of the proposed method. At the first clustering layer, the distance between every two nodes is computed based on the Minkowski distance measure. The whole network is clustered into eight clusters based on the Minkowski measure. The result of the clustering of the network to eight different clusters is shown in Figure 8.9.

At the second hierarchical clustering level, each of the eight clusters is considered as a parent cluster and in a similar clustering approach, the whole network is clustered into eight different smaller sub-clusters. The result of the second clustering layer is shown in Figure 8.10.

In this way, the whole water network is clustered into 64 different sub-clusters. For each of the sub-clusters, the multi-objective genetic algorithm is applied to find the best sensor placement. The hierarchical clustering helps in reducing the computational complexity and also increasing the convergence speed of the multi-objective genetic algorithm.

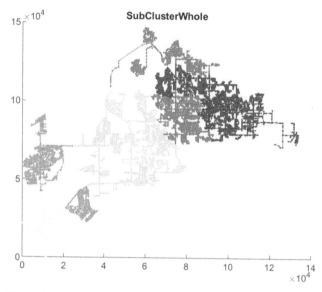

Figure 8.9 The Municipal Water Distribution Network Is Clustered into Eight Different Clusters in the First Layer.

8.5 Results and Discussion

The numerical experiment was performed using a Windows Server 2012 R2 with eight Intel(R) Xeon(R) CPU E5-2690 v2 with frequency 3.00 GHz. The amount of installed physical memory (RAM) was 40GB. There are 12527 nodes in the whole water network clustered into eight clusters, each of which approximately contains about 1400 nodes. In the second hierarchical clustering step, each of the clusters is again clustered into eight different sub-clusters that contain approximately about 150 nodes. Through hierarchical clustering, the resulted subclusters contain about 25 nodes in the end subclusters. The mean Euclidean distance between two nodes in the water network is 54742 meters.

In Figure 8.11 the histogram of Euclidian distance between every two nodes in the whole water monitoring network is shown. For placing the sensors in the links between the nodes, initially the distance between every two nodes is discretized with a predefined resolution. The discretizing resolution should be lower than the communication technology range, such that the communication between two sensors is possible over the links. The problem is to identify the best sensor location from these discretized sensor locations that concurrently optimize four objectives' functions: (1) the minimum number of sensors; (2) minimum observability radius; (4) connectivity measure of the network; and (5) the maximum number of arbitrary sensor failures that is tolerable and would not affect the connectivity of the network. After discretizing the locations of the sensors, there are many alternative locations for sensor placement. Considering the four objective functions, the computational cost and time to perform the multi-objective genetic algorithm on the whole discretized sensor's locations are very high. Hierarchical clustering helps to perform the multi-objective genetic algorithm on each sub-cluster and manage the computational cost.

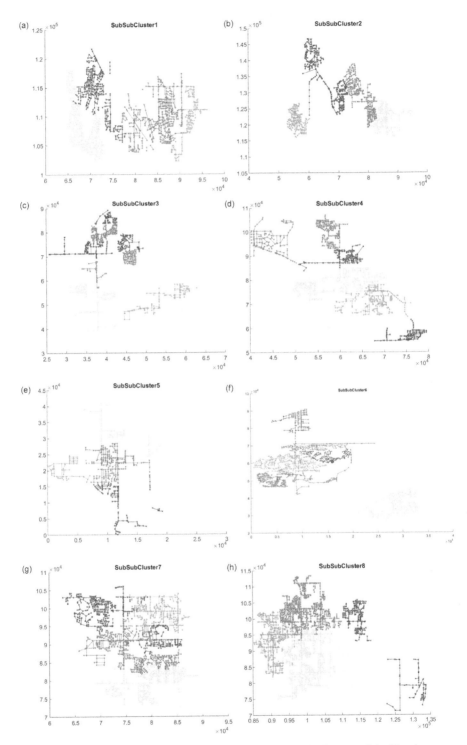

Figure 8.10 Second Layer Clustering, Hierarchically Each Sub-Clusters of the First Layer Clustering Is Clustered to Eight Different Sub-Clusters.

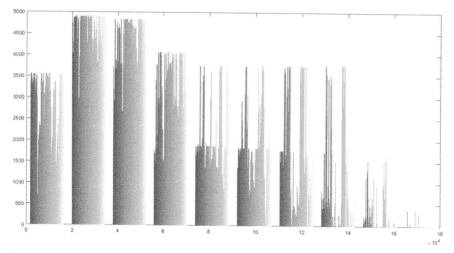

Figure 8.11 Histogram of the Euclidian Distance between Every Two Nodes in the Whole Water Monitoring Network.

The Pareto-front curve is the representative diagram that shows the optimum values of the objective functions and their trade-off. Each of the points in the Pareto-front curve is represented by selected sensor locations that are optimized considering a specific combination of weights for the four objective functions. For example, Figure 8.12 shows the Pareto-front curve when the technology communication range is set as 3700 meters. The horizontal axis shows the number of sensor failures, whereas the vertical axis corresponds to the number of sensors. All these points of the corresponding curve support that the network remained connected while such sensor failures occur. Figure 8.12 is the histogram of the distribution of the optimum points in the three-objective functions. As the figure shows, most of the optimum points have 898.601 meters as the observability radius that is plotted by the yellow color in the histogram.

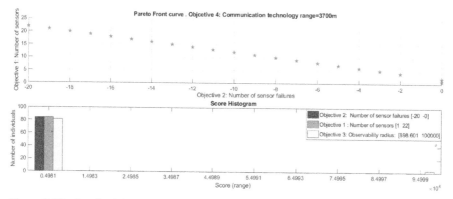

Figure 8.12 Result of the Multi-Objective Genetic Algorithm. The First Plot Is the Pareto-Front Curve When the Argument of the Multi-Objective Function Is As Follows: Objective 1: Number of Sensor Failures; Objective 2: Number of Sensors; Objective 3: Observability Measure; Objective 4: Communication Distance. The Bottom Plot Is the Score Histogram for 3 Objective Functions, Number of Sensors, Number of Sensor Failures, and the $r_{observability}$ Measure.

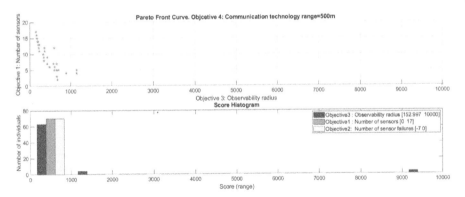

Figure 8.13 Result of Multi-Objective Genetic Function.

The multi-objective genetic algorithm returns the corresponding sensor locations of the optimum points and it is up to the designer to select which of the optimum sensor locations after performing all the simulations step over all sub-clusters. Such a decision could be made accounting for sensor failures, radius observability measures for fault detection, and network maintenance.

In Figure 8.13, the Pareto-front curve is displayed for the case that the communication range is 500 meters. The observability radius is the horizontal axis, whereas the vertical axis corresponds to the number of sensors. This Pareto-front curve displays the optimum points while the corresponding number of sensors are selected and reflects the corresponding observability radius. In Figure 8.13, the histogram of the optimum points is plotted for three objective functions. In Figures 8.12 and 8.13, Objective 2—the number of sensor failures—is plotted by a negative sign since the multi-objective genetic function wants to maximize this objective whereas for other objectives the goal is to minimize them.

8.6 Conclusion

Nowadays smart network management methods are crucial for continuous monitoring of the water network to ensure the cost-effective operability of the network and healthy conditions, with special emphasis on municipal water distribution networks. With the widespread application of Industry 4.0 technologies in water networks, there are significant opportunities to develop methods for monitoring water networks based on IoT, communication solutions, network, and cloud computing.

In this chapter, the municipal water network was analyzed, and for continuous monitoring of the network, a new multi-objective sensor placement method based on the genetic algorithm is proposed for finding the optimum locations of the sensors. Since the problem is multifaceted, four objective functions were considered. A hierarchical clustering algorithm was applied to reduce the computational cost and increase the convergence speed of the multi-objective genetic algorithm. The simulation result demonstrates that the proposed method can solve the problem with lower computational cost and more speed. The proposed method shows that monitoring municipal water networks is possible by installing sensors in appropriate locations and utilizing appropriate communication solutions. The case study investigated in the present research demonstrates that utilizing the communication ability of smart water sensors for sending and receiving data improves the robustness and efficiency of water networks.

References

1 Alabi, M., Telukdarie, A., and Van Rensburg, N.J. (2019). Industry 4.0: Innovative solutions for the water industry. in *Proceedings of the International Annual Conference of the American Society for Engineering Management*, 1–10.

2 AlMetwally, S.A.H., Hassan, M.K., and Mourad, M.H. (2020). Real Time Internet of Things (IoT) Based Water Quality Management System. *Procedia CIRP* 91: 478–485.

3 Wang, Y., Puig, V., and Cembrano, G. (2017). Non-linear economic model predictive control of water distribution networks. *J. Process Control* 56: 23–34.

4 Ostfeld, A. et al. (2008). The battle of the water sensor networks (BWSN): A design challenge for engineers and algorithms. *J. Water Resour. Plan. Manag.* 134 (6): 556–568.

5 Rathi, S. (2021). S-PLACE GA for optimal water quality sensor locations in water distribution network for dual purpose: regular monitoring and early contamination detection--a software tool for academia and practitioner. *Water Supply* 21 (2): 615–634.

6 Rathi, S. and Gupta, R. (2015). A critical review of sensor location methods for contamination detection in water distribution networks. *Water Qual. Res. J. Canada* 50 (2): 95–108.

7 Li, J., Wang, C., Qian, Z., and Lu, C. (2019). Optimal sensor placement for leak localization in water distribution networks based on a novel semi-supervised strategy. *J. Process Control* 82: 13–21.

8 Xie, X., Zhou, Q., Hou, D., and Zhang, H. (2018). Compressed sensing based optimal sensor placement for leak localization in water distribution networks. *J. Hydroinformatics* 20 (6): 1286–1295.

9 Adedoja, O.S., Hamam, Y., Khalaf, B., and Sadiku, R. (2018). A state-of-the-art review of an optimal sensor placement for contaminant warning system in a water distribution network. *Urban Water J.* 15 (10): 985–1000.

10 Yazdi, J. (2018). Water quality monitoring network design for urban drainage systems, an entropy method. *Urban Water J.* 15 (3): 227–233.

11 Casillas, M. V, Garza-Castañón, L. E., and Puig, V. (2015). Optimal sensor placement for leak location in water distribution networks using evolutionary algorithms. *Water* 7 (11): 6496–6515.

12 Demetillo, A.T., Japitana, M.V., and Taboada, E.B. (2019). A system for monitoring water quality in a large aquatic area using wireless sensor network technology. *Sustain. Environ. Res.* 29 (1): 1–9.

13 Ciaponi, C. *et al.* (2018). Optimal sensor placement in a partitioned water distribution network for the water protection from contamination. in *Multidisciplinary Digital Publishing Institute Proceedings*, 2 (11): 670.

14 Jha, M.K., Sah, R.K., Rashmitha, M.S., Sinha, R., Sujatha, B., and Suma, K.V. (2018). Smart water monitoring system for real-time water quality and usage monitoring. in

2018 International Conference on Inventive Research in Computing Applications (ICIRCA), 617–621.

15 Al-Khashab, Y., Daoud, R., Majeed, M., and Yasen, M. (2019). Drinking water monitoring in mosul city using IoT. in *2019 International Conference on Computing and Information Science and Technology and Their Applications (ICCISTA)*, 1–5.

16 Arduino. Getting Started with Arduino UNO. https://www.arduino.cc/en/Guide/ArduinoUno.

17 Espressif. ESP8266, A cost-effective and highly integrated Wi-Fi MCU for IoT applications. https://www.espressif.com/en/products/socs/esp8266.

18 Miry, A.H. and Aramice, G.A. (2020). Water monitoring and analytic based ThingSpeak. *Int. J. Electr. Comput. Eng.* 10 (4): 3588.

19 Adedoja, O.S., Hamam, Y., Khalaf, B., and Sadiku, R. (2018). Towards development of an optimization model to identify contamination source in a water distribution network. *Water* 10 (5): 579.

20 Ung, H., Piller, O., Gilbert, D., and Mortazavi, I. (2017). Accurate and optimal sensor placement for source identification of water distribution networks. *J. Water Resour. Plan. Manag.* 143 (8): 4017032.

21 Giudicianni, C., Herrera, M., di Nardo, A., Carravetta, A., Ramos, H.M., and Adeyeye, K. Zero-net energy management for the monitoring and control of dynamically-partitioned smart water systems. *J. Clean. Prod.* 252: 119745.

22 Liang, J., Xue, X., He, Z., and Leung, V.C.M. (2019). Low-delay and high-coverage water distribution networks monitoring using mobile sensors, *IEEE Access* 7: 107111–107128.

23 Perumal, T., Sulaiman, M. N., and Leong, C. Y. (2015). Internet of Things (IoT) enabled water monitoring system. in *2015 IEEE 4th Global Conference on Consumer Electronics (GCCE)*, 86–87.

24 Keum, J., Kornelsen, K. C., Leach, J. M., and Coulibaly, P. (2017). Entropy applications to water monitoring network design: a review. *Entropy* 19 (11): 613.

25 N. Vijayakumar, Ramya, and R. (2015). The real time monitoring of water quality in IoT environment. in *2015 International Conference on Innovations in Information, Embedded and Communication Systems (ICIIECS)*, 1–5.

26 Cassiolato, G., Carvalho, E.P., Caballero, J.A., and Ravagnani, M.A.S.S. (2021). Optimization of water distribution networks using a deterministic approach. *Eng. Optim.* 53 (1): 107–124.

27 Overmars, A. and Venkatraman, S. (2020). Towards a secure and scalable IoT infrastructure: A pilot deployment for a smart water monitoring system. *Technologies* 8 (4): 50.

28 Siddula, S.S. Babu, P., and Jain, P.C. (2018). Water level monitoring and management of dams using IoT. in *2018 3rd International Conference on Internet of Things: Smart Innovation and Usages (IoT-SIU)*, 1–5.

29 Krishna, S., Sarath, T.V., Kumaraswamy, M.S., and Nair, V. (2020).IoT based water parameter monitoring system. in *2020 5th International Conference on Communication and Electronics Systems (ICCES)*, 1299–1303.

30 E. Sonde. The EXO platform - Continuous monitoring for superior water quality data. https://www.ysi.com/exo.

31 Kamaruidzaman, N.S. and Rahmat, S.N. (2020). Water monitoring system embedded with Internet of things (IoT) device: A review. in *IOP Conference Series: Earth and Environmental Science* 498 (1): 12068.

32 Kamaludin, K.H. and Ismail, W. (2017). Water quality monitoring with Internet of things (IoT). in *2017 IEEE Conference on Systems, Process and Control (ICSPC)*, 18–23.

33 Daigavane, V.V. and Gaikwad, M.A. (2017). Water quality monitoring system based on IoT. *Adv. Wirel. Mob. Commun.* 10 (5): 1107–1116.

34 Olatinwo, S.O. and Joubert, T.-H. (2019). Enabling communication networks for water quality monitoring applications: A survey. *IEEE Access* 7, 100332–100362.

35 Brentan, B., Carpitella, S., Barros, D., Meirelles, G., Certa, A., and Izquierdo, J. (2021). Water quality sensor placement: A multi-objective and multi-criteria approach. *Water Resour. Manag.* 35 (1): 225–241.

36 Fiedler, M. (1973). Algebraic connectivity of graphs. *Czechoslov. Math* 2 (98): 298–305.

9

An Overview of the Evolution of Oil and Gas 4.0

Maryam Ghadrdan[a],* and David Cameron[b]

[a] Principal Engineer Innovation, Equinor ASA, Norway
[b] Center Coordinator, Sirius Centre for Research-based Innovation, University of Oslo, Norway
* Corresponding author. Equinor ASA, Norway

9.1 Oil and Gas 4.0: Sensors, Communicators, and Analysis

9.1.1 Optimizing Operations in the Petroleum Industry

This chapter presents an overview of Oil and Gas 4.0, the application of Industry 4.0 methods and standards to operations and facilities in the upstream petroleum industry. The significance of the subject is due to the increasing economic and technical challenges that this industry faces. Production of oil and gas must be safe, available, and have a low-carbon footprint. The capital cost of facilities must fall. This can be done by introducing standard and modular concepts from smart manufacturing. Standardization and interoperability of data in design and operations can also reduce costs of capital investment and operation.

The mixture of liquid hydrocarbons, gas, water, and solids are separated and sent to the proper line for sale or disposal. Figure 9.1 shows a three-dimensional model of an offshore oil and gas processing facility. This design is a floating production storage and offloading ship. It is a self-contained community, with living quarters and a production plant built on a ship's hull.

A production facility often handles production from more than one well. The produced fluids go through a subsea collection network and up a riser to the topside for processing. This is shown in Figure 9.2. The produced fluids consist of a mixture of hydrocarbons. They can be gas (e.g., methane, butane, propane), condensates (medium-density hydrocarbons), or crude oil. The fluids also contain unwanted components such as CO_2, water, salts, and sand. The purpose of the facility is to process the well fluids into clean marketable products: oil, natural gas, or condensates. The facility also includes utility systems that provide power, water, air, and chemicals.

The topside process consists of a series of separator drums, each of which separates oil from gas and water. A common process will have three stages of separation, each at a lower pressure. The produced gas is recompressed and either exported or reinjected into the oilfield. The produced water is reinjected into the oilfield.

Industry 4.0 Vision for the Supply of Energy and Materials: Enabling Technologies and Emerging Applications, First Edition. Edited by Mahdi Sharifzadeh.
© 2022 John Wiley & Sons, Inc. Published 2022 by John Wiley & Sons, Inc.

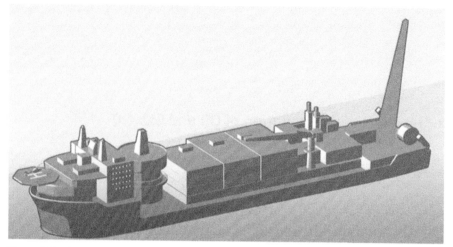

Figure 9.1 An Offshore Oil Processing Facility. Image courtesy Equinor.

This is the traditional way of constructing oil and gas facilities. Industry 4.0 offers new ways of working in oil and gas. It provides a way of introducing digital, end-to-end engineering for the whole life of a facility. Integrated value chains can be built. This will cut the cost of engineering, procurement, and construction and modifications and maintenance projects. Integrated value chains open for new business models and collaboration in the industry – based on data (data-based services). Standardization will allow a movement from tailor-made facilities toward modular solutions and services. The plants will be simpler and more robust. There is also the potential for using these methods to introduce robotics and autonomy in operations, leading to minimum staffing of facilities.

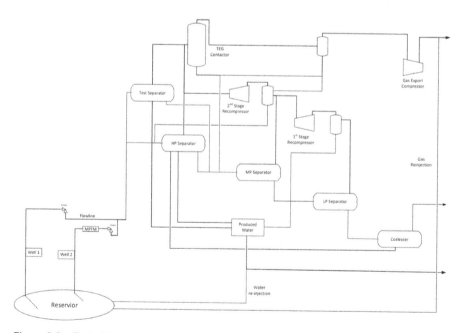

Figure 9.2 Typical Oil and Gas Processing Facility.

Most importantly, Industry 4.0 provides an effective vertical integration of the petroleum industry facilities. It makes it simpler to get data to the decision-maker. This allows better decisions to be made. Facilities that implement Industry 4.0 will realize the aims of the field of the future.

9.1.2 Data Access: Now and in Industry 4.0

The petroleum industry, like the wider process industry, has always been highly automated. Oil platforms are inherently dangerous. They process flammable materials at high pressures in isolated and difficult environments. This means that the safety and automation system (SAS) has been a critical component in the facility. It has also been the main source of information about how the facility behaves. Signals from the SAS are sent over to a process-historian database, where they are then available to support process improvement. This arrangement has several limitations. First, SAS sensors are expensive. This means that measurements are available only for variables that are used for control and alarms. Other sensors are seldom calibrated. Second, limitations in the SAS or its interfaces may limit the number and frequency of measurements sent. Furthermore, the process historian stores data in a way that can filter out information that is useful for analysis. Finally, data is stored in a flat data structure, referenced by tags, with no inherent structuring of the data.

Integration of data has followed the strictly hierarchical model proposed by ISA95 (internationalized as IEC62264 [1]). This is shown in Figure 9.3.

Overall, this means that it is difficult to obtain data for process improvement. Only a limited set of variables is available. Measurements needed to monitor the health of equipment are either not available or are locked inside a proprietary system. If they are available, they are often filtered or poorly calibrated. Applications that use the data must use custom-built cross-referenced lists of tags. This tagging is not consistent between facilities. This means that a solution for one facility needs a completely new data interface when it is applied to another facility.

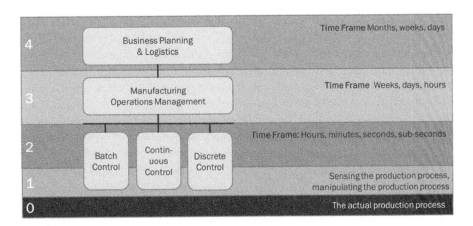

Figure 9.3   The ISA95, IEC62264, and Purdue Model.

Fortunately, we are seeing technical changes that can improve this situation:

- Wireless, 4G, and 5G networks offer cheaper and quicker ways of installing new sensors in plants.
- Instrument vendors offer stand-alone sensors for monitoring processes. These sensors are often noninvasive and can be simple and cheap to install.
- Equipment vendors offer "smart" equipment, which allows packaged access to operational variables and monitoring variables.
- The accelerating adoption of the so-called industrial Internet of things (IIoT) is offering alternative ways of integrating instrumentation and smart equipment, in parallel with the control system integration.
- Standard protocols for accessing and sending facility data are maturing and becoming common. Thus, open platform communication-unified architecture (OPC-UA) [2] and IoT protocols provide complementary ways of accessing operational data and making it available at other applications.
- Proprietary SAS systems and programmable logic controllers are moving toward open architectures. We will look at the open automation initiative in more detail later in this chapter.
- Cloud computing, novel databases, and platform developments offer new ways of distributing and working with data.

Industry 4.0 proposes a move to a network rather than a hierarchy. Here, automation and management of a facility are implemented as a network of independent, communicating equipment and applications. This arrangement is flexible and allows easy upgrading of the technology in a facility. However, this network is challenging to implement in safety-critical facilities. Responsibility for decisions is unclear in the network. Security can be compromised by weak nodes in the network. Evaluations and analyses of safety are complex. It is hard to get a coherent overview of the information in the networked system.

The IIoT is another manifestation of such "network thinking" approach. Its vision is to install networked sensors and smart devices that then support analysis and optimization calculations, either in the cloud or on the edge. Its usual value proposition is that inexpensive sensors and smart devices can be connected to networks in parallel to the control system for the plant. These networks can be wireless, that is, Wi-Fi, 4G, or 5G.

There is an overlap between the Industry 4.0 view of the world and that given by the Internet of things community. At the risk of generalization, the difference reflects the difference of view between the operating technology (OT) and information technology (IT) communities. Industry 4.0 builds on the background of automation and control standards such as ISA95 and OPC. The IIoT approach builds more on computer science and consumer electronics. These are complementary and necessary perspectives that need to work together as automation systems and enterprise IT systems interact, share technology, and compete for roles.

The landscape for Industry 4.0 and the IIoT is crowded, and it is difficult to get an overview. Paniagua and Delsing [3] gave a review of seven IIoT platforms, in addition to the IoT frameworks offered by the large cloud vendors: Amazon, Google, and Microsoft. Automation vendors offer solutions and frameworks (e.g., GE, ABB, Siemens, Schneider Electric, Honeywell, and Emerson), as do the product life cycle management vendors (e.g., Dassault, Bentley, PTC).

In this chapter, we will try to give a framework for implementing Industry 4.0 in an oil and gas facility. It is not possible to give an encyclopedic view of solutions, standards, and methods. Instead, we will use the ideas of a digital twin and the Reference Architecture Model for Industry (RAMI) 4.0 architecture to illustrate and introduce key ideas and standards.

9.1.3 The Digital Twin

Implementing Industry 4.0 in oil and gas involves building digital twins of the facility and using them to optimize design and operations. Cameron et al. [4] describe the idea of a digital twin and review its applications in oil and gas. They propose a conceptual model for a digital twin that combines three sources of data into a framework that allows analysis, reasoning, and decision support (Figure 9.4).

The three sources of data are:

1. The system's design and configuration. This is data about the facility as designed and built. It includes engineering data, product design data, and information from the construction, commissioning, and maintenance (material, method, process, QA historical usage data, and maintenance records).
2. Measurements and observations about the state of the facility.
3. Models for simulation and analysis. A digital twin may contain a variety of computational models. These may be based on first principles, data-oriented (machine learning, artificial intelligence), geometrical (computer-aided design, computer-aided engineering), or visualization-oriented (3D simulation, augmented reality and virtual reality).

The digital twin needs to be integrated. Interfaces must be provided for software to access the data, invoke commands, or run models. This enables connectivity interaction between digital twins and applications. Standardized application programming interfaces are key for interoperability.

The digital twin in the context of Industry 4.0 is implemented using an asset administration shell (AAS). This provides a standardized digital presentation of the asset. We will look at this in more detail later in this chapter, after we have put it into the context of the RAMI architecture.

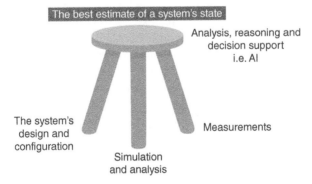

Figure 9.4 A Simple Conceptual Model of a Digital Twin.

9.1.4 Challenges in Introducing Industry 4.0

Introducing Industry 4.0 poses challenges to technology and work practices. The industry is experiencing an explosion of the amount and variety of data. This exposes missing links between data applications and models. It also exposes the need to build interoperability, the at-scale communication between many machines and data sources. This proliferation of data and interfaces also increases the attack surface for cyberattacks. Opening of systems and data sources must be accompanied by cybersecurity measures that address these risks. Further discussion of cybersecurity is beyond the scope of this chapter.

The petroleum industry also faces the challenge of brownfield facilities. These facilities are often highly profitable and can be made more so by the application of Industry 4.0 methods. However, they have much *technical debt* in the form of old systems with closed interfaces and a design basis that is based on paper archives. Industry 4.0 is not about discontinuing Industry 3.0. We must further improve what we do today, further automate the core processes, but at the same time start seizing opportunities in Industry 4.0. We need to find efficient ways of bringing these fields into the digital ecosystem while not wasting money on assets that will soon be decommissioned. Indeed, decommissioning may provide an incentive for digitalizing design information.

Scaling up implementations is also difficult. The technology of Industry 4.0 is still at the pilot and proof-of-concept stage. These pilots need to be scaled up to whole facilities and large projects to be able to succeed. What is easy to demonstrate for a single control loop or equipment package needs to be able to be used for a facility with 50,000 control system tags and hundreds of pieces of high-value equipment.

We should also recall that the biggest barriers to implementation are human and organizational. The move from hierarchical to networked technologies needs to be reflected in a change of skills and business practices in companies. A powerful but poorly packaged technology that does not support employee's perceived and actual productivity will not succeed. We must always remember that digitalization requires a balanced focus on humans, organizations, and technology.

9.2 The RAMI Architecture

9.2.1 Introduction to the RAMI Architecture

Interoperability is needed in the transformation from Industry 3.0 to Industry 4.0 [5]. Industry 4.0 is about the fusion between technologies like cloud computing, sensor-to-cloud connectivity, 5G and artificial intelligence, and a shared digital language/standard for semantics and services (RAMI 4.0). RAMI 4.0 is a reference architecture for systems that implement Industry 4.0. It was developed by a collaboration between government, industry, and academia in Germany. Work started in 2013 and has matured in the last three years. This discussion builds on the book by Heidel et al. [6]. This architecture combines three dimensions or ways of looking at systems:

1. A *life cycle* dimension, which covers the life of a product or facility, from design to disposal. The IEC62890 standard [7] describes these principles for control and automation systems.

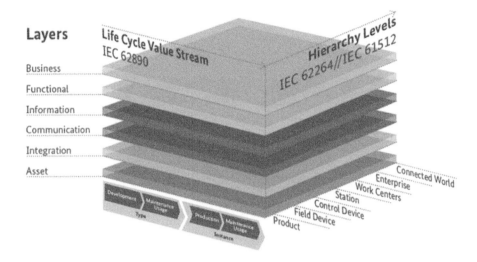

Layers

Business

Functional

Information

Communication

Integration

Asset

© Plattform Industrie 4.0, ZVEI

Figure 9.5 The RAMI Architecture.

2. A *hierarchy* or location in the company or factory. This is based on the Purdue model (ISA95, IEC62264 [1]), presented in Figure 9.5. The other standard mentioned in Figure 9.5 is IEC61512 [8], the ISA88 batch standard.
3. A *layering* that describes the information technology and computing architecture. This builds on and is comparable to the open systems interconnection layered model (ISO/IEC 7498).

We will briefly look at the life cycle dimension before using the layer model to structure the rest of the chapter.

9.2.2 The Life Cycle Dimension

The life cycle presented in the RAMI model is a *product* life cycle, where a vendor develops a piece of equipment, an *asset*, of a certain *type* (Figure 9.6). The vendor then sells an *instance* of this equipment to an end user. Consider a car. A pump manufacturer develops a type (or model) of the pump. They build this as a prototype, development model, and sales model. They then sell many thousand instances of this type.

We see that we can then organize the data about an asset into *type data* and *instance data*. Type data applies to all instances of a piece of equipment, whereas instance data is unique to a specific instance. Figure 9.7 shows examples of the two types of data.

Given that the type data is valid for all instances of an asset, it is costly and wasteful to acquire, manage, and store a copy of this with each asset that we own. A central repository of type data simplifies the management of information. This is the idea behind the EqHub system implemented in Norway [9]. This is a Web-based repository of type data for equipment used on the Norwegian Continental Shelf. This allows suppliers to hand over a reference to standard documentation to an operator, rather than,

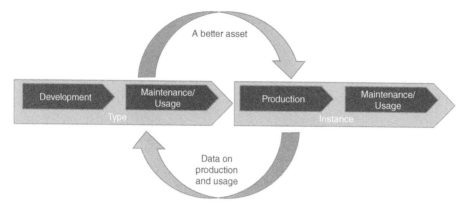

Figure 9.6 The Life Cycle Dimension of the RAMI Architecture. Adapted from ZVEI Plattform Industrie 4.0 and [6].

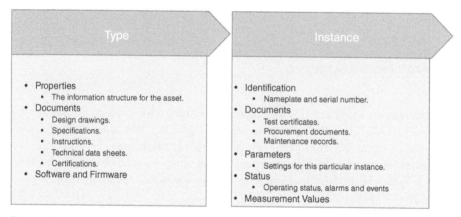

Figure 9.7 Type and Instance Data. Adapted from ZVEI Plattform Industrie 4.0.

as previously, sending copies of the same data for management in the operator's life cycle information system.

9.2.3 Tying Asset Information Together: The Asset Administration Shell

We have now seen that an asset is characterized by many types of type and instance data. Today, this data is spread across many databases, file systems, control systems, and physical archives. For this reason, Industry 4.0 proposes an extensible software framework that allows information for an asset (both types and instances) to be managed at a single point. This framework is the asset administration shell.

The AAS can be viewed as the digital counterpart of a physical asset. It also contains the design and measurement information needed to make a digital twin of the asset. The structure of the shell is shown in Figure 9.8.

The AAS consists of a header that contains identifiers and addresses for accessing the shell and a body that contains the data. The AAS data is divided into sub-models that divide the data according to logical groups. In practice it means that we can group type and instance data together. It also means that we can reuse standard sub-models

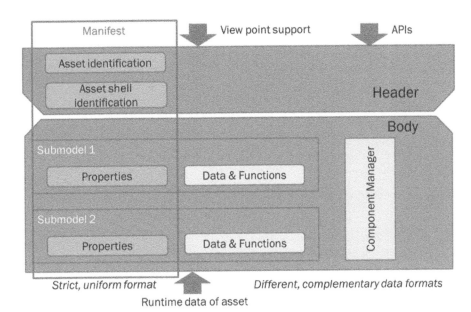

Figure 9.8 Simplified Structure of the Asset Administration Shell. Based on Plattform Industrie 4.0 and ZVEI Publications; see also [6].

to represent common groupings of information, such as nameplates, and explosive atmosphere classifications.

Each sub-model consists of a manifest in a strict, uniform format. The manifest is a list of properties. A property is an agreed, standard name for a data item. To allow interoperability, it is vital that properties use references from an agreed standard, such as the IEC 61360 Common Data Dictionary [10], eCl@ss [11], the IOGP JIP33 specification sheets [12], or the CFIHOS IOGP JIP36 property codes [13]. The property then points to data that is either stored in the AAS object itself or elsewhere. A component manager organizes the sub-models in the shell.

Sub-models can be added to and removed from the AAS as it is developed and handed over to a customer. Thus, an equipment supplier can have confidential sub-models of type data in their copies of the AAS. These are removed when the AAS is supplied to the customer and are replaced with sub-models that are needed for the operation of the asset. Figure 9.9 shows an example of an asset shell, in this case, a Siemens transmitter, viewed through the open-source AASX Package Explorer software.

Here you can see the organization into sub-models and properties. You can see that properties can also be grouped into sub-model components.

Asset administration shells can be grouped hierarchically. This allows an operator to build an asset shell for a package or system from the shells of the components in the system. The upper-layer shells contain system-level documentation and references to the lower-level sub-systems and components.

The AAS can be seen as providing a central, standardized window, on all the available information about an asset. The shell itself will store very little of this information. It rather provides references to relevant databases and data sources. This is an important feature. The AAS de-couples the information and functionality from the

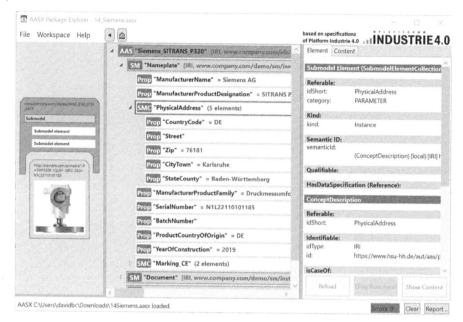

Figure 9.9 Using the AASX Package Explorer to View an Example File Supplied by Siemens.

application-specific representations and modeling of information. Without a digital twin framework like AAS, the use cases that need asset information and functionality will typically have to connect to all relevant application-specific interfaces, each holding a piece of the digital twin. Here they will meet application-specific data formats, interfaces, information models, and access policies.

9.2.4 The Implementation Layers

We have seen how to organize the information for an asset and will now look at the layers view of the RAMI architecture. This allows us to discuss the role of other technologies and standardization.

The RAMI architecture defines six layers, as shown in Figure 9.10.

The *asset* layer is in the physical work. It is the asset itself, the piece of equipment. An asset that is compliant with Industry 4.0 must be able to supply information about itself, in a digital form. This information is made available through the *integration* layer. This includes the sensors and actuators. Here the analog data is made digital, and digital signals are converted to physical control actions. The technology here is usually proprietary and closed. Actuators are linked to control and safety systems. This means that there is a need to convert from the Integration layer so that data is presented in an Industry 4.0 digital format. In practice, this means that data from this layer is exposed in an OPC-UA server.

The *communication* layer ensures the transmission of data about the asset. This layer deals with *how* (not *what*) information is transmitted. This layer organizes the used network, transport, and middleware protocols. This layer builds on the foundation of network protocols, such as ethernet, 4G, and 5G. On top of this, we then have the Internet layer, IP, protocol and transport protocols, transmission control protocol

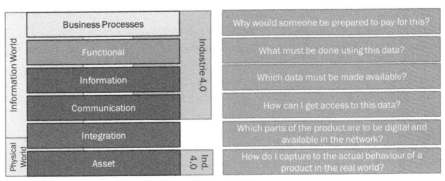

Source. ZVEI. RAMI 4.0 – Ein Orientierungsrahmen für die Digitalisierung. 2018

Figure 9.10 The RAMI Layer Model, Annotated by the Authors. Adapted from ZVEI [14].

(TCP), and user datagram protocol (UDP). This provides a foundation for the choice of middleware protocols: MQTT, advanced message queuing protocol (AMQP), hypertext transfer protocol (HTTP), and simple object access protocol (SOAP). The OPC-UA standard also provides two middleware communications protocols: one binary and the other based on XML (Extensible Markup Language).

The *information* layer defines the data that must be made available. Here we need to know what the data is called, where it can be found, and what its quality is. This is the layer where we want to implement shared and standard models of the data that free us from building one-to-one interfaces for each data source and application. It is in this layer that we want to agree on shared reference data (such as that supplied by ISO15926 Part 4) and asset models.

Once we have access to the data, we need to do business, we can implement a *functional* layer. This consists of the applications and services that support business using data. Heidel et al. [6] presented condition monitoring as an example of a generic function. The functional layer implements the calculations and analyses that justify the work done in the lower layers of the architecture. This approach is also taken by DNV, in their recent Recommended Practice on Digital Twins [15]. They proposed a model that breaks a digital twin into *functional elements*, each of which supports a business process or decision. The functional element then combines data streams, an asset information model, user interface components, and computational models to support a need.

Finally, we have the *business processes*, the actual work that drives the development of functional elements and provides the benefits from the architecture. The challenge of using the architecture is to find business processes that can be transformed—through increased efficiency or replacement—by implementing data-based functionality.

9.2.5 Industrial Internet of Things

This discussion so far has used the Industry 4.0 approaches and models. These approaches reflect the concerns and interests of the automation community. An alternative, and complementary, perspective is provided by the idea of an Internet of things, which in an industrial context becomes an industrial Internet of things [16, 17].

The IoT reflects the concerns and interests of the IT and telecommunications industry. This means that there is sometimes a divergence between RAMI and IIoT architectures. This is also demonstrated in the lack of interoperability between 4IR communication standards (open platform communications, or OPC) and IoT standards (Open Connectivity Foundation, or OCF) [18].

The Industrial Internet Consortium (IIC), which was founded by Dell, Purdue University, and Huawei, has published a reference architecture for IIoT. Lin and Mellor [19] discuss the relationship between IIC and Industry 4.0 and compare the RAMI architecture and the top-level IIoT architecture. The authors note that the IIC addresses interoperability across a broad range of applications. They list energy, health care, manufacturing, public administration, and transportation. 4IR addresses a manufacturing value chain. This means that the IIC architecture concentrates on cross-sectoral concepts and ideas, of which a thing is being monitored by sensors as the main element.

The IIC reference architecture, as shown in Figure 9.12, uses four viewpoints. Two of the viewpoints have the same name as a layer in RAMI, but the words mean different things [19]. So the term *business* in IIoT is about the business case, that is, why a system is to be built, whereas in RAMI, it means interaction with manufacturing life cycle business processes. The term *functional* in RAMI refers to specific applications and functions, whereas in IIoT it refers to a broader functional decomposition of the system into domains.

The IIoT functional view is a complex, three-dimensional structure where a breakdown into five functional domains is superimposed onto two additional axes: system characteristics and cross-cutting functions (Figure 9.11). System characteristics are generic desired features that a system must support if it is to work well. They are

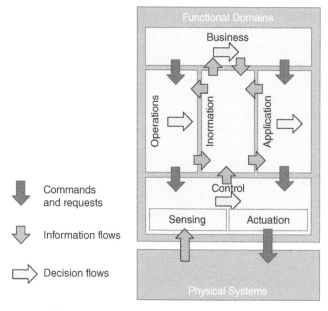

Figure 9.11 The Functional View of the Industrial Internet Reference Architecture. Adapted from [20].

safety, security, resilience, reliability, privacy, and scalability. Cross-cutting functions are functions or services that are needed irrespective of functional domain, such as connectivity, distributed data management, analytics, and control algorithms.

In the oil and gas industry, we believe that the deep manufacturing focus of the RAMI 4.0 is an advantage over the more comprehensive but more general industrial Internet reference architecture. We will therefore structure the rest of this chapter using the RAMI layers.

9.3 The Asset and Integration Layer: Novel Sensors

9.3.1 The Role of Novel Sensors in the Process Industries

Real-time sensors open for a new era of process control in several process industries. The pulp and paper industry gives a good example of how smart methods have changed production methods over the last decades. Similar progress is seen in the pharmaceutical, food, chemical, and metals industries.

The life span of a process system depends on several factors. Some of them are linked to the parameters in the manufacturing phase, and others depend on operational parameters. For example, an oil and gas facility can experience the following problems:

- High gas temperatures accelerate degradation in a compressor. Running a compressor outside its designed operation zone can damage the machinery.
- Foaming in separation drums can be caused by surfactants that are injected into the wells. They were added to improve oil recovery with high temperature and salinity in the reservoir.
- The subsea collection of production fluids is made complicated by flow assurance issues. Multi-phase flows can be unstable and form large slugs of fluid.
- Natural gas can form hydrates that block piping. Heavy components and wax in the oil can also block flow. It is therefore essential that we have good estimates (and, if possible, measurements) of the flow and chemical composition in these pipes.

Measurements are required to understand the physics and chemistry of the process systems, optimize system design, and enable feedback control. There are several hundred simple flow, temperature, and pressure sensors in a typical oil processing platform. More advanced sensors have been developed in recent years. Some of these are presented in this chapter. All these measurements (or a combination of them) are potential controlled variables. The optimal set of measurements can be chosen as control variables with the target of economically optimal operation of plants (see [21, 22] to gain insight about building an optimal control structure).

Explanations of some of the modern measurement tools are offered next as examples.

9.3.1.1 Spectrometer on a Chip

The spectrometer on a chip will enhance the portfolio of in situ technologies by providing a means for point detection of fluids and analysis of their content. Obtaining real-time hydrocarbon composition measurement data and using it for planning for the future can increase productivity, reduce operating and production costs, and enhance the profit.

Real-time composition measurement is normally done at the battery limits. It might be a bit complicated to find out the reason for an off-spec gas at the transport terminal. The traditional spectrometers are huge and expensive. It is not practical to have several of them in the plant. A miniaturized optical spectrometer technology as well as an advanced machine-learning-based data processing algorithm has pioneered the detection of a wide range of compounds exemplified by hydrocarbons and others. The device will be compatible with direct integration with chemical process instruments (e.g., a continuous-flow pipeline), offer real-time, multi-chemical sensing capability with a short sampling time (<1 s), and deliver the same detection capability as a benchtop spectrometer while featuring potentially significantly lower cost.

9.3.1.2 Lab on a Chip

A lab-on-a-chip is a device that integrates one or several laboratory functions on a single chip to achieve automation and high-throughput screening. A subset of microelectromechanical systems, these devices can handle extremely small fluid volumes down to less than pico-liters.

Advanced gas injection pressure-volume-temperature (PVT) data is the foundation for tuning an equation of state model, which will be extensively used for compositional reservoir simulation. For gas injection projects, such data is crucial for drainage strategy selection. PVT studies usually include swelling tests, multi-contact experiments, and minimum miscibility pressure. Microchip technology provides the opportunity to study reservoir flow and PVT studies.

9.3.1.3 Distributed Temperature Sensors

Distributed temperature sensing (DTS) systems are optoelectronic devices that measure temperatures utilizing optical fibers functioning as linear sensors. They assist reservoir engineers in optimizing the well lifetime. One example is to use DTS for flow profiling in the wells. See [23] for an example. An analytical-numerical model was used to predict wellbore temperature profiles as a function of flow rates and sand-face fluid-entry temperatures. These models can then be used in a reverse-modeling context for flowrate estimation from distributed temperature sensor measurements.

9.3.1.4 Trace Gas Sensors

Most gas sensor operation principles are normally highly versatile. For a variety of different analytes, a non-differentiable sensor response is expected. But one important objective of the manufacturer is high selectivity. Therefore, continuous research for sensitive and selective gas and vapor detection for a broad application field like process control, inspection, occupational safety, and environmental analysis are necessary. One of the important applications is to detect gas leakage in the pipeline. There have been reports of using metal organic frameworks (MOFs) as pre-treatment for trace gas sensing to increase the sensitivity of the sensors. Absorption of the analyte by a MOF will lead to sharper sensor signals (high sensitivity). Suitable MOF selection may give a high degree of selectivity.

9.3.2 Soft Sensors

Even though measuring many process variables is possible, implementing some of the sensors is costly and includes time delay. Some important variables are not accessible for measurement. These variables can be estimated by combining secondary variable measurements.

Some examples of soft-sensor development for the oil and gas industry are:

- *Dynamic estimation of model parameters*: For example, application of seismic data for estimation of reservoir parameters, such as porosity and permeability): This is done to have a model that describes the reservoir behavior and can be used for optimization purposes. A method based on the Gauss-Newton optimization technique was proposed by Dadashpour et al. [24] for continuous reservoir model updating with respect to production history and time-lapse seismic data.

- *Static estimation of process variables*: For example, compositions of the distillation products are estimated based on temperature measurements. Online composition measurement devices are expensive to be used directly in closed-loop control. Temperature measurements are fast and inexpensive and have been used for distillation column control instead of composition analyzers. A combination of all temperature measurements has been proposed by Ghadrdan [25].

- *Dynamic estimation of process* variables: For example, extended Kalman filters are used to estimate the states in a centrifugal compression system [26]. The estimator is dependent on enough information retrieved from the system through measurements to uniquely estimate all states. Active surge control or surge avoidance is needed to protect against compressor damage.

- *Data-based estimation of process variables*: For example, the syngas heating value is estimated using neural network-based nonlinear autoregressive with external input (NARX) by measuring flowrates involved in the mass and energy balance. Kabugo et al. [27] present a process data analytics platform built around the concept of Industry 4.0. The platform utilizes state-of-the-art IIoT platforms, machine-learning algorithms, and big data software tools.

- *Fault detection*: For example, statistical pattern recognition techniques such as least square estimation, autoregressive moving average, and Kalman filters were applied to build fault detection models for refinery data [28]. Sensor data are from temperature, pressure, flow rate, vibration, depth, and so forth of drills, turbines, boilers, pumps, compressors, and injectors. When a sensor starts malfunctioning, its data values must be discarded. Knowledge on the nature of processes and location and type of the measurements are important to improve predictive maintenance performance.

These examples suggest several types of estimations and different areas of application for the estimated values. The examples fall into three broad categories [29]: process monitoring, process control, and offline operation assistance. Process data analysis is the first step in the design of soft sensors. Investigation of operational data enables us to extract relevant information contained in data, select influential variables, and assess data quality, that is, reliability, accuracy, completeness, and representativeness.

Principal component regression [30] and partial least squares [31] are two of the most used data-based estimation methods. These estimators are based on mapping the measurement space to a lower-dimensional subspace, with a linear relationship between the data sets ($Y = XB + B0$, where B is the matrix of optimization variables). The least square solution to this problem is $B = YX$.

Mathematical models can be utilized to design estimators. The simplest model-based static estimator was introduced more than 40 years ago [32]. Ghadrdan et al. [33] developed an estimation method with the focus on application of the estimated values. The optimal estimators are derived for four cases [33]:

- **Case S1**. Predicting primary variables from a system with no control
- **Case S2**. Predicting primary variables from a system where the primary variables are measured and controlled
- **Case S3**. Predicting primary variables from a system where the control inputs are used to control the secondary variables
- **Case S4**. Predicting primary variables from a system where the primary variables are estimated and controlled

Case S1 is the direct extension of the Brosilow estimator, which includes measurement noise. The first three cases are relevant if the estimator is used for monitoring ("open-loop" estimation). Case S4 is the relevant case when we use the estimator for control ("closed-loop" estimation). The derivation of the estimators is based on results for optimal measurement combination for self-optimizing control [25].

Ghadrdan et al. [33] have shown that the prediction error is the lowest when the right scenario estimator is used (e.g., S3 should be used) when the estimated variable is used for monitoring and there are already some stabilizing control loops in place. The purpose of this section is to show that the quality of data is more important than quantity of data, when it comes to soft-sensor application. The interested reader is referred to [25] for more information.

The key challenge in adopting the sensors is to implement them reliably in a network (interoperability). It is important for sensors to be self-identifiable and exploitable in a network. They should also have the capability of self-observation to help identify when measurements are or will become wrong soon [34]. To communicate with sensors in a network, standardization of sensor and software interfaces across vendors is necessary.

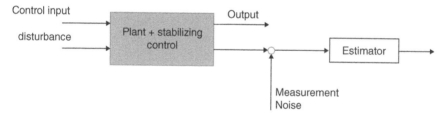

(a) "open-loop" estimation (three monitoring cases)

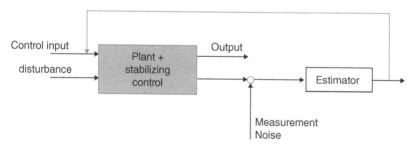

(b) "closed-loop" estimation

Figure 9.12 Block Diagrams for Different Estimation Cases.

9.4 The Open Integration Layer

9.4.1 Introduction

Successful digitalization requires access to the information provided by operational systems. This access must be secure, as the information contained in these systems is business critical and that cyberattacks can cause damage and material loss [35]. The integration layer is characterized by closed, proprietary systems and a variety of application programming interfaces and communication protocols. This makes building digital applications expensive and difficult.

Here, we provide a brief introduction to two activities that are trying to open the integration layer. The first of these is the Open Process Automation Framework (OPAF), which is trying to create an open ecosystem of interchangeable components for automation and manufacturing execution systems. The second is the NAMUR Open Architecture (NOA) [36], which is a German initiative started in the process industries to open and secure data access to operational data.

9.4.2 Open Process Automation Framework

The Open Process Automation Framework is a standard developed by the Open Group, following an initiative led by Exxon Mobil. The aim of OPAF is to build an "open, interoperable, secure-by-design process automation architecture" [37]. This will free operating companies from vendor lock-in, will increase the rate of innovation in process operations, and will improve the access to data from operations. Closed and proprietary control systems are expensive to upgrade and maintain, and it is costly to integrate new "best-in-class" third-party technology. High total cost of ownership is due to integration and maintenance tasks. The goal is to achieve interoperability, modularity, and portability of software and hardware components within the OPAF scope.

OPAF scope covers the following areas [37]: input and output (I/O), human–machine interface for distributed control systems, and programmable logic controllers. Advanced control and manufacturing execution systems are also in scope. Safety instrumented systems are *not* in scope. OPAF also looks at the linkage between its components and business systems and field devices.

An O-PAS (Open Process Automation Standard) system is built up of components that are connected through an O-PAS connectivity framework (Figure 9.13). The system consists of hardware components called distributed control nodes (DCN). This is typically a microprocessor-based controller or a gateway device [38]. Each DCN can host one or more distributed control frameworks (DCF). A DCF is a software system that runs applications. The standard opens for proprietary systems, such as instruments, analyzers, control systems, and programmable logic controllers, to host a DCF. This then enables them to exchange information with O-PAS components through the connectivity framework.

O-PAS also includes an advanced computing platform. The role of this component is to handle calculations and computations that are too resource-intensive for running on microprocessor DCN hardware. The advanced computing platform can host many DCFs and applications of optimization, user interface, advanced control, and scheduling.

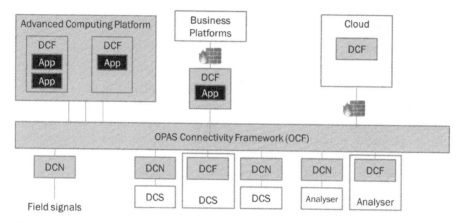

Figure 9.13 The O-PAS system architecture. Simplified from [37].

Krauss et al. [38] demonstrated the use of O-PAS (Open Process Automation™ Standard) in a chemical industry case. They realize the communication framework using two segregated virtual local area networks. This demonstrator shows the feasibility of the approach in a realistic chemical pilot plant. They conclude that this modular approach benefits both suppliers and users. Users obtain more flexible, cheaper solutions, and vendors can concentrate on their differentiating core capabilities.

9.4.3 NAMUR Open Architecture

Whereas O-PAS provides a framework for implementing the next generation of automation systems, NAMUR Open Architecture is designed to support digitalization in existing facilities. As noted by Krauss et al. [38], NOA provides a framework for opening automation systems so that they can supply data to monitoring and optimization calculations while leaving the existing infrastructure and core systems in place. The standard document states that NOA provides a way of connecting to existing systems at the site operations level. The approach is based on the existing systems offering an OPC-UA server that exposes the data that is to be used by applications in the NOA system. The data is to be structured using an information model for field instrumentation that was published in summer 2021 as NAMUR NE 176 [14].

An important contribution of NOA is the idea of an NOA diode. Recall that a diode is an electrical component that permits flow in only one direction. An NOA diode is a hardware–software system that ensures that there is no flow of data back from the monitoring and optimization layer to the core process control systems. This can be seen as a firewall that is tailored to be aware of the specific types of data flow that need to be blocked.

An NOA application can be realized using existing technologies. The architecture defines a set of good practices that ensure scalable, secure, and flexible access to data from control systems.

9.4.4 NAMUR Module Type Package (MTP)

A third integration technology has been proposed by the NAMUR organization to allow the integration of independent plant modules, each with their own control system, into an overall plant with its, orchestrated control system.

The NAMUR module type package (MTP) is described by the recommendation document NE 171. This implements the vision of modular production described by Lier et al. [39]. The MTP provides the specification for a framework in which a plant module is delivered together with its automation system. This system provides functionality for all automation aspects: alarms, human–machine interface, process control, maintenance diagnosis, and security. The MTP interface allows integration with an overlying control system or an orchestrating layer that builds an entire plant from MTP modules.

9.4.5 Conclusion

The industry still needs to work out how to best integrate these initiatives. Each of these addresses a specific part of the integration puzzle.

The MTP aims to increase the flexibility of production plants to cope with fast-changing market demands by modularizing process technology, mechanical construction, and automation technology. It facilitates the integration of units into the core process control domain since they are also equipped with their own automation systems. They need to be easily integrated into the distributed control systems or an overlying process orchestration level.

OPAF is being developed with the goal to bring an end to the era of proprietary automation vendor solutions and to define a standards-based, open, secure, and interoperable process control architecture, both hardware and software-wise. It is a disruptive change of the core process control domain.

NOA is being developed with the goal to enable new digitalization and Industry 4.0 use cases in existing and retrofitted plants. It has a clear focus on monitoring and optimization and leaves the core process control systems largely untouched.

9.5 The Communication and Information Layer

9.5.1 Industry 4.0 Communication Protocols and Middleware

The communication layer in Industry 4.0 deals with how we transmit information in a system. Here, we need to define the core IT networks, standards, and protocols that we will use. This requires us to choose the following elements:

- *Network*: Are we using ethernet, Wi-Fi, 4G, 5G, satellite, or some specialized industrial wireless network? The Internet protocol is used in this layer.
- *Transport*: Which transport protocol are we using? Usually, this is a choice between transmission control protocol and user datagram protocol.
- *Low-level application protocols*. Are we going to use HTTP, MQTT, SOAP, or OPC-UA?

Note that here we are considering transmission, not content. This means we are concerned not with what we send, but how we send it. Fieldbus protocols are viewed as legacy and part of the integration layer. They need to be translated to OPC-UA.

We begin to consider the content of the data we want to send in the Information layer. Here we define the data and properties that we want to make available.

It is beyond our scope to look at network, transport, and generic protocols here. We will, however, look at OPC-UA and messaging protocols, because of their importance in industrial applications.

9.5.2 Opc Ua

The OPC standards define an open and royalty-free set of standards for accessing and sharing real-time data from control systems and sensors. The standards are maintained by the OPC Foundation and were based on a set of standards, originally created in the 1990s, that provided the first vendor-neutral way of accessing control system data. These standards were based on the Microsoft COM technology. This was a barrier to further development, so that the standards were redesigned in the first decade of this century to a unified architecture: OPC-UA.

The types of services offered by OPC-UA are shown in Figure 9.14.

The fundamental idea is for OPC to be the interoperability standard for the secure and reliable exchange of data in the industrial automation space and in other industries. A producer and consumer of information shall be able to work together using the information being exchanged. OPC-UA is helping make data that is not only available but also in a context (information) that humans and machines are able to understand, use, and act upon.

An important innovation in OPC-UA was the introduction of support for information modeling. OPC-UA implements a framework for building graph data structures for information modeling [40]. This allows the definition of complex, tree-structured

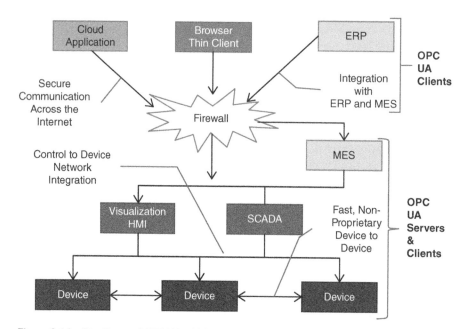

Figure 9.14   The Scope of OPC-UA within a Company [2].

object types and variable types, which can be used to create objects and variables with these types. Types and objects can refer to each other. The modelling also supports sub-typing of types. A modelling tool is needed to create and maintain these models. This information modeling capacity allows us to access structured data sets from a specific piece of equipment. It is also possible to use these so-called node sets to simplify interoperability with other semantic representations of the same data.

It is possible to develop standard or shared node sets for specific applications. The OPC Foundation is developing *Companion Specifications.*[1] Equinor has recently published a node-set that is based on their oil & gas platforms. This is freely available[2] and provides a model of equipment and control systems described as system control diagrams.

Historically, OPC supported a client–server approach, in which persistent connections are made between servers and clients. This approach is not scalable when bandwidth is limited or where there are many servers and clients that must be served. In these cases, messaging middleware offers an alternative approach.

9.5.3 Messaging Middleware: MQTT and AMQP

Figure 9.15 shows the difference between client–server and messaging middleware. A client–server system requires the use of data pipelines between each client and each server. This can be inefficient when there are many servers and clients in the system and when information changes only intermittently.

In a publish–subscribe system, a broker component is responsible for sending data from servers to clients. Servers publish data by telling the broker that a set of data is available. Clients can then subscribe to that data. When they do this, the broker will send any changes in that data to the clients as they are received from the server.

MQTT[3] is one such protocol. It was originally developed for satellite telemetry on pipelines and is designed to require low bandwidth. From this basis, it has been

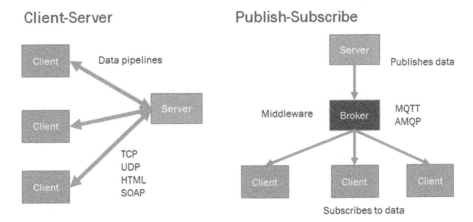

Figure 9.15 Client–Server and Publish–Subscribe (Messaging) Frameworks.

---

1 These can be downloaded from https://github.com/OPCFoundation/UA-Nodeset.

2 See https://github.com/equinor/opc-ua-information-models.

3 http://docs.oasis-open.org/mqtt/mqtt/v3.1.1/mqtt-v3.1.1.html.

developed into a widely applicable protocol, with an emphasis on industrial operations. MQTT can be viewed as the "OT" messaging middleware.

The alternative AMQP protocol was developed with financial applications, such as high-velocity stock trading, in mind. It has similar functionality to MQTT and is the middleware in Microsoft's Azure Service Bus and Event Hubs. AMQP is the "IT" messaging middleware.

9.6 The Functional Layer

9.6.1 Introduction

In this section, we will look at several use cases for Industry 4.0 in the petroleum industry. First, one of important applications needed to be put in the context of Industry 4.0, namely predictive maintenance, will be explained. Then, some examples of applying the open integration and communication frameworks in Equinor, as one of the frontrunner energy companies in the field of Industry 4.0 will be given. The purpose of this is to give motivations for further development and implementation.

9.6.2 Corrosion Under Insulation

Corrosion under insulation (CUI) is a known problem in process industry, and it makes preventive maintenance and CUI detection methods of critical importance.

Equipment and piping are typically constructed of steel (stainless steel, carbon steel with a stainless-steel layer, or epoxy coated). Insulation of pipes prevents pipes from contracting and expanding and heat loss due to the temperature difference with the environment because operational temperatures are typically also higher than the ambient temperature. Insulation helps protect the environment from high pressure and explosive gases. In addition, insulation provides a better barrier against moisture, helping pipes stay dry. However, keeping the pipes dry is not always easy. CUI happens in the event of penetration of moisture into the insulation.

A CUI program will have three elements:

1. CUI prevention: New installations should have the focus on insulating only when needed. It is believed that CUI can be prevented if robust and efficient solutions are adopted early enough during the design/engineering phase of the installations. This area of focus concerns green field developments.
2. CUI detection: For brown fields where the insulation systems are old and not necessarily optimal with regards to CUI prevention, it is important to be able to detect CUI degradations. There is potential for solutions based on novel sensors and data analytics.
3. Rectifications after CUI has occurred: Repairs and changes to limit or prevent CUI in the future

CUI is a challenge for the technical integrity of the installations and generates high maintenance costs. Figure 9.16 shows a typical timeline for prediction of failure. The required method to detect CUI is close visual inspection (CVI), which requires that the insulation must be removed. This method gives the highest probability of detection,

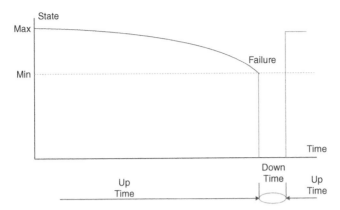

Figure 9.16 Timeline for Prediction of Failure and Maintenance Actions.

but it is time-consuming and labor intensive. It involves scaffolding, which contributes to 80% of the cost.

Alternative methods for inspection have been tested. The purpose is to find a more cost-efficient method without compromising with the probability of detection. New methods shall avoid building scaffolding and removing the insulations.

Some measurement tools used for inspection are a pulse eddy current, guided waves, X-ray backscattering, and digital radiography. None of these methods can replace CVI. Qualification of coating and insulation, monitoring of water under insulation, and air drying of insulation are some of the preventive measures. There is more weight on preventive methods and predictive maintenance. The highest business case is on water monitoring. This method can help narrow down the scope of the inspection, by segregating between dry and wet pipes. The pipes with dry insulations cannot corrode and can be excluded from the inspection scope. Predictive maintenance is explained in more detail in the next section. CUI is only one example to show the need of predictive maintenance.

9.6.2.1 Predictive Maintenance

According to NS-EN 13306: Maintenance terminology, maintenance is the "combination of all technical, administrative and managerial actions during the life cycle of an item intended to retain it in, or restore it to, as state in which it can perform the required function" [41].

Maintenance consists of two parts: preventive and corrective maintenance. Preventive maintenance is "maintenance carried out at predetermined intervals or according to prescribed criteria and intended to reduce the probability of failure or the degradation of the functioning of an item." There are two ways of executing preventive maintenance [42]:

1. Condition-based maintenance: "Preventive maintenance which include a combination of condition monitoring and/or inspection and/or testing, analysis and the ensuing maintenance actions." Condition monitoring, inspection, and/or testing may be scheduled, on request or continuous. Scheduled condition monitoring is regular inspection of components giving measures from a machine, or manual inspections. These are based on time, calendar, or known failure pattern.

Condition monitoring on request is inspections of the condition on machines based at suspicions or known failure development. Continuous condition monitoring is equal to real-time condition monitoring, defined later in the report.

2. Predetermined maintenance: "Preventive maintenance, opposed to corrective maintenance, carried out in accordance with established intervals of time or number of units of use but without previous condition investigation." Calendar-based scheduled maintenance actions are also common within predetermined maintenance. Predetermined maintenance regularly gets referred to as periodic maintenance.

Enabling the flow of information between all involved parties through data integration and assisting workflows with performant software is vital to reaching maintenance goals, with the goal of keeping production high and costs low. Maintenance engineers should have real-time access to relevant data and context information, allowing them to act in a timely way. Data from various sources are gathered and integrated to improve maintenance procedures. Figure 9.17 shows an example of the predictive maintenance planning framework based on a building information model and IoT.

9.6.3 Equinor Use Cases

In this section, some applications of open integration and communication platforms used by Equinor, an energy company in Norway, are explained.

9.6.3.1 Offshore wind
Offshore Wind

Equinor is determined to be a global offshore wind energy major. Offshore wind is at the center of revolutionary transition to green energy. In order to succeed, industrial and financial strength, and production capacity at a global level are important. In the context of Industry 4.0, the focus areas are broadly towards building an Operation & Maintenance decision support platform, failure prediction of structures and components, moving towards an industry standard data model and building a flexible, but fit for purpose OT/IT architecture.

Different data requirements across unmanned assets can be complicated when the size of wind farms are large, as it needs tedious mapping between tags. Common data model with standardization between equipment manufacturers should be developed. Plus, the solution needs to be fit for purpose for a low margin business.

An activity has been initiated to recommend a cloud-native technology stack and operating model that enables interoperability according to Industry 4.0 principles and supports the Offshore Wind portfolio for operation & maintenance, access design and wind farm optimization. It should provide us with a cost-effective solution that is

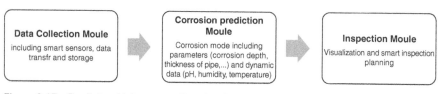

Figure 9.17 Predictive Maintenance Planning Framework Based on a Building Information Model and IoT. Modified from [43].

ready to iterate new product development on, but more importantly guarantees data consistency. Digital twins are being developed for individual wind turbines, and collection of them (wind farm), electrical systems infrastructure, foundations (also floaters) and substations (onshore and offshore).

9.6.3.2 Krafla/Spine Project

The Krafla oil and gas field was discovered in the North Sea in 2011. The field will be developed with an unmanned process platform. To assess Industry 4.0–based platforms, a pilot is initiated together with Equinor, AkerBP, Aker Solutions, Kværner, Cognite, and the SIRIUS Centre at the University of Oslo. The pilot uses READI IMF for the information model and Asset Administration Shell for the transfer of data between the companies.

9.6.3.3 OPC UA Implementation in Johan Sverdrup

Johan Sverdrup field came on stream in October 2019. Equinor and the partners tested the plant capacity in November 2020 to verify a possible production rise. Rates up to 535,000 barrels of oil per day were tested [44]. Johan Sverdrup selected ISA95, and its OPC-UA companion specification to create an enterprise OPC-UA–type library for main equipment and function blocks according to the old and well-known NORSOK I005, which now has become IEC63131.

The following are the main statistics for the Johan Sverdrup Phase 1 OPC-UA integration. There are 19 OPC-UA aggregating servers/data interface gateways/IACS gateways that provide information from the field center. Two OPC-UA OT-IT gateways, operating as redundant pairs, adapt the installation specific timeseries to enterprise usage.

The OPC-UA asset structure contains and maps more than 31,000 pieces of equipment, with more than a million items and references in the model. When the aggregating servers from SAS are expanded to handle the total facility load, the model will contain more than a million items, with more than 20 million references. The framework provides separate, read-only access to the entire alarm and events log, using OPC-UA alarm and conditions and the field device configuration.

The enterprise OPC-UA library contains the operational data required for operation and leap start of the digital journey. We have defined more than 50 different object types, more than 90 class types or interfaces, and more than 920 attributes. This OPC-UA enterprise model is automatically created from the TIE bus. We hope that when the next greenfield development project is done, the aggregating servers at the plant will serve standard models so no mapping is needed.

9.7 Conclusions and Way Forward

Lack of interoperability is no longer seen as merely being an inconvenience. It is now an obvious disadvantage that industries are no longer willing to accept. Today, this is addressed by "translating" and adapting the systems, which is inefficient. Interface standardization and semantic interoperability is at the core of Industry 4.0.

In this chapter, we briefly explained how Industry 4.0 revolution affects the oil and gas industry. Industry 4.0 provides integration between facilities, also between companies.

Standardization will allow a movement from tailor-made facilities toward modular solutions and services. This will cut the cost of engineering, procurement, and construction and modifications and maintenance projects. The plants will be simpler and more robust. On the other hand, there are challenges against implementing the concept of Industry 4.0 in the oil and gas business, some of which are still not fully solved. Adapting into an Industry 4.0 company is a challenging journey. Cross-industry collaboration is the key to its achievement in addition to the technology aspects. A combination of machine-to-machine communication, industrial Internet of things, cybersecurity, mobile technologies, big data analytics, and digital twins are some of the key technology components of Industry 4.0. Not only do multiple stakeholders in the industry need to coordinate with each other, legal frameworks, business models, enterprise solutions and policies at par with international standards are necessary for the Industry 4.0 ecosystem to function. Equinor as one of Norway's most influential companies have a unique responsibility and capability to guide other Norwegian companies. Some examples of applying Industry 4.0 concepts at Equinor were mentioned in the last section. The advantage with Industry 4.0 is that one company does not have to do the pulling alone anymore. The global industry is pulling. This is just the beginning.

References

1 IEC 62264-1 (2013). Enterprise-control system integration - Part 1: Models and terminology. International Organization for Standardization, Geneva, Switzerland, https://www.iso.org/standard/57308.html.

2 OPC Foundation (2021). OPC UA for Asset Administration Shell (AAS) (OPC UA Companion Specification No. OPC30270). [Online]. Available: https://reference.opcfoundation.org/v104/I4AAS/v100/docs.

3 Paniagua, C. and Delsing, J. (2020). Industrial Frameworks for Internet of Things: A Survey. *IEEE Syst. J.* 1–11: [Online]. Available: https://doi.org/10.1109/JSYST.2020.2993323.

4 Cameron, D. B., Waaler, A., and Komulainen, T. M. (2018). Oil and Gas digital twins after twenty years. How can they be made sustainable, maintainable and useful?. in *59th Conference on Simulation and Modelling (SIMS 59)* 9–16.

5 Sørenes, S. (2021). Energy Industry 4.0 - Why? What? How?. Linkedin, https://www.linkedin.com/pulse/energy-industry-40-why-what-how-steffan-sørenes.

6 Heidel, R., Hoffmeister, M., Hankel, M., and Döbrich, U. (2019). *The Reference Architecture Model RAMI 4.0 and the Industrie 4.0 component*, First Ed. Berlin, Germany: VDE Verlag.

7 IEC 62890 (2020). Industrial-process measurement, control and automation - Life-cycle-management for systems and components. https://webstore.iec.ch/publication/30583.

8 IEC 61512-1 (1997). Batch control - Part 1: Models and terminology. https://webstore.iec.ch/publication/5528.

9 Norsk Olje og Gas EPIM EqHub. https://collabor8.no

10 International Electrotechnical Commission (2017). IEC 61360-4 - Common Data Dictionary (CDD - V2.0014.0017). International Organization for Standardization, Geneva, Switzerland, https://cdd.iec.ch/cdd/iec61360/iec61360.nsf.

11 ECLASS (2021). Standard for master data and semantics for digitalization. https://www.eclass.eu/en/index.html [Online].

12 IOGP (2016). JIP33 Joint Industry Program Standardizing Procurement Specifications. www.iogp-jip33.org.

13 CFIHOS (2021). Capital Facilities Information HandOver Specification Website. https://www.jip36-cfihos.org/cfihos-standards [Online].

14 Joint NAMUR/ZVEI Working Group WG 2.8.2 Automation Architecture – Information Mode (2021). NAMUR Open Architecture NOA Information Model (NAMUR Recommendation No. NE 176 EN). [Online]. Available: https://www.namur.net/en/focus-topics/namur-open-architecture.html.

15 DNV GL RP A204 Qualification and Assurance of Digital Twins.

16 Behrendt, A., de Boer, E., Kasah, T., Koerber, B., Mohr, N., and Richter, G. (2021). Leveraging Industrial IoT and advanced technologies for digital transformation. [Online]. Available: https://www.mckinsey.com/~/media/mckinsey/business functions/mckinsey digital/our insights/a manufacturers guide to generating value at scale with iiot/leveraging-industrial-iotand-advanced-technologies-for-digital-transformation.pdf Accessed June 2021.

17 Boyes, LL H., Hallaq, B., Cunningham, J., and Watson, T. (2018). The industrial internet of things (IIoT): An analysis framework. *Comput. Ind.* 101: 1–12. doi: https://doi.org/10.1016/j.compind.2018.04.015.

18 Cavalieri, S., Salafia, M. G., and Scroppo, M. S (2019). Towards interoperability between OPC UA and OCF. *J. Ind. Inf. Integr.* 15: 122–137.

19 Lin, S.-W. and Mellor, S. (2017). Architecture Alignment and Interoperability, An Industrial Internet Consortium and Plattform Industrie 4.0 Joint Whitepaper (No. IC: WHT:IN3:V1.0:PB:20171205).

20 IIC (2019). The Industrial Internet of Things Volume G1: Reference Architecture. [Online]. Available: https://www.iiconsortium.org/pdf/IIRA-v1.9.pdf.

21 Skogestad, S. and Postlethwaite, I. *Multivariable Feedback Control: Analysis and Design.* Wiley.

22 Ghadrdan, M., Skogestad, S., and Halvorsen, I. J. (2012). Economically optimal control of kaibel distillation column: Fixed boilup rate. in *IFAC Proceedings Volumes (IFAC-PapersOnline)* 8: no. 15. 744–749. doi: 10.3182/20120710-4-SG-2026.00126.

23 Ouyang, Liang-Biao and Belanger, D. L. (2006). Flow Profiling by Distributed Temperature Sensor (DTS) System - Expectation and Reality. *SPE Prod Oper.* 21 (2): 269–281.

24 Dadashpour, M., Echeverría-Ciaurri, D., Kleppe, J., and Landrø,M. (Dec, 2009). Porosity and permeability estimation by integration of production and time-lapse near and far offset seismic data. *J. Geophys. Eng.* 6 4: 325–344. doi: 10.1088/1742-2132/6/4/001.

25 Ghadrdan, M. (2004). *Optimal operation of Kaibel columns.* Norwegian University of Science and Technology.

26 Hovd, T. H. (2007). Modeling, state observation and control of Compression System.

27 Kabugo, J. C., Jämsä-Jounela, S.-L., Schiemann, R., and Binder, C. (2020). Industry 4.0 based process data analytics platform: A waste-to-energy plant case study. *Int. J. Electr. Power Energy Syst.* 115: 105508. doi: https://doi.org/10.1016/j.ijepes.2019.105508.

28 Khodabakhsh, A., Ari,I., and Bakir,M. (2017). Cloud-based Fault Detection and Classification for Oil & Gas Industry.

29 Khatibisepehr,S., Huang, B., and Khare, S. (2013). Design of inferential sensors in the proess industry: A review of bayesian methods. *J. Process Control* 23: 1575–1596.

30 Massy, W. F. (1965). Principal component regression in exploratory statistical research. *J. Am. Stat. Assoc.* 60: 234–246.

31 Wold, H. (1975). 11-Path Models with Latent Variables: The NIPALS Approach. In: *Quantitative Sociology International Perspectives on Mathematical and Statistical Modeling* 307–357. Academic Press.

32 Joseph, B. and Brosilow, C. B. (1978). Inferential control of processes: Part i. steady state analysis and design. *AIChE J.* 24 (3): 485–492.

33 Ghadrdan, M., Grimholt, C., and Skogestad, S., (2013). A new class of model-based static estimators *Ind. Eng. Chem. Res.*, 52 (35): 12451–12462. doi: 10.1021/ie400542n.

34 EY Sensors as drivers of Industry 4.0 - A study on Germany, Switzerland and Austria. https://www.oxan.com/insights/client-thought-leadership/ey-sensors-as-drivers-of-industry-40.

35 Corallo, A., Lazoi, M., and Lezzi, M. (2020). Cybersecurity in the context of industry 4.0: A structured classification of critical assets and business impacts. *Comput. Ind.* 114: 103165. doi: https://doi.org/10.1016/j.compind.2019.103165.

36 NAMUR Working Group WG 2.8 Automation Architecture (2020). NE 175 'NAMUR Open Architecture – NOA Concept' (NAMUR Recommendation No. NE175). NAMUR. [Online]. Available: https://www.namur.net/en/focus-topics/namur-open-architecture.html.

37 Brandl, D. et al. (2018). Requirements for an Open Process Automation™ Standard (No. W184).

38 Krauss, M. et al. (2020). Open Automation Demonstrator für die chemische Industrie: Synergien von O-PAS™, ein Standard der Open Group, MTP und NOA. 62: 60–67. doi: https://doi.org/10.17560/atp.v62i6-7.2476.

39 Lier, M., Wörsdörfer, S., and Grünewald, D. (2015). Wandlungsfähige Produktionskonzepte: Flexibel, Mobil, Dezentral, Modular, Beschleunigt. *Chemie Ing. Tech.* 87: 1147–1158. [Online]. Available: https://doi.org/10.1002/cite.201400191.

40 Mahnke, W. Informationsmodellierung mit OPC Unified Architecture: Die Basis für standardisierte Informationsmodelle. *atp Mag. 3/2020* 62: 58–65.

41 Manzini, R., Regattieri, A., Pham, H., and Ferrari, E. (Eds). (2010). *Introduction to Maintenance in Production Systems BT - Maintenance for Industrial Systems.* London: Springer London. 65–85.

42 Berg, T. (2017). Predictive maintenance and Industry 4.0.

43 Cheng, J. C. P., Chen, W., Chen, K., and Wang, Q. (2020). Data-driven predictive maintenance planning framework for MEP components based on BIM and IoT using machine learning algorithms. *Autom. Constr.* 112: 103087. doi: https://doi.org/10.1016/j.autcon.2020.103087.

44 Equinor ASA (2021). Expecting a third capacity increase at Johan Sverdrup. https://www.equinor.com/en/news/20210128-expecting-third-capacity-increase-johan-sverdrup.html accessed September 5th 2021 (accessed Sep. 05, 2021).

10

Electrification of Transportation

Transition Toward Energy Sustainability

Hossein Ranjbar and Mahdi Sharifzadeh*

Sharif Energy, Water and Environment Institute (SEWEI), Sharif University of Technology, Tehran, Iran

*Corresponding author. Sharif Energy, Water and Environment Institute (SEWEI), Sharif University of Technology, Tehran, Iran

10.1   The Trend of Transportation Electrification

A sustainable and clean transportation system is the most important parameter to improve the standard of living. However, the current transportation system is rarely sustainable due to its high dependency on fossil fuels. Besides, it is not environmentally friendly considering its contribution to nearly 25% of the total greenhouse gas (GHG) emissions [1].

Electrification of transportation is an effective solution for decreasing the high dependence on fossil fuels and moving toward sustainable and environmentally friendly energy systems. Thus, numerous countries around the world have been progressively setting targets for moving toward transportation electrification. For instance, the European Commission has targeted to reduce 90% of the generated GHG emissions in the transportation part [2]. To achieve this goal, almost all vehicles including light- and heavy-duty cars, buses, and vans should be zero-emission electric vehicles (EVs) [2].

There were several waves of electrification of transportation in history during the 19th century [3]. However, since the emergence of lithium-ion batteries about 20 years ago, the strongest wave of electrification of vehicles has been started. Besides, due to supporting policies and incentives executed in numerous countries globally, the number of EVs on the road has been continuously growing meeting more than 10 million EVs at the end of 2020 [4]. However, despite all the efforts made by policymakers, the transportation sector accounts for a small share of electricity consumption. For instance, Figure 10.1 illustrates the share of world electricity consumption for each sector in 2018, which indicates a negligible share of used electricity by transportation (only 2%) [5].

Figure 10.2 illustrates the number of EVs on the road worldwide. China accounts for 48% of total EVs in the world, which is followed by overall Europe (30%), the United States (16%), and other countries (6%) [4]. The trend of EV utilization is rising in which more than 10 million EVs are on the road at the end of 2020, which is 41% higher than in 2019.

Industry 4.0 Vision for the Supply of Energy and Materials: Enabling Technologies and Emerging Applications, First Edition. Edited by Mahdi Sharifzadeh.
© 2022 John Wiley & Sons, Inc. Published 2022 by John Wiley & Sons, Inc.

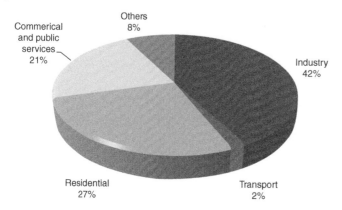

Figure 10.1 Share of World Electricity Consumption by Sectors Based on [5].

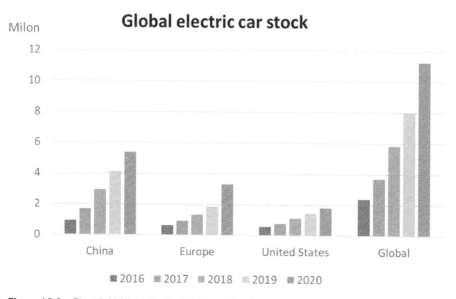

Figure 10.2 Electric Vehicles in the World and Leading Countries Based on [4] [5].

Europe experienced 1.4 million new registered EVs in 2020, which is more than other countries and regions in the world. Among the European countries, Germany, France, and the United Kingdom are leading countries that, respectively, registered 295,000, 185,000, and 176,000 new electric cars. From the view of sales share, Norway reached the astonishing record of 75% EVs' sales share, up about one-third from 2019. The following countries are Iceland, Sweden, and the Netherlands with sales shares of EVs exceeded 50%, 30%, and 25%, respectively [4].

The increase in the number of EVs in Europe regardless of the pandemic-related economic crisis is associated with two main support policies introduced by many European governments [6]. First, the European Union's CO_2 emission standard limits the average emitted carbon dioxide to 95 g CO_2/km for 2020–2021, and even tighter targets are provisioned in 2025 and 2030 [4, 7]. Second, an increasing number of subsidies for EVs as several incentive packages by many European governments to reduce the negative impacts of the pandemic [4, 7].

Unlike Europe, the overall EV markets in China and the United States were impacted by the pandemic in which the total new EV registration in China and the United States were down about 9% and 17%, respectively. In other countries, Japan, Korea, and Canada have the highest EV stock market such that the new registered EVs in Canada, Korea, and Japan are about 68,000, 44,000, and 30,000 in 2020, which in total covers 90% of new EVs in other countries except for Europe, China, and the United States[4, 8].

Various support policies were adopted by countries worldwide for EVs to motivate car manufactures for major extensions in EV models. These support policies include subsidies on the price of EVs. In addition, the tax credit for purchasing or registration of EVs is employed to decrease the price difference with conventional cars. Examples of such policies are the Energy Improvement and Extension Act of 2008 in the United States [9], the exemption of new energy vehicle purchase tax in China [10], and the European Union's CO_2 emissions standards [4].

10.2 Transportation Electrification Opportunities

Transportation electrification provides a significant number of benefits to people, utility infrastructure, and government. These benefits can be categorized as economic, environmental, and societal benefits.

10.2.1 Environmental and Health Benefits

The current transportation system, which is mainly based on fossil-fuel vehicles, accounts for about 31% of petroleum product utilization and almost a fourth of the world's energy-related CO_2 outflows (Figure 10.3 [11, 12]). According to the World Energy Outlook 2021 [5], the annual CO_2 emissions in energy sectors globally were increasing since 2000 (except for 2020 due to the pandemic) and the transport sector contributes to approximately 20% of that, regardless of implemented CO_2 reduction policies. Thus, the obvious benefit of transportation electrification is the reduction of GHG emissions by replacing fossil fuel-based cars with EVs. However, reducing the

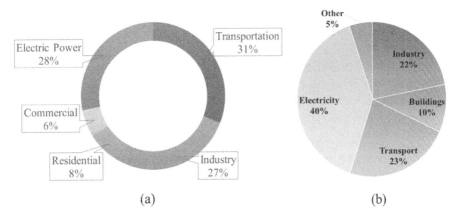

(a)   (b)

Figure 10.3 Fossil Fuel Consumption and GHG Emissions by Sector: (a) Fuel Consumption. (b) GHG Emission Based on [12] [14].

emissions produced in the transport sector will only solve one part of the emissions problem. In other words, the decarbonization of energy generation must also take place, simultaneously. Because electric vehicles need electricity for their operation which is generated using power plants that produce emissions. Thus, electrification of the transport sector will not significantly improve the environmental issues without the integration of more renewable energy resources [13].

In addition to GHG emission, which is mostly a global issue, the transportation sector is responsible for significant effects on air quality, including photochemical exhaust cloud, metropolitan and territorial cloudiness, aerosol particle organic matters (that affect lung work), acid rain and acid deposition, and airborne effects on the environment, as well as stratospheric ozone depletion [11]. These environmental issues will endanger human health and create indirect costs to society. According to a report conducted by the European Environment Agency, most citizens in Europe are exposed to harmful air pollutants generated by automobiles [13, 15]. Moreover, transport-resulted air pollution in the United States is responsible for approximately 5 to 10% of premature mortality every year [16]. To this end, transportation electrification can significantly reduce air pollutions and its subsequent impacts on human health. For instance, the study in [17] illustrated that deployment of 30% EVs can reduce the harmful pollutants in China, especially nitrogen oxides (NO_x), carbon monoxide (CO), and particulate matter (PM), for about 25% in 2025. Moreover, it is expected that global NO_x- and PM-based emissions generated by road vehicles will decline by 65% and 53%, respectively, between 2010 and 2050 using transportation electrification policies [11].

Furthermore, EVs provide more comfort because of the significantly low vibrations and motor noises. They can reduce noise pollution, which is a serious issue in big cities and capitals. Noise pollution impacts millions of people daily and causes health problems such as high blood pressure, heart disease, sleep disturbances, and stress [18].

10.2.2 Economic and Social Benefits

Transportation electrification provides various socioeconomic benefits for utilities, customers, and the whole society. For instance, a study by the University of California (UC), Berkeley, illustrates that achieving 100% EV sales by 2030–2035 along with 90% clean and renewable energy resources by 2035 result in major benefits as [19]:

- Customer cost saving of $2.7 trillion by 2050;
- Avoiding 150,000 premature deaths followed by nearly $1.3 trillion indirect costs for health and environmental issues by 2050;
- Over 2 million created net jobs by 2035 in the various sectors such as EV development, EV technologies, clean energy transition, and infrastructure development; and
- A dependable grid and achievable investments in renewable energy, batteries, and charging infrastructure.

Moreover, another study by Advanced Energy Economy [20] has shown that significant economic advantages can be created by motivative investments in electrified transportation systems across the United States. Examples of these economic advantages include enhancing the economy, making a huge number of jobs, and providing extra income for the governments. This study illustrated that investment of $275 billion in US transportation electrification can provide the following economic benefits [20]:

- Boosting national GDP by $1.3 trillion, which means nearly five-times of return on public investment
- Creating 10.7 million jobs considering job positions in vehicle- and battery-related industries
- $231 billion in extra tax income for the governments
- $19 billion in annual customer, administrative, and business investment funds from changing to EVs

Furthermore, such a federal investment will stimulus private investors to invest in EVs for at least 2.6 times the amount invested by the government [20].

From the car owners' point of view, in particular, EVs can save significant costs in comparison with ICE-based vehicles. Although the purchase price of EVs is more than that of traditional gasoline-based cars, the maintenance and the required electricity costs for EVs are significantly lower than the maintenance and fuel costs of ICE-based cars. Generally, the fuel cost for EVs is approximately 50 to 75% lower than traditional ICE-based cars and the maintenance expenses of EVs are almost negligible compared to ICE vehicles. Moreover, the efficiency of EVs is more than traditional ICE vehicles. The average energy efficiency of ICE vehicles is between 10 and 20% for converting the fuel energy into forwarding motion. However, EVs can travel the same distance using approximately 24% less energy compared with ICE vehicles [21]. Table 10.1 compares the cost breakdown of two sample electric and ICE-based vehicles in Germany [22].

According to Table 10.1, the total costs of EVs and ICE-based vehicles are almost the same. The total fixed cost of EVs is more than that of ICE-based vehicles even considering subsidies. However, the development of EV technologies especially battery and charging technologies will reduce the purchase price of EVs and make them comparable with respect to the fixed as well as variable costs.

Traffic congestion is another issue that EVs can contribute to resolve or reduce. Generally, the substantial cause of a traffic congestion issue and the solution to resolve it are unknown. Figure 10.4 illustrates a summary of sources of traffic congestions. According to this figure, about 50% of traffic congestion is caused by traffic-influencing

Table 10.1 Cost Comparison of Two Sample Electric and ICE-Based Vehicles [22].

	Hyundai IONIQ Elektro	Hyundai i30 1.4 T-GDI DCT
Fixed costs		
Buying price	33,300 €	24,550 €
Charging foundations	1,100 €	0 €
Subsidy/purchase premium	-9,000 €	0 €
Variable costs (per year)		
Fuel cost	662 €	1,170 €
Vehicle tax	0 €	98 €
Protection/insurance	969 €	1,260 €
Maintenance/servicing	552 €	744 €
Leftover worth	7,100 €	6,070 €
Total Cost	34,213 €	34,840 €

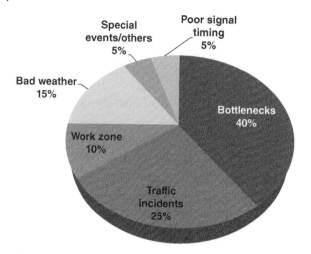

Figure 10.4 Summary of the Traffic Congestion Sources in the United States [24].

events or temporary disruptions such as traffic incidents, bad weather, and work zone. The other half is related to deficient roadway capacity that happens when the capacity of a roadway is less than the number of vehicles in it [23]. Traffic congestion creates huge societal issues and imposed costs such as increasing shipping costs and vehicle crashes, wasting fuel, and increasing air and noise pollution [23]. According to a study by Texas Transportation Institute in 2009, each US highway traveler wastes nearly 40 hours of a year stuck in traffic congestion and wastes $808 fuel [23]. Electrification of transportation with the development of Intelligent Transportation Systems (ITS) and Connected Autonomous Vehicles (CAVs) can potentially decrease and manage traffic congestion. This technology enables vehicle to vehicle (V2V) and vehicle to infrastructure (V2I) communications that can potentially improve roadway congestion by determining the best routes for each vehicle [23].

10.3 Barriers and Challenges of Transportation Electrification

Despite the significant benefits of the EV industry in transportation electrification, there are some challenges for consumers, utilities, and societies. These challenges and barriers are mainly related to EV technologies, infrastructure, power grid, economic, education, and environment [25–27].

The technological challenges of EVs are mainly associated with batteries performance characteristics, such as driving range, charging time, battery weight, and lifetime. Typically EVs can travel in the range of 200 to 350 km with one-time full charging, although scientists are continuously working to improve this issue [27]. The charging time of EVs is generally long, and 4 to 8 hours is required for full charging a typical EV. Even using the current fast charging technology, it can take at least 30 min for 80% capacity charging. For example, the Tesla Model S cars can be charged up to 50% and 80%, respectively, in 20 min and half an hour using the Tesla superchargers [27, 28]. Moreover, suitable batteries for EVs are weighty and need a large space in the vehicles.

The weight of a typical battery is approximately 200 kg that can vary based on the capacity [29].

In addition to the mentioned challenges, transportation electrification faces several technological challenges regarding cybersecurity, reliability, privacy, big data management, and modernization levels [30]. These challenges are mainly related to advanced connected and autonomous EV technologies. For instance, the FBI released a Public Service Announcement on vehicle vulnerabilities in March 2016 to cyberattacks and vehicle hacking incidents [30, 31].

EV infrastructure is of the most important challenges for EV adaptations. There are many countries without proper EV infrastructures. Moreover, customers are facing anxiety coming from the fear of not completing a trip before the battery runs out due to the lack of charging stations. To overcome this barrier, more charging infrastructures should be established in optimal locations that employ fast and ultra-fast charging technologies. For instance, based on a recent study, up to $11 billion investments are required in the United States for establishing charging stations and infrastructures by 2030 to ensure charging stations be as accessible as gas stations [32, 33].

Electrification of transportation might have huge impacts on power grid infrastructure, grid operation, and planning activities [34]. Studies illustrated that the large integration of EVs will impact electricity demand and grid activities. Thus, grid reinforcement and capacity expansion are inevitable to meet the demand in the presence of EVs [26, 34]. However, there are some non-wire solutions such as the utilization of EVs flexibility through the vehicle to grid technologies, demand-side management programs, and incorporating energy storage systems for efficient and cost-effective EV integration [35–39].

Despite the financial incentives and subsidies imposed by policymakers, the upfront cost of EVs is still high for a large portion of customers [25]. Moreover, it is expected that the incentives and subsidies will be lifted in the near future, which might introduce several financing challenges for customers [33]. For instance, a research study by Morning Consult has illustrated that only one-third of possible customers would consider EVs as a purchase option due to high purchase prices and required charging infrastructure. Moreover, it has been shown that 84% of car buyers have concerns about the purchase incentives regarding EVs [33, 40].

Another important challenge of EV integration is the need to prepare qualified professionals of EVs considering differences between the electric and ICE vehicles. These professionals should be qualified in both the administration and maintenance of the electrified transportation sector, particularly charging and fast charging technology as well as the maintenance and repair of EVs. Moreover, security forces, health-related crews, and firefighters need to be trained in response to the EVs' risks, especially EV-related accidents and fires [26].

With respect to environmental concerns, EVs help to reduce GHG emissions as well as noise pollutions. However, current battery energy storage systems contain non–eco-friendly materials that make their disposal a big challenge for the environment. Currently, lithium-ion batteries is the most widely utilized technology in EVs that uses cobalt or nickel materials for cathode material, both of which are costly and highly toxic [39, 41]. To reduce this risk, battery storage systems should be designed eco-friendly, with reuse, remake, and recycling capability [26, 42]. Therefore, some countries such as China and European Commission countries have developed the technical procedures for industrial recycling of batteries [26, 43, 44].

10.4 Electrification Technologies

Transportation electrification refers to the utilization of more electrical energy to control impetus and non-drive loads in vehicles. Traditionally, internal combustion engines (ICEs), which have an average efficiency of less than 30%, are responsible for powering these loads. However, electrical systems provide much higher efficiencies. For instance, electric motors are intended to work with an efficiency of over 90% [1, 45]. Besides, electrical energy can be generated by renewable and emission-free energy resources such as wind and solar [46, 47].

10.4.1 Degree of Electrification

There are various levels of electrification provided by automakers based on the proportion of the total required power of the vehicle generated by electrical sources [1]. Table 10.2 illustrates the various levels of electrification for the same vehicle platform [1, 3]. According to this figure, the first level of electrification is related to the existing ICE-based vehicles that include up to 10% of electrification. These vehicles use electrical systems for non-propulsion loads such as steering, cooling system, pumps, and so on [1]. Moreover, these vehicles utilize the belt-driven starter generator for restarting the motion after the engine stops when the vehicle is braked out of gear. This system offers a low-cost and easily adaptable electrification option for automakers, that can enhance the efficiency of fuel consumption (up to 10%) and decrease CO_2 emissions by about 5% based on the fact that almost 17% of vehicle inefficiency is associated with the time that the car is in idle mode [48].

The second category of electrification is related to hybrid electric vehicles (HEVs). Powertrain system in HEVs consists of an ICE (primary system) and an electric motor (supporting system) in which the electric motor is only responsible for short distances or to support the primary system in some situations such as at stoplights [3, 49, 51, 52]. Similar to conventional vehicles, the HEVs cannot be plugged into the electricity grid and consumes fossil fuels as the main energy source. Generally, HEVs are designed as series or parallel powertrain types. In a series configuration, the ICE is utilized as an electricity generator for charging the batteries, which is responsible for powering the wheels. Thus, in this design, the ICE does not deliver power to wheels directly. On the other hand, parallel powertrains employ both the electric motor and the ICE for powering the wheels [49, 51].

HEVs provide a higher degree of electrification between 8% and 50% relying upon the configuration of the powertrain [1, 48, 49]. The configuration of the powertrain characterizes the degree of fuel-efficiency enhancement in highway and city driving

Table 10.2 Levels of Electrification of Vehicles Based on [1], [48]–[50].

Vehicle type	Fuel Efficiency Improvement	CO$_2$ Reduction
ICE	up to 20%	about 5%
HEV	8–50%	10–22%
PHEV	40–90%	30–60%
BEV	100%	about 100%

situations. HEVs provide a significantly appealing cost-worth ratio for automakers such that each vehicle can produce 10–22% less CO_2 emissions at a cost of $1,000 to $1,200 [48].

Higher fuel efficiency can be achieved by increasing the degree of electrification. The powertrain design of plug-in HEVs (PHEVs) is almost the same as that of HEVs with larger battery packs. Moreover, PHEVs can be utilized external electricity energy sources to charge their battery and have a longer all-electric driving range [1, 3, 51]. Also, batteries in PHEVs can be charged by the regenerative braking system. which uses the generated electricity by electric motor in generator mode.

Another type of PHEVs is the extended range EVs (EREVs), which are just one step behind fully electric vehicles. These vehicles look more like series-type HEVs that the ICE generates electrical power for the electric motor. Although PHEVs consist of ICE, their main energy source is electricity and the ICE is only used for sustainable supply of electricity to the battery, and extending the driving range [1, 48, 51]. The CO_2 reduction of PHEVs is between 30% and 60% compared to a conventional ICE vehicle [50]. However, the cost of moving from the conventional ICE to PHEV is considerable (i.e., $3,200 to $3,600 on average for each vehicle) [48].

The battery electric vehicle (BEV) is the actual zero-emission vehicle that does not utilize any ICE and is completely powered by large-capacity batteries. In BEVs, the batteries are recharged by connecting to the power grid through wall sockets or at a dedicated charging station. BEVs face several challenges for growth including high battery costs, limited and unreliable driving range, and the lack of accessible charging infrastructures everywhere. Therefore, a few automakers (e.g., Tesla Motors, Faraday Futures) are dedicated to this technology. However, BEV market share is expected to grow due to the increasing stimulation measures and policies introduced by countries worldwide.

10.4.2 Electric Motors

The core part of EVs is the electric motor, which generates mechanical motion for the wheels using the stored electrical energy in the batteries. Moreover, the electric motor works as a generator unit to driven back the electric energy to the battery during braking or when the speed of the vehicle is reduced using a regenerative braking system. Electric motors have serious impacts on fuel utilization, speed increase, fast execution, and driving solace. Electric motors with higher efficiency and performance will enhance the utilization of electrical status and increase the efficiency of the engine in hybrid EVs as well as a higher range for all-electric EVs [1].

There are various types of electric motors with distinct design and construction technologies to be utilized in EVs. The most common types of electric motors are induction motors, permanent-magnet motors, and switched reluctance motors [39, 53]. The main requirements of electric motors for EVs, regardless of the machine type, are durability, small size, high power, energy, and torque characteristics as well as low cost and noise [54, 55].

Induction motors (IMs) have been successfully employed in some of the EV models such as the General Motor EV1 as well as Tesla Roaster and Model S and provide acceptable performances [56, 57]. The main advantages of IMs are their simplicity, low cost, robustness, less maintenance, mature technology, high peak torque, and good dynamic response [57, 58]. However, the low efficiency at light loads can be considered as their main drawback [53]. Moreover, the operation of IMs is with a significant amount of inherent rotor copper losses. These losses produce a lot of heat especially during high-torque operation, which limits the torque-density of IMs [1, 59].

The most utilized electric motor in the EV industry is the permanent magnet synchronous motor (PMSM). Among various topologies and designs for PMSM, surface permanent magnet (SPM) and interior permanent magnet (IPM) motors are the two main types that are employed in EVs [53, 60]. The SPM machines have a relatively simple design/structure, lower rotor inertia, and simpler control [61, 62]. However, the larger air gap between the rotor and stator due to locating the magnet outside of the rotor reduces the performance and efficiency of the machine. Thus, the application of SPM in the EV industry is now quite limited [53]. On the other hand, IOM motors can provide more efficiency and higher torque density compared with SPM, particularly at low and medium speed ranges due to embedding the permanent magnet inside the rotor [1].

Switched reluctance motors (SRMs) have a simpler design, lower costs, and more robustness compared to IMs and PM machines. Moreover, they do not utilize rare-earth materials for constructing permanent magnets [39]. The structure of the rotor in SRM is a common salient pole design that is composed of only thin silicon steel laminations with no windings or magnets [1, 53]. Also, SRMs contain a salient pole type stator in which each pole is wounded with concentrated coils. The main drawbacks of SRMs are the significant torque ripples and strong radial forces, which result in vibration and acoustic noises that limit the application of SRMs in EVs. However, these factors can be reduced using advanced design and control techniques and make this technology a suitable candidate in the long run [1, 63–66].

10.4.3 Battery Technology of EVs

The advances in battery technology have the greatest impact on the EV industry. EV batteries should provide continuous power to the vehicle that necessitates being high capacity in terms of power, energy, and energy density [67–70]. Lead-acid, lithium-ion, Zebra, and nickel-metal hydride (NiMH) batteries are the most utilized rechargeable battery types in EVs [39, 71]. Lead-acid is the oldest rechargeable technology used on EVs that was invented in 1860 [27]. The energy density and specific energy ratios of lead-acid batteries are significantly low, which enables them to mainly use in low-speed EVs. However, it is utilized in some EVs such as GM EV1 and the Toyota RAV4 EV because of some of its key advantages such as low cost and high reliability. The main drawback of lead-acid batteries for utilization in EVs is the negative environmental issues with manufacturing, disposal, and recycling. However, there is a significant improvement in the collection and recovery of lead-based batteries in the European Union and United States, recently.

The sodium nickel chloride or "Zebra" battery is another relatively mature technology of utilized batteries in EVs. Zebra battery has better specific energy, power, and energy density characteristics than lead-acid batteries. However, its electrolyte must be heated to about 300°C, which dissipates some amount of energy, creates challenges for long-term charging, and increases the risk of explosion [39, 71].

Table 10.3 Comparison of the Most Commonly Used Battery Technology in EVs.

	Lead-Acid	Zebra	Nickel-Metal Hydride	Lithium-Ion
Working temperature (°C)	-20 – 60	300 – 350	0 – 50	-20 – 60
Energy density (Wh/L)	60 – 100	110 – 120	100 – 300	200 – 735
Specific energy (Wh/kg)	30 – 50	160	60 – 120	100 – 275
Specific power (W/kg)	75 – 100	150 – 200	250 – 1000	350 – 3000
Efficiency (%)	70 – 90	80 – 90	50 – 80	90 – 95
Cycle durability	500 – 800	2500 – 4500	500	400 – 3000

Another battery technology used in EVs is NiMH, which utilizes nickel hydroxide as the anode, a hydrogen-absorbing alloy as cathode, and an electrolyte made of potassium hydroxide. NiMH batteries provide the least efficiency (50–80%) in charging and discharging than other technologies and suffers from high self-discharging problem [27, 39]. However, the main advantage of NiMH batteries is their long lives, which is used in the first-generation of Toyota RAV4 EVs and can remain operational after 100,000 km [39, 72].

Lithium-ion batteries have simultaneously small size and large power and energy capacity, which provides higher energy density and specific energy compared to other battery technologies [39]. Additionally, they have other advantages such as higher energy efficiency, an ignorable memory effect, and long life, which make them the first choice for energy storage of EVs [73, 74]. However, this technology suffers from instability problems caused by its structure and chemical process, which limits its operation within a restricted working area based on voltage and working temperature. Exceeding the restricted area will cause a significant reduction in the battery performance and in some cases increases the risk of fire and explosion.

Table 10.3 compares the main characteristics of the aforementioned battery technologies. According to the table, the lead-acid and lithium-ion batteries have better working temperature ranges that cover low temperatures. However, the low-temperature issue significantly affects the capacity of lithium-ion batteries and increases self-discharging, for which the operating temperature is usually set near 40°C [27]. Regarding the specific energy and the energy density, the lithium-ion technology is significantly better than others especially the lead-acid and NiMH batteries. Lead-acid and Zebra technologies offer the worst specific power (up to 200 W/kg) in comparison with other battery technologies. On the other hand, lead-acid and NiMH provide the worst life cycle specifications, whereas lithium-ion and Zebra batteries provide up to 3000 and 4500 cycles, respectively [27, 39]. Therefore, considering all these parameters, lithium-ion technology is the optimal choice for current electric vehicles.

10.4.4 Charging Technology and Infrastructures

In addition to the battery-related challenges, charging technology and infrastructure is another challenge for plug-in EVs, that is, PHEVs and BEVs. Nowadays EV charging is becoming more convenient and quicker with the fast improvement in charging technologies and infrastructures. There are various types of battery charging technologies for EVs, which can be categorized as conductive charging, inductive charging, and battery swapping [39, 75].

Conductive charging technology requires a physical connection between EVs and the charger for transferring power. This technology is the most widely used charging technology that can be performed by on-board and off-board chargers [75]. The charging activity in on-board chargers is carried out inside the EVs and mainly applied for slow charging. On the other hand, off-board chargers provide fast charging at the expense of taking the vehicles to a charging station [39, 75].

In inductive charging technology, the power is transferred to EVs through an electromagnetic field. This technology provides three charging systems known as stationary, semi/quasi-dynamic, and dynamic systems [39, 75]. Stationary charging systems work similarly to current off-board conductive chargers, but no physical connection is required between EV and power supply. This technology requires an on-board receiving coil in the EV and a transmitting coil on the road surface. Semi/quasi-dynamic charging systems give a brief span charging activity at a certain situation and places like bus stations, taxi stops, and traffic signals. Dynamic charging systems can charge the EV during moving through the surface. Overall, although inductive charging does not require plugging and unplugging, its major drawbacks are low efficiency and energy transfer capability as well as high power loss [76, 77].

In battery swapping, which is a new and under developing concept for EV charging, the depleted battery is replaced by a fully charged one without waiting for charging. In this method, service providers are responsible for the swapping process and the drivers do not worry about battery life or maintenance [75, 78]. However, the main disadvantage of the battery swapping method is the standardization of battery types considering the fact that each EV model utilizes a different battery size and design. Moreover, swappable batteries imposed more cost to end users by energy companies [75, 79].

Several standards for charging EVs are mainly determined based on regions. In particular, the SAE-J1772 standard is employed in North America and the Pacific zone for loading EVs. The GB/T 20234 and the IEC-62196 are other standards in Europe and China, respectively. Table 10.4 shows the charging specifications based on the IEC-62196 standard, which is the well-accepted standard by most countries [27]. According to this standard, clients are given four charging modes for EVs:

1) Mode 1 (Slow charging): This mode is introduced for charging EVs at home with a single-phase or three-phase power outlet at the maximum current of 16 A.

Table 10.4 Charging Characteristics of the IEC-62196.

Charging Method	Type	Max. Current	Max. Voltage	Max. Power	Specific Connector
Mode 1 (slow)	AC – Three	16 A	230 – 240 V	3.8 kW	No
	AC – Single		480 V	7.6 kW	
Mode 2 (semi-fast)	AC – Three	32 A	230 – 240 V	7.6 kW	No
	AC – Single		480 V	15.3 kW	
Mode 3 (fast)	AC – Three	32 – 250 A	230 – 240 V	60 kW	Yes
	AC – Single		480 V	120 kW	
Mode 4 (ultra-fast)	DC	250 – 400 A	600 -1000 V	400 kW	Yes

2) Mode 2 (Semi-fast charging): In this mode, the maximum current is defined as 32 A and can be employed by users at home or in public locations. Similar to the previous mode, it needs standard single- or three-phase power outlets.

3) Mode 3 (Fast charging): This mode requires a specific power supply called EV Supply Equipment (EVSE) to provide electrical current between 32 and 250 A. The EVSE is located in charging stations and can communicate with other vehicles, monitor the charging process, and stop energy flow after failing the vehicle connection to protect the system.

4) Mode 4 (Ultra-fast charging): In this mode, EVs are connected to a DC supply network with a maximum current of 400 A and up to 1000 V of DC voltage. This mode can be employed in charging stations using only a special charger to provide protection and control over charging and enable communication between the vehicle and the charging point.

Transportation electrification and EV development require a very strong investment in charging infrastructures. Countries with a high share of EVs have established policies and support measures for the development of charging infrastructures [39, 75]. For instance, the American Recovery and Reinvestment Act in 2009 was the first established supporting measure in the United States for charging infrastructures, which results in building approximately 18,000 public charging stations between 2010 and 2013. After that, other private companies and governmental agencies have started to invest in building charging stations [39, 80]. In Europe, the total number of 165,000 EV charging stations were exited in 2019, which shows a more than 300% increase since 2015 [75, 81]. Moreover, new investments in charging infrastructures have focused on fast-charging technology such that the number of fast-charging stations in European countries increased from 3396 in 2015 to 17,056 in 2019 [75].

10.4.5 Developing Technologies of the Future Transportation

EVs have the potential to provide more benefits to society in addition to reducing greenhouse gas emissions and oil dependency. Vehicle-to-everything (V2X) is referred to technologies that employ the energy in the batteries of EVs for other purposes than transportation [36]. Vehicle to grid (V2G), vehicle to vehicle (V2V), vehicle to infrastructure (V2I), vehicle to building (V2B), and vehicle to cloud (V2C) technologies are examples of V2X that can significantly improve both energy efficiency and traffic safety [82]. For instance, V2I and V2V communications could enhance driving quality by choosing optimal velocity based on received data from other vehicles and infrastructures such as distance to traffic lights and waiting time. Moreover, these technologies can provide smoother and steadier driving by decreasing the number of starts/stops during driving activities [39, 83]. V2P communication system is another V2X technology that has attracted increasing attention due to the safety or convenience it can provide for society [84]. V2X technologies are mostly motivated for improving road safety, traffic efficiency, and energy savings through communication with grids, vehicles, and infrastructures. However, there is a long road for real-world applications of these technologies since they are still in experimental stages with several challenges to overcome.

10.4.5.1 V2G technology

V2G technology, which is of the most important V2X technologies, refers to using the energy and the power capability of EVs to provide various grid management services such as frequency regulation, peak shaving, and shifting, spinning reserve, voltage support, and transmission & distribution upgrade deferral [36]. In this technology, EVs can play the role of electrical loads or distributed energy resources that can exchange energy with the grid through the market environment. The main challenge of V2G is the EV owner acceptance due to the high degradation costs of EVs [39, 85–87]. Moreover, it needs several communication infrastructures for exchanging energy, power, and economic information between EVs and the market. Although V2G has not been implemented widely, several studies illustrate that V2G enables various economic profits for EV owners, parking lots, and grid operators [88, 89].

10.4.5.2 Connected and Autonomous Vehicle Technology

Connected and autonomous vehicle (CAV) technologies are expected to play an important role in the future of transportation electrification [90]. CAVs refer to different types of vehicles such as light- and heavy-duty vehicles as well as drones that depend on artificial intelligence for operation using various human inputs. CAVs include two types of technologies: (1) connecting vehicles; and (2) automating vehicles [91]. For connecting vehicles, various types of communication technologies are utilized in vehicles to enable the car for communication with the driver, other cars on the road (V2V), roadside infrastructure (V2I), and the pedestrian (V2P). This technology can be used to improve not only vehicle safety but also vehicle efficiency, traffic congestion, and commute times [23, 39, 91].

With respect to automated vehicles, the US Department of Transportation's National Highway Traffic Safety Administration (NHTSA) defines the fully automated, autonomous, or self-driving vehicles as the vehicle that operates without any driver intervention to control the car. However, to reach this level of automation, there are different automotive levels based on the human interface and "who does what, when" as illustrated in Table 10.5 [91–93].

10.4.5.3 Communication Technologies

Communication technology in the electrification of transportation refers to a communication system that enables bidirectional information and energy exchange between multiple devices such as other vehicles, infrastructures, pedestrians, buildings, and clouds [94]. To establish communication between devices, each device must employ the same protocols. For instance, several devices in a network can be connected by a specific platform and standard (e.g., Wi-Fi or Ethernet) or in a broader sense, they can exchange information through different types of technologies such as the Internet [94]. In this sense, there need suitable standards for the communication network platform and its respective protocols to satisfy several criteria such as compatibility, user-friendliness, and cost. The IoT architecture is of the most promising and novel concepts for the communication of electrified vehicles. The IoT concept enables the utilization of the Internet for data exchange, processing, storing, and analysis by devices, either automatically or with the user's assistance. This architecture is different than the current Internet utilization in which humans produce and consume most of the data [95, 96].

Table 10.5 Different Levels of Autonomous Driving. Based on [91]–[93].

	Driver	Vehicle
Level 0 (no automation)	Driver is in charge of all the driving all the time.	Vehicle responses only to inputs from the driver.
Level 1 (driver assistance)	Driver must do all the driving with some basic help from the vehicle.	Vehicle provides some basic help such as automatic emergency braking and lane control support.
Level 2 (partial automation)	Driver can assign some basic driving tasks to the vehicles but must be fully alert.	Vehicle can automatically both accelerate/decelerate and perform basic steering functions.
Level 3 (conditional automation)	Driver must be always ready to act when the self-driving system is unable to continue.	Vehicle is able to perform real-time monitoring of the driving situation and control acceleration, turning, or brake actions with the expectations of human intervention when alerted.
Level 4 (high automation)	Driver can be a passenger that should take over the car control in with notice of self-driving system some situation.	Vehicle can do all the driving operations without driver attention.
Level 5 (full automation)	No human driver is required.	Vehicle is in charge of all the driving tasks and operates in all conditions with no human intervention.

Several studies have proposed various IoT architectures for EVs and transportation systems [96–99], but there is still a long way to reach a consensus and standard model in this area. However, according to the proposed approaches, the basic IoT architecture can be categorized into three layers known as sensing layer, network layer, and application layer [94]. The sensing layer, which is also called as perception layer or object layer, is equipped with several sensors to gather information for processing. The network layer is the middle layer in the IoT architecture, which connects sensing and application layers through communication infrastructure. The application layer provides specific services such as automation of processes for the users according to the gathered information by the sensing layer through the network layer [94].

10.4.5.4 Blockchain Technology

Blockchain is a novel technology that can create an environment for secure, privacy-reserved, and anonymous transactions, using a decentralized network of devices [100, 101]. This technology was first introduced in financial markets for the cryptocurrency Bitcoin in 2008 [102]. This technology has found many applications in several industries outside the financial world such as healthcare [103–105], travel and hospitality [106–108], energy trading [109–111], and transportation systems [112–114].

The future of transportation systems is an intelligent automotive architecture based on connected and autonomous electric vehicles. In this context, EVs are connected with each other and other devices using communication links such as an IoT platform. The main challenge toward the widespread deployment of IoTs is the security of exchanging data [100, 115]. Blockchain can provide a secure and privacy-reserved environment for exchanging information between connected vehicles and devices

[116–119]. As mentioned before, the IoT platform includes three sensing, network, and application layers. For instance, blockchain technology can be utilized in the network layer by exchanging data of vehicles and infrastructures to each other in a secure manner. Examples of such exchanged information are the state of charge of the vehicle, distance to charging station, and type of batteries that can be utilized in blockchain-based EV charging management [120, 121]. Moreover, in the application layer, blockchain can be utilized for making contracts, payments, settlements, and so forth [119, 122, 123] [124, 125]. This technology has achieved popularity in several industries such as health care [126, 127], transportation [128, 129], power system [130–132], and banking [133–135] due to its ability to employ underutilized computational resources and to reduce the computational complexity for large-scale cases [125].

Cloud computing provides different types of services in which infrastructure as a service (IaaS), platform as a service (PaaS), and software as a service (SaaS) are its main three services [124, 136]. An example of IaaS is the Amazon Web Services (AWS) that share its computing, network, and storage resources for customers. PaaS (e.g., Google AppEngine, and Microsoft Azure) provides an environment for platform developers to use the hosted platforms on the cloud without installing any applications. In SaaS, the expensive license-required software such as IBM CPLEX can be utilized by customers without a need for buying the whole software through only paying for a subscription, in a "pay-as-you-go" model [124, 137].

Future transportation systems equipped with a huge number of CAVs and communicable devices require a platform for processing data and making coordinated decisions. The vast number of CAVs and infrastructures makes the processing and decision-making computationally expensive and interactable [137, 138]. Cloud computing offers a remarkable opportunity for data processing and computation for large-scale cases and in a real-time manner [139]. Traffic management and transportation coordination [140–143] as well as EVs' energy management and monitoring [139, 144–146] are examples of cloud computing applications in the transportation sector.

Moreover, the huge number of connected EVs on the roads, buildings, and parking lots as well as available charging infrastructures provide abundant computational resources that can be employed for cloud computing services [124, 137]. In this vein, the underutilized resource capacity of EVs such as storage, Internet access, sensors data, and computing devices can be shared in the cloud for utilization of other customers and authorities [147].

10.4.5.5 Artificial Intelligence and Robotic Technology

Artificial intelligence (AI) enables the utilization of machine capabilities to perform human tasks like perceiving, reasoning, learning, and problem-solving using available historical data [122, 148]. AI determines a complex pattern between input data and output decisions based on historical data and contains several types of algorithms and techniques such as machine learning algorithms and meta-heuristic optimization methods [149, 150]. AI methods have been successfully employed in various industries such as power systems [46, 151–153], transport sector [154–157], healthcare [158–160], education, social media, and marketing.

In the transportation sector, AI can be employed in different ways including the application of AI (1) in traffic coordination and incident prediction; (2) to enhance public transportation performance; and (3) to improve the productivity of CAVs [148].

With regards to the first category, AI methods determine the optimal decision for the vehicle routes, traffic lights, and other traffic control systems according to the prediction of the traffic volume, traffic conditions, and incidents [148, 154]. The quality of the public transportation sector can be improved by utilizing AI methods in long-term and short-term forecasting of transportation situations for future expansion of infrastructures and daily scheduling of existing vehicles [148, 150].

Naturally, CAVs utilize AI algorithms for real-time decision-making of self-driving systems, energy management, and communication with other devices. The self-driving system is taught how to drive the car safely using the receiving data from the sensors and processing them in a real-time manner. Moreover, AI methods help the vehicle for improving its energy management and efficiency [161]. In addition, AI can be utilized in CAVs for monitoring passenger's health conditions using the available sensors in the vehicles and the connected devices to make appropriate decisions such as going to a hospital, calling emergency units, and stopping the vehicle.

Moreover, various types of robots work in the transportation sector like robot chargers that are located in charging stations and provides the selected charging technology for the connected EV. For example, the Volkswagen group is working to develop a mobile charging robot that can drive and connects to the EV autonomously, and communicates with it for opening the charging socket, plugging into the power, and disconnecting the charger [162]. Furthermore, a CAV can be considered as a robot that can connect to other devices, process data, and make decisions.

10.4.5.6 Cybersecurity in Electrified Transportation

As mentioned before, intelligent transportation systems (ITS) include a huge number of connected vehicles (CVs). Although integration of CVs into the transportation sector provides a variety of opportunities and improvements, security and privacy concerns will emerge that need to be resolved [163]. In [164, 165], some of the real-world cyberattacks in the United States between 2007 and 2017 have been listed. Although the occurred attacks can be considered as simple mischief, they reveal the vulnerability of transportation systems and draw considerable cybersecurity concerns for ITS. The significance of cyberattacks at ITSs is that they often results in catastrophic safety consequences.

The main security issues of ITS are associated with the internal communication environment, that is, vehicle-to-sensor (V2S) and external communication technologies such as V2V and V2I. In this vein, the US Department of Transportation categorizes the main security requirements in the three core elements as confidentiality, integrity, and availability [166, 167]. Confidentiality concern necessitates that the exchange information between vehicles and infrastructures must not be accessed by unauthorized users. With regard to integrity, any security solution must protect communication data from alternations and modifications by unauthorized users. However, availability-based security issues are related to latency and accessibility problems in V2X communications [167–169].

Various types of attacks are possible in ITS based on the aforementioned security requirements. Denial of service, spoofing, tampering, message alternations, malware and spam, and eavesdropping are the main types of cyberattacks that can highly impact ITS operations [164, 167, 168]. In [170–172], the effects of these cyberattacks are discussed and the proper preventive measures are introduced.

10.5 Summary and Conclusion

In this chapter, an overview of the trends, opportunities, challenges, and technological requirements of transportation electrification was presented. According to the reviewed data, transportation electrification provides significant economic, environmental, and societal benefits and opportunities for different sectors of society such as people, utilities, and government. Thus, numerous countries around the world have been progressively setting targets for moving toward transportation electrification such that the global number of EVs on the road was increased almost five times between 2016 and 2020. However, despite all the efforts made by policymakers, the transportation sector accounts for a small share of electricity consumption. The reason is the unsolved challenges and barriers in electrified transportation systems mainly related to EV technologies, infrastructures, power grid, high purchase costs, and education.

The various levels of transportation electrification were also discussed in this chapter. Moreover, different types of EVs were introduced along with the key technologies of the EVs such as battery energy storage, electric motors, charging methods and standards, and communication systems. Besides, the V2X application, connected and autonomous vehicles, blockchain technologies, cloud computing services, AI and robotic technologies, and cybersecurity concerns were presented at the end of this chapter as the emerging technologies for the future development of EVs.

References

1 Bilgin, B., et al. (2015). Making the case for electrified transportation. *IEEE Trans. Transp. Electrif.* 1 (1): 4–17.

2 E.-E. Commission. (2020). Sustainable and Smart Mobility Strategy–putting European transport on track for the future, SWD (2020) 331 final (COM (2020) 789 final). Brussels.

3 Ajanovic, A., and Haas, R. (2019). On the environmental benignity of electric vehicles. *J. Sustain. Dev. Energy, Water Environ. Syst.* 7 (3): 416–431.

4 International Energy Agency. (2021). Global EV Outlook 2021—analysis—IEA. https://www.iea.org/reports/global-ev-outlook–2021.

5 International Energy Agency. (2020). World electricity final consumption by sector, 1974–2018 IEA, Paris. https://www.iea.org/commentaries/how-global-electric-car-sales-defied-covid-19-in-2020.

6 Virta Liikennevirta Oy (Ltd.). (2021). The Global Electric Vehicle Market Overview in 2022 –Statistics & Forecasts. https://www.virta.global/global-electric-vehicle-market.

7 International Energy Agency. (2021). How global electric car sales defied Covid-19 in 2020. IEA, Paris. https://www.iea.org/commentaries/how-global-electric-car-sales-defied-covid-19-in-2020.

8 International Energy Agency. (2021). Global EV Data Explorer. IEA, Paris. https://www.iea.org/articles/global-ev-data-explorer.

9 H.R.6049-110th Congress (2007–2008): Energy Improvement and Extension Act of 2008. https://www.congress.gov/bill/110th-congress/house-bill/6049.

10 Ministry of Finance of the People's Republic of China. (2021). Announcement on the exemption of new energy vehicle purchase tax. http://www.lawinfochina.com/display.aspx?id=35596&lib=law#.

11 Samaras, Z., and Vouitsis, I. (2013). Transportation and Energy. *Clim. Vulnerability Underst. Addressing Threat. To Essent. Resour.* 3: 183–205.

12 U.S. Energy Information Administration. (2020). U.S. fossil fuel consumption by source and sector, 2020. https://www.eia.gov/totalenergy/data/monthly/pdf/flow/fossil_fuel_spaghetti_2020.pdf.

13 Gaspari, E. (2020). Intelligent mobility for energy transition: Accelerating towards more sustainable societies. *Eur. Innov. Partnersh. Smart Cities Communities.*

14 International Energy Agency. (2021). Global energy-related CO2 emissions by sector, IEA, Paris. https://www.iea.org/data-and-statistics/charts/global-energy-related-co2-emissions-by-sector.

15 Bruyninckx, H. (2018). Air pollution still too high across Europe. European Environmental Agency. https://www.eea.europa.eu/highlights/air-pollution-still-too-high.

16 Al-Qadi, D. (2021). Equitable Transportation Electrification. AECOM Engineering Company. https://publications.aecom.com/transportation-electrification/article/equitable-transportation-electrification.

17 Hu, X., Chen, N., Wu, N., and Yin, B. (2021). The potential impacts of electric vehicles on urban air quality in Shanghai city. *Sustainability* 13 (2): 496.

18 Gabrielli, A. (2021). Noise Pollution. National Geographic Society. https://www.nationalgeographic.org/encyclopedia/noise-pollution.

19 Baldwin, S., Myers, A., O'Boyle, M., and Wooley, D. (2021). Accelerating clean, electrified transportation by 2035: policy priorities. Policy

20 Hibbard, P. and Darling, P. (2021). Economic impact of stimulus investment in transportation electrification. Advanced Energy Economy.

21 Salisbury, M. (2014). Economic and Air Quality Benefits of Electric Vehicles in Nevada. Southwest Energy Efficiency Project. http://swenergy.org/data/sites/1/media/documents/publications/documents/Economic_and_AQ_Benefits_of_EVs_in_NV-Sept_2014.pdf.

22 Mobility House. (n.d.). Cost Comparison: Electric Car vs. Petrol (TCO analysis). https://www.mobilityhouse.com/int_en/knowledge-center/cost-comparison-electric-car-vs-petrol-which-car-costs-more-annually.

23 Boucher, M. (2019). Transportation Electrification and Managing Traffic Congestion: the role of intelligent transportation systems. *IEEE Electrif. Mag.* 7 (3): 16–22.

24 Federal Highway Administration. (2020). *Traffic Congestion and Reliability: Trends and Advanced Strategies for Congestion Mitigation.* U.S. Department of Transportation.

25 Romero Lankao, P., et al. (2019). Urban Electrification: knowledge Pathway Toward an Integrated Research and Development Agenda. *SSRN Electron. J.* 10: 1–20.

26 Pereirinha, P.G., González, M., Carrilero, I., Anseán, D., Alonso, J., and Viera, J.C. (2018). Main trends and challenges in road transportation electrification. *Transp. Res. Procedia* 33: 235–242.

27 Sanguesa, J.A., Torres-Sanz, V., Garrido, P., Martinez, F.J., and Marquez-Barja, J.M. (2021). A Review on Electric Vehicles: Technologies and Challenges. *Smart Cities* 4 (1): 372–404.

28 Tesla. (2019). Tesla Official Website. https://www.tesla.com/en_EU/supercharger.

29 Berjoza, D., and Jurgena, I. (2017). Effects of change in the weight of electric vehicles on their performance characteristics. *Agron. Res.* 15 (1): 952–963.

30 Graham, R.L., Francis, J., and Bogacz, R.J. (2017). *Challenges and Opportunities of Grid Modernization and Electric Transportation.* US Department of Energy and Allegheny Science & Technology.

31 Valasek, C., and Miller, C. (2016). Motor vehicles increasingly vulnerable to remote exploits. https://www.ic3.gov/Media/Y2016/PSA160317#fn1.

32 Eisenstein, P.A. (2019). VW's $2 billion penalty for diesel scam, Electrify America, builds electric charging network across US to boost EV market. *CNBC.*

33 Rogotzke, M., Eucalitto, G., and Gander, S. (2019). *Transportation Electrification: States Rev Up.* National Governors Association Center for Best Practices.

34 Blonsky, M., Nagarajan, A., Ghosh, S., McKenna, K., Veda, S., and Kroposki, B. (2019). Potential impacts of transportation and building electrification on the grid: A review of electrification projections and their effects on grid infrastructure, operation, and planning. *Curr. Sustain. Energy Reports* 6 (4): 169–176.

35 Camus, C., Esteves, J., and Farias, T. (2011). Integration of electric vehicles in the electric utility systems. In *Electr. Veh. Benefits Barriers* (ed. S. Soylu), 135–158. Intechopen.

36 Pearre, N.S., and Ribberink, H. (2019). Review of research on V2X technologies, strategies, and operations. *Renew. Sustain. Energy Rev.* 105: 61–70.

37 Arteaga, J., Zareipour, H., and Thangadurai, V. (2017). Overview of lithium-ion grid-scale energy storage systems. *Curr. Sustain. Energy Reports* 4 (4): 197–208.

38 Cetin, K.S. and O'Neill, Z. (2017). Smart meters and smart devices in buildings: a review of recent progress and influence on electricity use and peak demand. *Curr. Sustain. Energy Reports* 4 (1): 1–7.

39 Sun, X., Li, Z., Wang, X., and Li, C. (2020). Technology development of electric vehicles: A review. *Energies* 13 (1): 90.

40 Toth, J. (2019). For Widespread Adoption of Electric Vehicles, Many Roadblocks Ahead. Morning Consult. https://morningconsult.com/2019/05/22/for-widespread-adoption-of-electric-vehicles-many-roadblocks-ahead.

41 Helmers, E., and Marx, P. (2012). Electric cars: Technical characteristics and environmental impacts. *Environ. Sci. Eur.* 24 (1): 1–15.

42 Chen, F., Taylor, N., and Kringos, N. (2015). Electrification of roads: Opportunities and challenges. *Appl. Energy* 150: 109–119.

43 Daly, T. (2018). *Chinese Carmaker BYD Close to Completing Battery Recycling Plant.* Reuters.

44 Beaudet, A., Larouche, F., Amouzegar, K., Bouchard, P., and Zaghib, K. (2020). Key challenges and opportunities for recycling electric vehicle battery materials. *Sustainability* 12 (14): 5837.

45 Zhang, M., Suntharalingam, P., Yang, Y., Jiang, W., and Emadi, A. (2014). Fundamentals of hybrid electric powertrains. In: *Advanced Electric Drive Vehicles* (ed. A. Emadi), 369–410. Taylor & Francis.

46 Chen, Y., Wang, Y., Kirschen, D., and Zhang, B. (2018). Model-free renewable scenario generation using generative adversarial networks. *IEEE Trans. Power Syst.* 33 (3): 3265–3275.

47 Gubman, J., Pahle, M., Steinbacher, K., and Burtraw, D. (2016). Transportation Electrification Policy in California and Germany. http://joannagubman. com/2016EVStudy.pdf.

48 Eichenberg, P. (2021). The Multiple levels of Vehicle Electrification. https://www. pauleichenberg.com/blog/multi-level-electrification-explained/.

49 Wirasingha, S.G., Khan, M., and Gross, O. (2014). 48-V electrification: belt-driven starter generator systems. In: *Advanced Electric Drive Vehicles* (ed. A. Emadi), 331–368. Taylor & Francis.

50 Elgowainy, A., Burnham, A., Wang, M., Molburg, J., and Rousseau, A. (2009). Well-to-wheels energy use and greenhouse gas emissions of plug-in hybrid electric vehicles. *SAE Int. J. Fuels Lubr.* 2 (1): 627–644.

51 Ajanovic, A. (2015). The future of electric vehicles: prospects and impediments. *Wiley Interdiscip. Rev. Energy Environ.* 4 (6): 521–536.

52 Hensley, R., Knupfer, S., and Pinner, D. (2009). Electrifying cars: how three industries will evolve. *McKinsey Q.* 3 (2009): 87–96.

53 Kumar, M.S., and Revankar, S.T. (2017). Development scheme and key technology of an electric vehicle: an overview. *Renew. Sustain. Energy Rev.* 70: 1266–1285.

54 Lee, H.-K., and Nam, K.-H. (2016). An overview: Current control technique for propulsion motor for EV. *Trans. Korean Inst. Power Electron.* 21 (5): 388–395.

55 Rajashekara, K. (2013). Present status and future trends in electric vehicle propulsion technologies. *IEEE J. Emerg. Sel. Top. Power Electron.* 1 (1): 3–10.

56 Sultana, U., Khairuddin, A.B., Sultana, B., Rasheed, N., Qazi, S.H., and Malik, N.R. (2018). Placement and sizing of multiple distributed generation and battery swapping stations using grasshopper optimizer algorithm. *Energy* 165: 408–421.

57 Agamloh, E., Von Jouanne, A., and Yokochi, A. (2020). An overview of electric machine trends in modern electric vehicles. *Machines* 8 (2): 20.

58 Sutikno, T., Idris, N.R.N., and Jidin, A. (2014). A review of direct torque control of induction motors for sustainable reliability and energy efficient drives. *Renew. Sustain. Energy Rev.* 32: 548–558.

59 Bilgin, B., and Emadi, A. (2014). Electric motors in electrified transportation: a step toward achieving a sustainable and highly efficient transportation system. *IEEE Power Electron. Mag.* 1 (2): 10–17.

60 Shao, L., Karci, A.E.H., Tavernini, D., Sorniotti, A., and Cheng, M. (2020). Design approaches and control strategies for energy-efficient electric machines for electric vehicles—A review. *IEEE Access* 8: 116900–116913.

61 Lara, J., Xu, J., and Chandra, A. (2016). Effects of rotor position error in the performance of field-oriented-controlled PMSM drives for electric vehicle traction applications. *IEEE Trans. Ind. Electron.* 63 (8): 4738–4751.

62 Fodorean, D., Sarrazin, M.M., Marțiș, C.S., Anthonis, J., and Van Der Auweraer, H. (2016). Electromagnetic and structural analysis for a surface-mounted PMSM used for light-EV. *IEEE Trans. Ind. Appl.* 52 (4): 2892–2899.

63 Ye, J., Bilgin, B., and Emadi, A. (2014). An extended-speed low-ripple torque control of switched reluctance motor drives. *IEEE Trans. Power Electron.* 30 (3): 1457–1470.

64 Azer, P., Bilgin, B., and Emadi, A. (2019). Mutually coupled switched reluctance motor: Fundamentals, control, modeling, state of the art review and future trends. *IEEE Access* 7: 100099–100112.

65 Bostanci, E., Moallem, M., Parsapour, A., and Fahimi, B. (2017). Opportunities and challenges of switched reluctance motor drives for electric propulsion: A comparative study. *IEEE Trans. Transp. Electrif.* 3 (1): 58–75.

66 Gan, C., Wu, J., Sun, Q., Kong, W., Li, H., and Hu, Y. (2018). A review on machine topologies and control techniques for low-noise switched reluctance motors in electric vehicle applications. *IEEE Access* 6: 31430–31443.

67 Hannan, M.A., Lipu, M.S.H., Hssain, A., and Mohamed, A. (2017). A review of lithium-ion battery state of charge estimation and management system in electric vehicle applications: challenges and recommendations. *Renew. Sustain. Energy Rev.* 78: 834–854.

68 Pelegov, D.V., and Pontes, J. (2018). Main drivers of battery industry changes: Electric vehicles—a market overview. *Batter.* 4 (4): 65.

69 Saxena, S., Floch, C.L., Macdonald, J., and Moura, S. (2015). Quantifying EV battery end-of-life through analysis of travel needs with vehicle powertrain models. *J. Power Sources* 282: 265–276.

70 Alshahrani, S., Khalid, M., and Almuhaini, M. (2019). Electric vehicles beyond energy storage and modern power networks: challenges and applications. *IEEE Access* 7: 99031–99064.

71 Nykvist, B., and Nilsson, M. (2015). Rapidly falling costs of battery packs for electric vehicles. *Nat. Clim. Chang. 2014 54* 5 (4): 329–332.

72 Toyota RAV4 EV (2021). Wikipedia. https://en.wikipedia.org/w/index. php?title=Toyota_RAV4_EV&oldid=952112719.

73 Bresser, D., et al. (2018). Perspectives of automotive battery R&D in China, Germany, Japan, and the USA. *J. Power Sources* 382: 176–178.

74 Ding, Y., Cano, Z.P., Yu, A., Lu, J., and Chen, Z. (2019). Automotive Li-ion batteries: current status and future perspectives. *Electrochem. Energy Rev.* 2 (1): 1–28.

75 Gönül, Ö., Duman, A.C., and Güler, Ö. (2021). Electric vehicles and charging infrastructure in Turkey: an overview. *Renew. Sustain. Energy Rev.* 143: 110913.

76 Niu, S., Xu, H., Sun, Z., Shao, Z.Y., and Jian, L. (2019). The state-of-the-arts of wireless electric vehicle charging via magnetic resonance: Principles, standards and core technologies. *Renew. Sustain. Energy Rev.* 114: 109302.

77 Machura, P., and Li, Q. (2019). A critical review on wireless charging for electric vehicles. *Renew. Sustain. Energy Rev.* 104: 209–234.

78 Ji, Z., and Huang, X. (2018). Plug-in electric vehicle charging infrastructure deployment of China towards 2020: Policies, methodologies, and challenges. *Renew. Sustain. Energy Rev.* 90: 710–727.

79 Andwari, A.M., Pesiridis, A., Rajoo, S., Martinez-Botas, R., and Esfahanian, V. (2017). A review of Battery Electric Vehicle technology and readiness levels. *Renew. Sustain. Energy Rev.* 78: 414–430.

80 Hall, D., and Lutsey, N. (2017). Emerging best practices for electric vehicle charging infrastructure. *Washington, DC Int. Counc. Clean Transp.*

81 Zhang, Y., Liu, X., Wei, W., Peng, T., Hong, G., and Meng, C. (2020). Mobile charging: a novel charging system for electric vehicles in urban areas. *Appl. Energy* 278: 115648.

82 Cacciato, M., Nobile, G., Scarcella, G., and Scelba, G. (2016). Real-time model-based estimation of SOC and SOH for energy storage systems. *IEEE Trans. Power Electron.* 32 (1): 794–803.

83 Bo, Z., Di, W., MinYi, Z., Nong, Z., and Lin, H. (2019). Electric vehicle energy predictive optimal control by V2I communication. *Adv. Mech. Eng.* 11 (2): 1687814018821523.

84 Sewalkar, P., and Seitz, J. (2019). Vehicle-to-pedestrian communication for vulnerable road users: survey, design considerations, and challenges. *Sensors* 19 (2): 358.

85 Saber, H., Ehsan, M., Moeini-Aghtaie, M., Fotuhi-Firuzabad, M., and Lehtonen, M. (2020). Network-constrained transactive coordination for plug-in electric vehicles participation in real-time retail electricity markets. *IEEE Trans. Sustain. Energy* 12 (2): 1439–1448.

86 Qiu, T., Xu, B., Wang, Y., Dvorkin, Y., and Kirschen, D.S. (2017). Stochastic multistage coplanning of transmission expansion and energy storage. *IEEE Trans. Power Syst.* 32 (1): 643–651.

87 Su, W., Eichi, H., Zeng, W., and Chow, M.Y. (2012). A survey on the electrification of transportation in a smart grid environment. *IEEE Trans. Ind. Informatics* 8 (1): 1–10.

88 Habib, S., Kamran, M., and Rashid, U. (2015). Impact analysis of vehicle-to-grid technology and charging strategies of electric vehicles on distribution networks–a review. *J. Power Sources* 277: 205–214.

89 Yilmaz, M., and Krein, P.T. (2012). Review of the impact of vehicle-to-grid technologies on distribution systems and utility interfaces. *IEEE Trans. Power Electron.* 28 (12): 5673–5689.

90 Schiller, P.L. (2021). The future of road transport. *Int. Encycl. Transp.* 306–314.

91 Autocaat.org. (2021). Connected and Automated Vehicles. http://autocaat.org/Technologies/Automated_and_Connected_Vehicles.

92 Underwood, S. (2014). Automated, connected, and electric vehicle systems: expert forecast and roadmap for sustainable transportation. Technical report. Graham Inst. Sustain. Univ. Michigan, Ann Arbor.

93 Thorn, E., Kimmel, S.C., Chaka, M., and Hamilton, B.A. (2018). A framework for automated driving system testable cases and scenarios. United States. Department of Transportation. National Highway Traffic Safety.

94 Monteiro, V., Afonso, J.A., Sousa, T.J.C., Cardoso, L.L., Pinto, J.G., and Afonso, J.L. (2019). Vehicle electrification: technologies, challenges, and a global perspective for smart grids. *Innov. Energy Syst. - New Technol. Chang. Paradig.* Nov.

95 Gubbi, J., Buyya, R., Marusic, S., and Palaniswami, M. (2013). Internet of Things (IoT): A vision, architectural elements, and future directions. *Futur. Gener. Comput. Syst.* 29 (7): 1645–1660.

96 Al-Fuqaha, A., Guizani, M., Mohammadi, M., Aledhari, M., and Ayyash, M. (2015). Internet of things: A survey on enabling technologies, protocols, and applications. *IEEE Commun. Surv. Tutorials* 17 (4): 2347–2376.

97 Xu, L.D., He, W., and Li, S. (2014). Internet of things in industries: A survey. *IEEE Trans. Ind. Informatics* 10 (4): 2233–2243.

98 Khan, R., Khan, S.U., Zaheer, R., and Khan, S. (2012). Future Internet: The Internet of things architecture, possible applications and key challenges. In: *2012 10th international conference on frontiers of information technology*, 257–260.

99 Krčo, S., Pokrić, B., and Carrez, F. (2014). Designing IoT architecture (s): A European perspective. *2014 IEEE World Forum on Internet of Things (Wf-iot)*, 79–84.

100 Florea, B.C., and Taralunga, D.D. (2020). Blockchain IoT for smart electric vehicles battery management. *Sustainability* 12 (10): 3984.

101 Florea, B.C. (2018). Blockchain and Internet of Things data provider for smart applications. In: *2018 7th Mediterranean Conference on Embedded Computing (MECO)*, 1–4.

102 Nakamoto, S. (2008). Bitcoin: A peer-to-peer electronic cash system. *Decentralized Bus. Rev.* 21260.

103 Gordon, W.J., and Catalini, C. (2018). Blockchain technology for healthcare: Facilitating the transition to patient-driven interoperability. *Comput. Struct. Biotechnol. J.* 16: 224–230.

104 Zhang, P., Schmidt, D.C., White, J., and Lenz, G. (2018). Blockchain technology use cases in healthcare In: *Advances in Computers* (ed. P. Raj and G.C. Deka), Vol. 111, 1–41. Elsevier.

105 Siyal, A.A., Junejo, A.Z., Zawish, M., Ahmed, K., Khalil, A., and Soursou, G. (2019). Applications of blockchain technology in medicine and healthcare: Challenges and future perspectives. *Cryptography* 3 (1): 3.

106 Ozdemir, A.I., Ar, I.M., and Erol, I. (2020). Assessment of blockchain applications in travel and tourism industry. *Qual. Quant.* 54 (5): 1549–1563.

107 Rashideh, W. (2020). Blockchain technology framework: Current and future perspectives for the tourism industry. *Tour. Manag.* 80: 104125.

108 Filimonau, V., and Naumova, E. (2020). The blockchain technology and the scope of its application in hospitality operations. *Int. J. Hosp. Manag.* 87: 102383.

109 Esmat, A., De Vos, M., Ghiassi-Farrokhfal, Y., Palensky, P., and Epema, D. (2021). A novel decentralized platform for peer-to-peer energy trading market with blockchain technology. *Appl. Energy* 282: 116123.

110 Andoni, M., et al. (2019). Blockchain technology in the energy sector: A systematic review of challenges and opportunities. *Renew. Sustain. Energy Rev.* 100: 143–174.

111 Hamouda, M.R., Nassar, M.E., and Salama, M.M.A. (2020). A novel energy trading framework using adapted blockchain technology. *IEEE Trans. Smart Grid* 12 (3): 2165–2175.

112 Krishna, A.M., and Tyagi, A.K. (2020). Intrusion detection in intelligent transportation system and its applications using blockchain technology. In: *2020 International Conference on Emerging Trends in Information Technology and Engineering (ic-ETITE)*, 1–8.

113 Singh, M., and Kim, S. (2017). Blockchain based intelligent vehicle data sharing framework. *arXiv Prepr. arXiv1708.09721.*

114 Narbayeva, S., Bakibayev, T., Abeshev, K., Makarova, I., Shubenkova, K., and Pashkevich, A. (2020). Blockchain technology on the way of autonomous vehicles development. *Transp. Res. Procedia* 44: 168–175.

115 Kim, M., et al. (2019). A secure charging system for electric vehicles based on blockchain. *Sensors* 19 (13): 3028.

116 Son, B., Lee, J., and Jang, H. (2020). A scalable IoT protocol via an efficient DAG-based distributed ledger consensus. *Sustainability* 12 (4): 1529.

117 Li, Z., Kang, J., Yu, R., Ye, D., Deng, Q., and Zhang, Y. (2017). Consortium blockchain for secure energy trading in industrial Internet of things. *IEEE Trans. Ind. Informatics* 14 (8): 3690–3700.

118 Knirsch, F., Unterweger, A., and Engel, D. (2018). Privacy-preserving blockchain-based electric vehicle charging with dynamic tariff decisions. *Comput. Sci. Dev.* 33 (1): 71–79.

119 Huang, X., Xu, C., Wang, P., and Liu, H. (2018). LNSC: A security model for electric vehicle and charging pile management based on blockchain ecosystem. *IEEE Access* 6: 13565–13574.

120 Ochoa, J.J., Bere, G., Aenugu, I.R., Kim, T., and Choo, -K.-K.R. (2020). Blockchain-as-a-Service (BaaS) for Battery Energy Storage Systems. In: *2020 IEEE Texas Power and Energy Conference (TPEC)*, 1–6.

121 Javed, M.U., and Javaid, N. (2019). Scheduling charging of electric vehicles in a secured manner using blockchain technology. In: *2019 International Conference on Frontiers of Information Technology (FIT)*, 351–3515.

122 Sadiq, A., Javed, M.U., Khalid, R., Almogren, A., Shafiq, M., and Javaid, N. (2020). Blockchain based data and energy trading in iIternet of electric vehicles. *IEEE Access* 9: 7000–7020.

123 Dubois, A., Wehenkel, A., Fonteneau, R., Olivier, F., and Ernst, D. (2017). An app-based algorithmic approach for harvesting local and renewable energy using electric vehicles. In: *Proceedings of the 9th International Conference on Agents and Artificial Intelligence (ICAART 2017)*.

124 Olariu, S., Hristov, T., and Yan, G. (2013). The next paradigm shift: from vehicular networks to vehicular clouds. In: *Mobile Ad Hoc Networking: Cutting Edge Directions*, 2nd ed. (ed. S. Basagni, M. Conti, S. Giordano, and I. Stojmenovic), 645–700. Wiley.

125 Mell, P., and Grance, T. (2017). The National Institute of Standards and Technology [NIST] definition of cloud computing. *NIST Spec. Publ.* 800 (145): 1–4.

126 Chauhan, R., and Kumar, A. (2013). Cloud computing for improved healthcare: Techniques, potential and challenges. In: *2013 E-Health and Bioengineering Conference (EHB)*, 1–4.

127 Ali, S., et al. (2020). Towards pattern-based change verification framework for cloud-enabled healthcare component-based. *IEEE Access* 8: 148007–148020.

128 Abbasi, M., Rafiee, M., Khosravi, M.R., Jolfaei, A., Menon, V.G., and Koushyar, J.M. (2020). An efficient parallel genetic algorithm solution for vehicle routing problem in cloud implementation of the intelligent transportation systems. *J. Cloud Comput.* 9 (1): 1–14.

129 Abbasi, M., Yaghoobikia, M., Rafiee, M., Jolfaei, A., and Khosravi, M.R. (2020). Energy-efficient workload allocation in fog-cloud based services of intelligent transportation systems using a learning classifier system. *IET Intell. Transp. Syst.* 14 (11): 1484–1490.

130 Song, Y., Chen, Y., Yu, Z., Huang, S., and Shen, C. (2020). CloudPSS: a high-performance power system simulator based on cloud computing. *Energy Reports* 6: 1611–1618.

131 Talaat, M., Alsayyari, A.S., Alblawi, A., and Hatata, A.Y. (2020). Hybrid-cloud-based data processing for power system monitoring in smart grids. *Sustain. Cities Soc.* 55: 102049.

132 Hashmi, S.A., Ali, C.F., and Zafar, S. (2021). Internet of things and cloud computing-based energy management system for demand side management in smart grid. *Int. J. Energy Res.* 45 (1): 1007–1022.

133 Aruna, D.N. (2020). Cloud computing technology–A digital transformative solution for banking industry-an overview. *Eur. J. Mol. Clin. Med.* 7 (2): 5305–5310.

134 Tiwari, S., Bharadwaj, S., and Joshi, S. (2021). A study of impact of cloud computing and artificial intelligence on banking services, profitability and operational benefits. *Turkish J. Comput. Math. Educ.* 12 (6): 1617–1627.

135 Hajipour, V., Niaki, S.T.A., and Rahbarjou, M. (2021). An optimisation model for cloud-based supply chain network design: Case study in the banking industry. *Int. J. Commun. Networks Distrib. Syst.* 27 (2): 119–146.

136 Foley, J. (2008). Private clouds take shape. *InformationWeek* 1104.

137 Whaiduzzaman, M., Sookhak, M., Gani, A., and Buyya, R. (2014). A survey on vehicular cloud computing. *J. Netw. Comput. Appl.* 40: 325–344.

138 Hussain, M.M., Beg, M.M.S., Alam, M.S., Krishnamurthy, M., and Ali, Q.M. (2018). Computing platforms for big data analytics in electric vehicle infrastructures. In: *2018 4th International Conference on Big Data Computing and Communications (BIGCOM)*, 138–143.

139 Hou, J., and Song, Z. (2020). A hierarchical energy management strategy for hybrid energy storage via vehicle-to-cloud connectivity. *Appl. Energy* 257: 113900.

140 Ziani, A., Sadouq, Z.A., and Medouri, A. (2019). Use of cloud computing and GIS on vehicle traffic management. *Int. J. Intell. Enterp.* 6 (2–4): 382–392.

141 Wang, Y., Wang, X., Li, H., Dong, Y., Liu, Q., and Shi, X. (2020). A multi-service differentiation traffic management strategy in SDN cloud data center. *Comput. Networks* 171: 107143.

142 Ahmad, I., Noor, R.M., Ali, I., Imran, M., and Vasilakos, A. (2017). Characterizing the role of vehicular cloud computing in road traffic management. *Int. J. Distrib. Sens. Networks* 13 (5): 1550147717708728.

143 Kaur, N., Singh, J., Goyal, S., and Duhan, B. (2020). Load balancing in cloud computing: the online traffic management. *J. Nat. Remedies* 21 (2): 202–209.

144 Zhang, Y., Liu, H., Zhang, Z., Luo, Y., Guo, Q., and Liao, S. (2020). Cloud computing-based real-time global optimization of battery aging and energy consumption for plug-in hybrid electric vehicles. *J. Power Sources* 479: 229069.

145 Kong, C., Rimal, B.P., Reisslein, M., Maier, M., Bayram, I.S., and Devetsikiotis, M. (2020). Cloud-based charging management of heterogeneous electric vehicles in a network of charging stations: Price incentive vs. capacity expansion. *IEEE Trans. Serv. Comput.*

146 Shao, H., Zhang, G., and Xia, M. (2020). An intelligent charging navigation scheme for electric vehicles using a cloud computing platform. *Energy Sources, Part A Recover. Util. Environ. Eff.* 1–17.

147 Guerrero-Ibanez, J.A., Zeadally, S., and Contreras-Castillo, J. (2015). Integration challenges of intelligent transportation systems with connected vehicle, cloud computing, and Internet of things technologies. *IEEE Wirel. Commun.* 22 (6): 122–128.

148 Abduljabbar, R., Dia, H., Liyanage, S., and Bagloee, S.A. (2019). Applications of artificial intelligence in transport: An overview. *Sustainability* 11 (1): 189.

149 Machin, M., Sanguesa, J.A., Garrido, P., and Martinez, F.J. (2018). On the use of artificial intelligence techniques in intelligent transportation systems. In: *2018 IEEE Wireless Communications and Networking Conference Workshops (WCNCW)*, 332–337.

150 Iyer, L.S. (2021). AI enabled applications towards intelligent transportation. *Transp. Eng.* 5: 100083.

151 Hajibandeh, N., Faghihi, F., Ranjbar, H., and Kazari, H. (2017). Classifications of disturbances using wavelet transform and support vector machine. *Turkish J. Electr. Eng. Comput. Sci.* 25 (2): 832–843.

152 Ranjbar, H., Behrouz, M., and Deihimi, A. (2015). Neural network based global maximum power point tracking under partially shaded conditions. In *ICEE 2015 - Proceedings of the 23rd Iranian Conference on Electrical Engineering*, vol. 10.

153 Yousuf, H., Zainal, A.Y., Alshurideh, M., and Salloum, S.A. (2021). Artificial intelligence models in power system analysis. In: *Artificial Intelligence for Sustainable Development: Theory, Practice and Future Applications* (ed. A.E. Hassanien, R. Bhatnagar, and A. Darwish), 231–242. Springer.

154 Lee, M. (2020). An analysis of the effects of artificial intelligence on electric vehicle technology innovation using patent data. *World Pat. Inf.* 63: 102002.

155 Lee, H., Kang, C., Park, Y.-I., Kim, N., and Cha, S.W. (2020). Online data-driven energy management of a hybrid electric vehicle using model-based Q-learning. *IEEE Access* 8: 84444–84454.

156 Boukerche, A., Tao, Y., and Sun, P. (2020). Artificial intelligence-based vehicular traffic flow prediction methods for supporting intelligent transportation systems. *Comput. Networks* 182: 107484.

157 Fatemidokht, H., Rafsanjani, M.K., Gupta, B.B., and Hsu, C.-H. (2021). Efficient and secure routing protocol based on artificial intelligence algorithms with UAV-assisted for vehicular ad hoc networks in intelligent transportation systems. *IEEE Trans. Intell. Transp. Syst.* 22 (7): 4757–4769.

158 Yu, K.-H., Beam, A.L., and Kohane, I.S. (2018). Artificial intelligence in healthcare. *Nat. Biomed. Eng.* 2 (10): 719–731.

159 Jiang, F., et al. (2017). Artificial intelligence in healthcare: past, present and future. *Stroke Vasc. Neurol.* 2 (4).

160 Davenport, T., and Kalakota, R. (2019). The potential for artificial intelligence in healthcare. *Futur. Healthc. J.* 6 (2): 94.

161 Chitra, A., Holm-Nielsen, J.B., Sanjeevikumar, P., and Himavathi, S. (2020). *Artificial Intelligent: Techniques for Electric and Hybrid Electric Vehicles*. Wiley Online Library.

162 Volkswagenag.com. (2021). Charging robots: Revolution in the Underground Parking Garage. https://www.volkswagenag.com/en/news/stories/2019/12/volkswagen-lets-its-charging-robots-loose.html#.

163 Thuraisingham, B., (2020). Cyber security and artificial intelligence for cloud-based Internet of transportation systems. in *2020 7th IEEE International Conference on Cyber Security and Cloud Computing (CSCloud)/2020 6th IEEE International Conference on Edge Computing and Scalable Cloud (EdgeCom)*, 8–10.

164 Huq, N., Vosseler, R., and Swimmer, M. (2017). Cyberattacks against intelligent transportation systems. *TrendLabs Res. Pap.*

165 Kelarestaghi, K.B., Heaslip, K., Khalilikhah, M., Fuentes, A., and Fessmann, V. (2018). Intelligent transportation system security: hacked message signs. *SAE Int. J. Transp. Cybersecurity Priv.* 1 (11-01-02–0004): 75–90.

166 Sumra, I.A., Ahmad, I., and Hasbullah, H. (2011). Behavior of attacker and some new possible attacks in vehicular ad hoc network (VANET). In: *2011 3rd* *international congress on ultra modern telecommunications and control systems and workshops (ICUMT)*, 1–8.

167 Islam, M., Chowdhury, M., Li, H., and Hu, H. (2018). Cybersecurity attacks in vehicle-to-infrastructure applications and their prevention. *Transp. Res. Rec.* 2672 (19): 66–78.

168 Mejri, M.N., Ben-Othman, J., and Hamdi, M. (2014). Survey on VANET security challenges and possible cryptographic solutions. *Veh. Commun.* 1 (2): 53–66.

169 Raw, R.S., Kumar, M., and Singh, N. (2013). Security challenges, issues and their solutions for VANET. *Int. J. Netw. Secur. Its Appl.* 5 (5): 95.

170 Alnasser, A., Sun, H., and Jiang, J. (2019). Cyber security challenges and solutions for V2X communications: A survey. *Comput. Networks* 151: 52–67.

171 Lu, Y., and Da Xu, L. (2018). Internet of Things (IoT) cybersecurity research: a review of current research topics. *IEEE Internet Things J.* 6 (2): 2103–2115.

172 Acharya, S., Dvorkin, Y., Pandžić, H., and Karri, R. (2020). Cybersecurity of smart electric vehicle charging: a power grid perspective. *IEEE Access* 8: 214434–214453.

11

Computer-Aided Molecular Design

Accelerating the Commercialization Cycle

Tohid N. Borhani

Center for Engineering Innovation and Research, School of Engineering, Computing and Mathematical Sciences, University of Wolverhampton, UK

11.1 Introduction

The term Industry 4.0 is applied to express the Fourth Industrial Revolution in which the concept of connectivity will be considered in the production and manufacturing environment. It was predicted that in Industry 4.0 different machines in the smart factories can be connected, transfer information, and collaborate with each other through a global network. Hence, in the future, the smart factory can be operated automatically that the machine will recognize the manufacturing process of production of a chemical product and finally the product will be produced with tailor-made features [1].

In the process of production of unique and optimized chemical products, computer-aided molecular design (CAMD) can play a very important role. CAMD is the method to design and select the best candidates of molecules or components by considering the process conditions and constraints of the applications [2]. Therefore, this book chapter will address the concept of CAMD and also its application in the production of different chemicals. In addition to reviewing the recent progress and status of CAMD, this study will emphasize the importance of CAMD in Industry 4.0 and the chemical industry in the future.

Before arriving in CAMD with detail it will be useful to talk about the role of Industry 4.0 in chemical industries. The process of defining new and desirable chemicals for a specific application can be called *chemical product design* [3]. Products and chemicals can be categorized into three main types, namely, commodity or bulk products, fine products, and functional or specialty products. Commodity or bulk products are produced in large volume, and they are sold based on composition and purity, which determine their price. Some examples of this type are sulfuric acid, nitric acid, ethylene, and chloride. Fine chemicals are produced in smaller volumes than commodity products. Their production sale is based on composition and price. Some examples of this type are chloropropylene oxide, irnoxy resin, and dimethylformamide. Functional chemicals are manufactured for a particular or specific application. Here, the special functions of these chemicals determine their price, rather than their composition and purity. Some examples of this type are perfumes, flavorings, pharmaceutical products and simple chemicals such as mosquito repellents and domestic cleaners [4, 5]. It should be noted that these diverse classes of chemical products are different in many

Industry 4.0 Vision for the Supply of Energy and Materials: Enabling Technologies and Emerging Applications, First Edition. Edited by Mahdi Sharifzadeh.

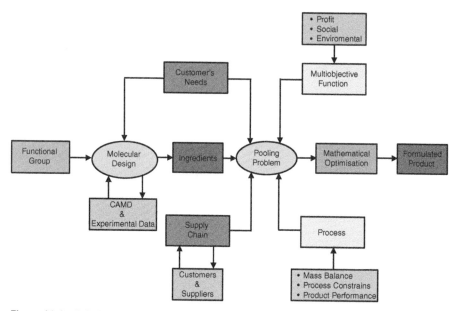

Figure 11.1 Relationship between Molecular Design, Its Parameters, and SCM [9].

aspects such as key steps of design, selling points and possible risk encountered during the product design, but the procedures of design for these chemical products are identical and similar [2].

Due to global competition, innovative products and processes should be developed fast [6]. Therefore, using a systematic method to search for new desirable molecules is necessary as there are numerous molecules (around 10^{200} organic molecules) in the chemical design space [7]. To meet this necessity, various methods utilizing CAMD have been developed.

It must be mentioned that the supply chain management (SCM) also has a very important role in product design and formulation [8]. There are different types of SCMs, namely, centralized, decentralized, cooperative, non-cooperative, global, or national. This will have an impact on the pooling problem either directly or indirectly, so it will be required to identify what form of the SCM will be considered. The relationship between molecular design, process, and SCM is illustrated in Figure 11.1.

11.2 Computer-Aided Molecular Design

CAMD allows for the systematic investigation of chemical space, which opens up a world of possibilities for efficient chemical product design [10]. CAMD is a reverse engineering method that creates molecules that meet a set of prespecified physical property parameters [11–13]. The schematic diagram of a CAMD problem is illustrated in Figure 11.2. As can be seen, a CAMD problem contains an exploration of chemical space, property estimation methods, and process optimization [10]. Therefore, molecular modelling methods, thermodynamics, and optimization techniques are combined in the CAMD to design useful or optimum molecular structures, many of them entirely novel.

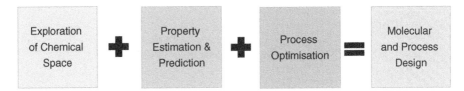

Figure 11.2 The Schematic Diagram of the CAMD Problem. Based on Gertig C et al., 2020.

11.3 Different Classes of CAMD

Papadopoulos et al. [13] provided a classification for CAMD problems. The authors mentioned that different objectives can be considered to categories molecular design problems: (1) a single or multiple performance criteria; (2) the determination of only optimum molecular structures; (3) integration of the process structural and operating decisions as well; (4) design of single or multiple components in the form of mixtures or blends; (5) accounting for the uncertainty in the employed models, internal and external process or application conditions; (6) consideration of inert or reactive systems; and (7) including models that directly account for matter behavior at the atomic or molecular level and can all be formulated through the optimization objective function and constraints. Based on all these variations, different types of CAMD problems are developed in the literature such as single-objective optimization (SOO) CAMD; computer-aided molecular and process design (CAMPD); multi-objective optimization (MOO) CAMPD [CAM(P)D]; computer-aided mixture design [CAM^b(P)D]; computer-aided molecular design under uncertainty [CAM^u(P)D]; computer-aided molecular, process, and control design [CAMPCD]; and computer-aided molecular design for reactive systems [CAM^R(P)D].

For instance, the single solvent selection and design problem can be formulated as follows:

$$\min_{n} C(n,p)$$

$$p = f(n) \quad \text{Solvent predictive model}$$
$$h_1(p,n) \leq 0$$
$$h_2(p,n) = 0$$
$$s_1(n) \leq 0$$
$$s_2(n) = 0$$
$$p_{min} \leq p \leq p_{max}$$
$$n_{min} \leq n \leq n_{max} \tag{11.1}$$

where C is objective or cost function which depends on two sets of variables, namely, vector n and vector p. Vector n demonstrates relevant structural information of the designed molecules and vector p is the vector of properties. The constraints h_1 and h_2 are related to property values and desired structural features. s_1 and s_2 ensure structural feasibility. A similar formulation can be considered for mixture selection and design:

$$\min_{n,x} C(n,p,q)$$

$q = f(x,n,p)$ Solvent, property, and mole fraction relation model

$$\sum_i x_i = 1$$

$$h_1(p,q,n) \leq 0$$

$$h_2(p,q,n) = 0$$

$$s_1(n) \leq 0$$

$$s_2(n) = 0$$

$$p_{min} \leq p \leq p_{max}$$

$$n_{min} \leq n \leq n_{max}$$

$$q_n \leq q \leq q_{max} \tag{11.2}$$

where here C is a function of n, p, and q, which q is the mixture property variables.

11.4 Implementation of a CAMD Problem

In [5] five hierarchical steps are considered to implement the CAMD problems: (1) problem formulation; (2) method and constraint selection; (3) solution of CAMD problem; (4) result analysis and verification; and (5) final selection. There are several tools developed to implement the CAMD models. SolventPro is a toolbox in ICAS [14] and can be used to design solvents for different applications [15]; AMODEO [16] can be used to design solvents in a three-stage process. The first stage is based on group contribution models and is composition design, the second step is structure determination, and the final stage is based on problem-specific models to recognize molecules [17]. ProCAPD [18] is another software tool that gained wide applications [9], with a built-in molecular-mixture design toolbox called OptCAMD [19]. The techniques for solving the molecular design problems can be classified as the following subsections.

11.4.1 Generate and Test Methods

Gani and Brignole [20] proposed this method for the first time. To assist the synthesis of feasible molecular structures, the "generate" process includes three primary stages: group selection, group characterization, and molecular feasibility rules. On the other hand, the group contribution methods for estimation of properties, calculated properties, property constraints, and evaluation are the primary elements of the "test" step.

11.4.2 Decomposition Methods

These methods include the formulation and consecutive solution of smaller subproblems, as well as the subsequent transmission of data, with the goal of facilitating the identification of an optimal solution. The subproblems might be solved at the same time or in successive parts inside the same software implementation. The decomposition method is applicable to different types of CAMD problems such as single-molecule design, mixture molecule design, and integration of process and product design [21].

11.4.3 Optimization Methods (Deterministic, Stochastic, and Heuristic)

In various circumstances, optimization techniques can be used to solve CAMD problems directly. With a minor change to the formulation or by utilizing the problem structure, optimization procedures become viable in additional cases. In molecular, mixture, or process models, optimization procedures are best suited to problems with a large number of alternative descriptors that often suffer from hard non-linearities or non-convexities.

11.5 Application of Predictive Models in CAMD

The essential ingredients for molecular design require (1) generation of a stable molecule structure; (2) generation of chemically feasible molecules; and (3) estimation of the required properties. The third part means the estimation of the required properties can be done using different methods such as quantitative structure–property relationships (QSPRs) and quantitative structure–activity relationships (QSARs) [22], quantum mechanical based models [10], group contribution models, and molecular dynamics.

Property estimation models or predictive models are one of the most important constraints in the CAMD optimization models, as they relate the molecular structure to the property, which is known as structure–property relationships [23]. Different theoretical/mechanics (such as quantum mechanical models and molecular simulation), semi-empirical (e.g., group contribution methods [GCM], UNIversal Functional-group Activity Coefficients [UNIFAC] model), and empirical models (such as QSPR models and chemometrics) can relate the structure of molecules to their properties. The accuracy of the property estimation methods has a significant effect on the components/molecules designed by CAMD.

Quantum mechanical models, molecular dynamics, and quantitative structure–property relationships (QSPRs) can be incorporated into CAMD formulation to design and select desirable molecules for a special application. It should be noted that in some studies, the authors considered the term QSPR as a general term to describe GCMs [21]. However, in this study QSPR and GCM models are classified separately.

11.5.1 Quantitative Structure–Property Relationships

Between predictive models that can be applied in the field of CAMD, QSPRs have widespread applications [21]. The basics of the QSPR models include finding a relationship/function between a target value (a property of chemical molecules) and the molecular structure (descriptors):

$$\text{Property} = f\left(\text{descriptor}\right) \tag{11.3}$$

Different types of descriptors can be used in QSPR modeling, such as quantum mechanical based descriptors, topological descriptors, and autocorrelation descriptors [24]. It should be noted that descriptors can be selected in the hybrid format, namely, the combination of all types of descriptors and also the application of experimental properties such as boiling point, critical points, dipole moments, and other properties of components [25]. The molecular descriptors of any chemical structure can be calculated using a variety of software tools. Todeschini and Consonni presented a review of various software packages [26]. They also discussed different types of descriptors.

Different modeling (regression) approaches are applicable in QSPR and QSAR studies, linear techniques such as multivariate linear regression, partial least squares (PLS) regression, and principal component regression [25] and nonlinear machine learning techniques such as artificial neural networks, genetic programming, support vector machines, and adaptive neuro-fuzzy inference systems [22, 27]. In QSPR studies, especially when dealing with the multivariate linear regression method, different types of algorithms such as classical algorithms (e.g., stepwise forward selection) and evolutionary or metaheuristic algorithms (e.g., genetic algorithm particle swarm optimization, simulated annealing, and ant colony) have been used in descriptor selection step to reduce the number of descriptors and keep the most influential ones in the prediction of property under study [28]. Different types of properties can be estimated using this method [29]. The developed models can be used in other levels of modeling jobs. The general flowchart of QSAR and QSPR methods is illustrated in Figure 11.3.

11.5.2 Group Contribution Methods

The underlying premise behind this semi-empirical technique is that a molecule's thermophysical behavior is defined by its intermolecular forces, which are dictated by the molecular structure. As a result, the properties of a molecule are determined as the sum of contributions from different atomic groups and bond types.

This method assumes that a molecule's properties may be anticipated by assuming multiple occurrences of various molecular sub-structures known as *groups*. For instance, the n-butanol molecule can be presented as a combination of the groups $-CH_3$, $-CH_2-$, and $-OH$, which (−) signify bonds to other groups. n-Butanol would no longer be thought of as the linked alcohol molecule but rather as a collection of its constituent groups in its group representation. Figure 11.4 shows the n-butanol group representation.

The group representation of a molecular structure is converted into an estimate for properties using group contribution methods. Group contribution approaches do this by defining a vector n that indicates the number of occurrences of each of the groups. By assuming that there are only three groups similar to the case of n-butanol, a vector can be considered for n-butanol as n = [1, 3, 1], where the digits in this array signify the number of occurrences of the groups $-CH_3$, $-CH_2-$, and $-OH$, respectively. Each of these groups (g) would also be accompanied by a coefficient (c_g), which counts its effect or "contribution" to a property. Therefore, the properties are estimated as follows:

$$\text{Property} = \sum_g c_g n_g \tag{11.4}$$

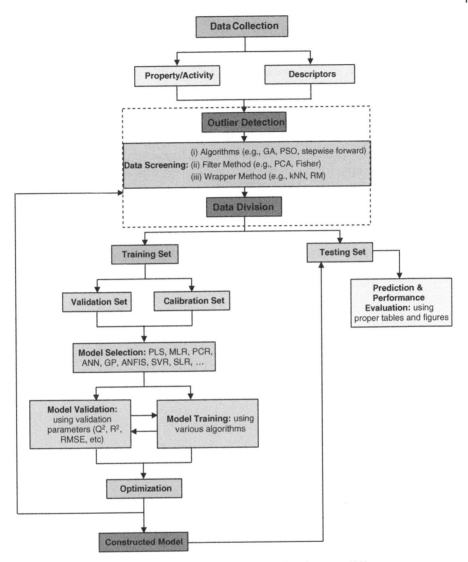

Figure 11.3 A Flowchart of the QSPR and QSAR Model Development [30].

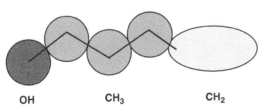

Figure 11.4 Group Representation of n-Butanol.

The coefficients vector c is derived through regression of the property over a large data set of various compounds. Identification of all the groups must be established a priori to regress these parameters. In the example of butanol illustrated in Figure 11.4, multiple sets of groups might readily be applied to describe butanol. For example, the groups $-CH_3$, $-CH_2-$, and $-CH_2OH$ also completely account for the atoms in butanol, and these may provide a better fit for the regression problem, with respect to the experimental data. As a result, different approaches for estimating different attributes using group contribution do not always use the same collection of groups. It must be mentioned that, because many group contribution methods include 50–100 groups, the vector n is usually much larger. Many group contribution methods also include the assumption that groups cannot overlap, implying that n is usually a sparse vector. Figure 11.5 depicts a visual example of the use of group contribution methods.

In this example, a hypothetical GCM is used for a hypothetical molecule that illustrates how a molecular structure is disintegrated into its fundamental groups and provides a count of each of these groups. The elements of the vector n are made up of these counts. An example property is calculated by pairing the n vector with a hypothetical c vector. Great discussion about GCM, their history, and different types of GC methods are provided by Austin et al. [21].

According to Figure 11.5, the following equations can be written:

$$n = \begin{bmatrix} 0, \ldots, 1, \ldots, 2, \ldots, 4, \ldots, 1, \ldots, 1, \ldots \end{bmatrix} \tag{11.5}$$

$$c = \begin{bmatrix} \ldots, 3.2, \ldots, -2.4, \ldots, 0.6, \ldots, 1.2, \ldots, 2.3, \ldots \end{bmatrix} \tag{11.6}$$

$$P = \sum_g c_g n_g = 3.2(1) - 2.4(2) + 0.6(4) + 1.2(1) + 2.3(1) \tag{11.7}$$

where n is group composition vector and c is coefficient vector.

GCM are beneficial since they are simple to utilize. They evaluate and analyze chemical structures in terms of their functional components, in a similar way as chemists do. Because the groups can be joined in a variety of ways to form a vast range of distinct structures, GCM can easily represent a huge and diversified chemical space.

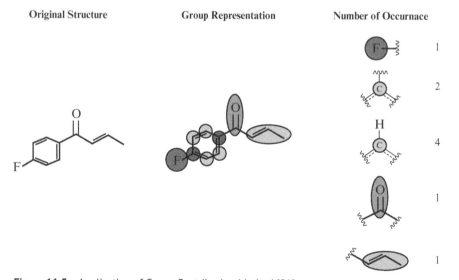

Original Structure	Group Representation	Number of Occurnace

Figure 11.5 Application of Group Contribution Method [21].

This is particularly useful from the perspective of the CAMD. Finally, because the inclusion and count of groups (the vector n) are simply expressed in the framework of mathematical optimization, GCM often result in large-scale nonlinear mathematical formulations for various CAMD problems.

There are a few limitations with the current GCM. One issue is that many of them have trouble distinguishing between various isomers with the same closed formulation. This represents a gap in the prediction capability of GCM since isomers might have highly diverse characteristics. Due to the inclusion of big groups, some GCM, such as GCM +, can differentiate between multiple isomers. The lack of consistency in the groups used to anticipate certain properties is a second issue with GCM. Though this does not influence estimating these properties for a given structure, it makes mathematical representations of the CAMD problem more difficult. Last, before regressing the GCM coefficients, the methods necessitate specifying the set of groups. Despite the fact that many GCM are extremely precise, there is no guarantee that the selection of groups utilized best reflects the property being modeled. The prediction power of a GCM model can be substantially altered by using different groups.

It should be noted that statistical associating fluid theory (SAFT) and UNIFAC are two important and powerful models that can be used in CAMD studies. These two predictive models are also categorized as group contribution methods is some references. They have been applied in many studies to predict different properties, and hence they have a high potential to be used in CAMD problems [31].

11.5.3 Quantum Mechanical Models

Quantum chemistry or quantum mechanics (QM) is based on Schrödinger's equation, also known as the wave function, which theoretically solves to a probability of finding a particle in a point in space [32]:

$$\mathcal{H}\,\Psi_i = E_i \Psi_i \tag{11.8}$$

where $\mathcal{H}$ is Hamiltonian operator, Ψ_i is the wave function corresponding to the ith electronic state and E_i is the energy of the ith electronic state. To calculate energy and other related properties of compounds, this equation must be solved. The exact solution of the equation is possible only for helium atom or for H_2^+ molecule and for other molecules, the approximate methods must be used. In general, there are three approximate methods to solve Schrödinger's equation: (1) semi-empirical; (2) ab initio (first principle); and (3) density functional theory. These types of calculations can be done using commercial programs such as Gaussian or using free programs, such as MOPAC (http://openmopac.net/manual) and GAMESS (http://www.msg.ameslab.gov/gamess).

QM-based predictive models showed considerable applications in recent years. The COSMO (COnductor-like Screening MOdel) solvation model [33] is a variant of continuum solvation models used for QM calculations. These models allow for the calculation of the required solvation properties considering the liquid environment of a molecule as a continuum. Two important continuum solvation–based models applied in CAMD problems are COSMO-RS and COSMO-SAC [10]. COSMO-RS (COnductor-like Screening MOdel for Real Solvents) [34] and COSMO-SAC (COnductor-like

Screening MOdel for Segment Activity Coefficient) [35], based on COSMO solvation, are two basic thermodynamic property estimation models. The solvent molecular orbital continuum model underpins COSMO-RS. COSMO-SAC is a model based on the COSMO-RS framework that has been proposed and regularly refined in recent years to increase the accuracy of activity coefficient estimates. The COSMO-RS/SAC models can predict the activity coefficients of novel molecules without interaction parameters and become popular in various fields in chemical engineering. Meanwhile, CAMD integrated COSMO-RS/SAC has been developed continuously. Although QM-based models are reliable and accurate and have shown good performance in combination with CAMD problems, they are computationally expensive.

11.6   Different Applications of CAMD

Since the emergence of high-performance computing in the last 20 years, many academic and industrial studies are devoted to the application of CAMD. Some important applications of CAMD that can improve the effectiveness of the SCM have been reviewed in the following subsections. It should be noted that CAMD application for pharmaceutical products is known as computer-aided drug design, which is out of the scope of the current chapter.

11.6.1   Fuel- and Oil-Related Applications

Zhang et al. [36] presented a computer-aided methodology integrated with experimental verification. They first used model-based computer-aided methods to design mixtures and blends and then examined the properties of the most promising product candidates using experiments or rigorous models. The authors mentioned that the starting point in their method is to analyze the product requirements and interpret them into target property constraints. Using a well-known computer-aided design molecular design approach, a list of molecules that serve as ingredient-chemicals to add to the blended product, as well as their pure compound qualities, was developed. To select the ingredient-chemicals and their compositions in the blended product, a mixed-integer nonlinear programming (MINLP) model was developed. The authors discussed the MINLP model's solution approaches. Phase equilibrium properties (e.g., liquid solution activity coefficients) were estimated using the UNIFAC model. The optimization results were verified using experimental data and rigorous models. The authors also highlighted the applications for tailor-made surrogate fuel designs of a gasoline blend and a jet-fuel blend.

Mah et al. [37] studied the addition of a solvent to bio-oil to improve its properties. The criteria to select the solvent was simplicity and low processing cost. The immiscibility of the final blend due to the difference in polarity of the solvent and bio-oil molecules is one of the primary issues in solvent design. To this end, a CAMD framework was developed to find viable solvent candidates that would allow bio-oil to meet desired properties with little solvent addition. Moreover, a model for phase stability analysis was developed based on Gibbs' tangent plane distance, which covers the miscibility check of the solvent-oil blend, generation of the phase diagram, and addition of binding agents to homogenize the blend.

11.6.2 Solvent Design and Selection for Separation

Solvent selection and design have wide applications in chemical studies. One of the main applications of CAMD is the absorption of gases from gas streams and separation technologies [38]. Ahmed et al. [39] presented a CAMD model to design new alternative solvents as replacements for commonly used solvents during the post-combustion carbon capture process in power plants. The authors showed that by using the proposed solvent, significant savings on regeneration energy (up to 31.4%) could be achieved compared with the conventional solvent, monoethanolamine. The authors considered five steps in their study: (1) problem formulation; (2) generation of the single reactive solvent candidate using ProCAMD tools in ICAS software; (3) prediction of the reaction mechanism between the potential amine solvent candidate and CO_2 using the zwitterion mechanism and base-catalyzed hydration mechanism; (4) evaluation of process performance by calculating the heat required for the solvent regeneration process; and (5) the selection of the best solvent candidate based on the desired process performance. Fang et al. [40] used the COSMO-SAC model to screen the promising ionic liquids for extractive distillation. The authors considered selectivity and solubility as the selection index. They also performed vapor–liquid equilibrium on the selected candidates and showed that the experimental and design results were consistent with each other. Papadopoulos et al. [41] proposed a systematic approach for simultaneous CAMD and sustainability assessment that deals efficiently with data gaps. The developed framework supports the life cycle assessment and environmental health and safety impacts from cradle-to-gate chemicals designed through CAMD. They investigated the effects of using sustainability objective functions during CAMD. They examined their framework on the design of phase-change solvents for chemisorption-based post-combustion CO_2 capture.

11.6.3 Design of Heat Transfer Fluids

The design of fluids for heat transfer applications includes refrigeration cycles, heat pumps, and organic Rankine cycle (ORC) [13]. Van Kleef et al. [42] developed a framework based on CAMD for designing working fluid (heat-transfer medium) for optimal ORC power systems that simultaneously considers both thermodynamic and economic objectives. The requisite thermodynamic parameters of the proposed working fluids are determined using the SAFT-Mie equation of state, with critical and transport properties approximated using empirical group-contribution methods. During a set of MINLP optimizations of the ORC systems with heat source temperatures of 150°C and 250°C, it was found that 1,3-butadiene and 4-methyl-2-pentene are the best performing working fluids, respectively, with SICs of 9640 per kW and 4000 per kW. Tillmanns et al. [43] developed continuous-molecular targeting computer-aided molecular design (CoMT-CAMD) framework. They used a physically based perturbed-chain SAFT (PC-SAFT) as a thermodynamic predictive model. They also integrated dynamic models for the ORC equipment into the process model to capture the ORC behavior under dynamic conditions. They reported an optimal control problem, which resulted in the design of optimal working fluid and the corresponding optimal process control for a given dynamic input. The authors showed that the dynamic CoMT-CAMD method can be used in an ORC for waste-heat recovery on a heavy-duty vehicle.

11.6.4 Reaction and Reactor Applications

Due to the diversity of materials and chemical reactions, the selection of solvents for reactions is a critical point. CAMD showed interesting application in reactions. Gertig et al. [44] presented a COSMO-CAMD model to design and select solvent to enhance the reaction kinetics of a Menschutkin reaction and a chain propagation reaction for the production of polymers and microgels. The solvent candidates obtained by this method showed substantial enhancement of reaction rates. The authors have mentioned that the COSMO part of the model does not require experimental data to predict the solvent performance because the prediction of reaction kinetics are based on transition-state theory and advanced quantum chemical methods [44].

11.6.5 Molecular Catalysts

Catalyst design is another field of study that utilizes the CAMD methodology and has attracted many attentions in recent years. Chang et al. [45] introduced an inverse design method to guide the design of catalysts based on a tight-binding-linear combination of atomic potential and model Hamiltonian. The approach necessitates the use of a reference catalyst with well-defined reaction intermediates as well as the catalytic process rate-limiting phase. In the space of alchemical structures defined by synthetically viable alterations in the reference complex, molecular candidates with improved performance were sought. The method was illustrated as applied to the search of a NiII catalyst for CO/CO_2 conversion using a biomimetic Ni^{II} _iminothiolate complex as the reference catalyst. Dittner and Hartke [46] presented a framework known as the globally optimal catalyst framework to select the best catalyst of chemical reactions. They used genetic algorithm and computational chemistry in their framework. They applied their framework to the prototype Menshutkin SN_2 reaction, using the semiempirical PM7 level. Gertig et al. [47] suggested a method for designing catalysts and processes using CAMPD. The essential part, according to the authors, is using modern quantum chemistry approaches to forecast reaction dynamics efficiently and accurately. As the demonstrating case, they presented the design method for catalytic carbamate cleavage, demonstrating that only an integrated catalyst and process design can identify catalysts that maximize the process' performance.

11.7 Conclusion

In emerging Industry 4.0 technologies, it is very crucial to automate the cycle of product design, testing, and commercialization. CAMD plays an enabler that offers opportunities to accelerate the product design and reduces the costs associated with physical search, screening, and testing. The present chapter demonstrates the application of CAMD for a number of applications including the systematic design of fuels, solvents, pharmaceuticals, heat transfer medium, and catalysts. Therefore, it can be said that CAMD will be useful method in the near future of SCM. The accuracy and speed of CMAD is not comparable with trial-and-error experimental methods. By this method, millions of molecules can be investigated, and the best candidates can be selected and designed. This will have a huge impact of the cost and progress of the product and process design. In aspect of predictive methods that must be used in CAMD problems, it can be said that QM-based methods are the most accurate ones but computationally expensive. On the other hand, QSPR and GCM are faster but have less accuracy compared with QM-based methods.

References

1 Tjahjono, B., Esplugues, C., Ares, E., and Pelaez, G. (2017). What does Industry 4.0 mean to supply chain? *Procedia. Manuf.* 13: 1175–1182.

2 Ng, L.Y., Chong, F.K., and Chemmangattuvalappil, N.G. (2015). Challenges and opportunities in computer-aided molecular design. *Comput. Chem. Eng.* 81: 115–129.

3 Gani, R. (2004). Chemical product design: challenges and opportunities. *Comput. Chem. Eng.* 28: 2441–2457.

4 Smith, R. (2005). *Chemical Process: Design and Integration*. John Wiley & Sons.

5 Seider, W.D., Seader, J.D., Lewin, D.R., and Widagdo, S. (2010). *Product and Process Design Principles: Synthesis, Analysis, and Evaluation*. Wiley.

6 Grützner, T., Schnider, C., Zollinger, D., Seyfang, B.C., and Künzle, N. (2016). Reducing time to market by innovative development and production strategies. *Chem. Eng. Technol.* 39: 1835–1844.

7 Reymond, J.-L. (2015). The chemical space project. *Acc. Chem. Res.* 48: 722–730.

8 Taifouris, M., Martín, M., Martínez, A., and Esquejo, N. (2020). On the effect of the selection of suppliers on the design of formulated products. *Comput. Chem. Eng.* 141: 106980.

9 Taifouris, M., Martín, M., Martínez, A., and Esquejo, N. (2020). Challenges in the design of formulated products: Multiscale process and product design. *Curr. Opin. Chem. Eng.* 27: 1–9.

10 Gertig, C., Leonhard, K., and Bardow, A. (2020). Computer-aided molecular and processes design based on quantum chemistry: Current status and future prospects. *Curr. Opin. Chem. Eng.* 27: 89–97.

11 Achenie, L., Venkatasubramanian, V., and Gani, R. (2002). *Computer Aided Molecular Design: Theory and Practice*. Elsevier Science.

12 Kontogeorgis, G.M., and Gani, R. (2004) *Computer Aided Property Estimation for Process and Product Design: Computers Aided Chemical Engineering*. Elsevier Science.

13 Papadopoulos11 AI, Tsivintzelis, I., Linke, P., and Seferlis, P. (2018). Computer aided molecular design: fundamentals, methods and applications. *Chem., Mol. Sci. and Chem. Eng.* Reference Collection. Elsevier. https://doi.org/10.1016/B978-0-12-409547-2.14342-2.

14 Gani, R., Hytoft, G., Jaksland, C., and Jensen, A.K. (1997). An integrated computer aided system for integrated design of chemical processes. *Comput. Chem. Eng.* 21: 1135–1146.

15 Harper, P.M., and Gani, R. (2000). A multi-step and multi-level approach for computer aided molecular design. *Comput. Chem. Eng.* 24: 677–683.

16 Samudra, A.P., and Sahinidis, N.V. (2013). Optimization-based framework for computer-aided molecular design. *AIChE J.* 59: 3686–3701.

17 Chemmangattuvalappil, N.G. (2020). Development of solvent design methodologies using computer-aided molecular design tools. *Curr. Opin. Chem. Eng.* 27: 51–59.

18 Kalakul, S., Eden, M.R., and Gani, R. (2017). The Chemical Product Simulator—ProCAPD. In: 27 Eur. Symp. Comput. Aided Process Eng., Vol. 40 (eds. A. Espuña, M. Graells and Puigjaner LBT-CACE), 979–984. Elsevier.

19 Liu, Q., Zhang, L., Liu, L., Du, J., Tula, A.K., Eden, M. et al. (2019). OptCAMD: An optimization-based framework and tool for molecular and mixture product design. *Comput. Chem. Eng.* 124: 285–301.

20 Gani, R., and Brignole, E.A. (1983). Molecular design of solvents for liquid extraction based on UNIFAC. *Fluid Phase Equilib.* 13: 331–340.

21 Austin, N.D., Sahinidis, N.V., and Trahan, D.W. (2016). Computer-aided molecular design: An introduction and review of tools, applications, and solution techniques. *Chem. Eng. Res. Des.* 116: 2–26.

22 Borhani, T.N.G., Saniedanesh, M., Bagheri, M., and Lim, J.S. (2016). QSPR prediction of the hydroxyl radical rate constant of water contaminants. *Water Res.* 98: 344–353.

23 Zhang, L., Pang, J., Zhuang, Y., Liu, L., Du, J., and Yuan, Z. (2020). Integrated solvent-process design methodology based on COSMO-SAC and quantum mechanics for TMQ (2,2,4-trimethyl-1,2-H-dihydroquinoline) production. *Chem. Eng. Sci.* 226: 115894.

24 Puzyn, T., Leszczynski, J., and Cronin, M.T. (2010). *Recent Advances in QSAR Studies: Methods and Applications*. Netherlands: Springer.

25 Borhani, T.N., García-Muñoz, S., Vanesa Luciani, C., Galindo, A., and Adjiman, C.S. (2019). Hybrid QSPR models for the prediction of the free energy of solvation of organic solute/solvent pairs. *Phys. Chem. Chem. Phys.* 21: 13706–13720.

26 Todeschini, R., Consonni, V., Mannhold, R., Kubinyi, H., and Folkers, G. (2009). *Molecular Descriptors for Chemoinformatics, 2 Volume Set: Volume I: Alphabetical Listing/Volume II: Appendices, References*. Wiley.

27 Borhani, T.N.G., Afzali, A., and Bagheri, M. (2016). QSPR estimation of the auto-ignition temperature for pure hydrocarbons. *Process Saf. Environ. Prot.* 103: 115–125.

28 Bagheri, M., Borhani, T.N.G., Gandomi, A.H., and Manan, Z.A. (2014). A simple modelling approach for prediction of standard state real gas entropy of pure materials. *SAR QSAR Environ. Res.* 25: 695–710.

29 Katritzky, A.R., Kuanar, M., Slavov, S., Hall, C.D., Karelson, M., Kahn, I. et al. (2010). Quantitative correlation of physical and chemical properties with chemical structure: Utility for prediction. *Chem. Rev.* 110: 5714–5789.

30 Yan, Y., Borhani, T.N., and Clough, P.T. (2020). Chapter 14 Machine Learning Applications in Chemical Engineering. In: Mach. Learn. Chem. Impact Artif. Intell., The Royal Society of Chemistry 340–371.

31 Papaioannou, V., Adjiman, C.S., Jackson, G., and Galindo, A. (2011). Group contribution methodologies for the prediction of thermodynamic properties and phase behavior in mixtures. In: Process Syst. Eng. 135–172. Wiley.

32 Foresman, J.B., Frisch, A.E., and Gaussian, I. (1996). *Exploring Chemistry with Electronic Structure Methods*. Gaussian Incorporated.

33 Klamt, A., and Schuurmann, G. (1993). COSMO: A new approach to dielectric screening in solvents with explicit expressions for the screening energy and its gradient. *J. Chem. Soc. Perkin. Trans.* 2: 799–805.

34 Klamt, A. (1995). Conductor-like screening model for real solvents: A new approach to the quantitative calculation of solvation phenomena. *J. Phys. Chem.* 99: 2224–2235.

35 Lin, S.-T., and Sandler, S.I. (2002). A priori phase equilibrium prediction from a segment contribution solvation model. *Ind. Eng. Chem. Res.* 41: 899–913.

36 Zhang, L., Kalakul, S., Liu, L., Elbashir, N.O., Du, J., and Gani, R. (2018). A computer-aided methodology for mixture-blend design. Applications to tailor-made design of surrogate fuels. *Ind. Eng. Chem. Res.* 57: 7008–7020.

37 Mah, A.X.Y., Chin, H.H., Neoh, J.Q., Aboagwa, O.A., Thangalazhy-Gopakumar, S., and Chemmangattuvalappil, N.G. (2019). Design of bio-oil additives via computer-aided molecular design tools and phase stability analysis on final blends. *Comput. Chem. Eng.* 123: 257–271.

38 N.Borhani, T., and Wang, M. (2019). Role of solvents in CO_2 capture processes: The review of selection and design methods. *Renew. Sustain. Energy Rev.* 114.

39 Ahmad, M.Z., Hashim, H., Mustaffa, A.A., Maarof, H., and Yunus, N.A. (2018). Design of energy efficient reactive solvents for post combustion CO_2 capture using computer aided approach. *J. Clean. Prod.* 176: 704–715.

40 Fang, J., Zhao, R., Su, W., Li, C., Liu, J., and Li, B. (2016). A molecular design method based on the COSMO-SAC model for solvent selection in ionic liquid extractive distillation. *AIChE J.* 62: 2853–2869.

41 Papadopoulos, A.I., Shavalieva, G., Papadokonstantakis, S., Seferlis, P., Perdomo, F.A., Galindo, A., et al. (2020). An approach for simultaneous computer-aided molecular design with holistic sustainability assessment: Application to phase-change CO_2 capture solvents. *Comput. Chem. Eng.* 135: 106769.

42 van Kleef, L.M.T., Oyewunmi, O.A., and Markides, C.N. (2019). Multi-objective thermo-economic optimization of organic Rankine cycle (ORC) power systems in waste-heat recovery applications using computer-aided molecular design techniques. *Appl. Energy* 251: 112513.

43 Tillmanns, D., Petzschmann, J., Schilling, J., Gertig, C., and Bardow, A. (2019). ORC on tour: Integrated design of dynamic ORC processes and working fluids for waste-heat recovery from heavy-duty vehicles. In: 29th Eur. Symp. Comput. Aided Process Eng. 46 (eds. A.A. Kiss, E. Zondervan, R. Lakerveld, and L. Özkan), Elsevier.

44 Gertig, C., Kröger, L., Fleitmann, L., Scheffczyk, J., Bardow, A., and Leonhard, K. (2019). Rx-COSMO-CAMD: Computer-aided molecular design of reaction solvents based on predictive kinetics from quantum chemistry. *Ind. Eng. Chem. Res.* 58: 22835–22846.

45 Chang, A.M., Rudshteyn, B., Warnke, I., and Batista, V.S. (2018). Inverse design of a catalyst for aqueous CO/CO_2 conversion informed by the NiII–iminothiolate complex. *Inorg. Chem.* 57: 15474–15480.

46 Dittner, M., and Hartke, B. (2018). Globally optimal catalytic fields—Inverse design of abstract embeddings for maximum reaction rate acceleration. *J. Chem. Theory. Comput.* 14: 3547–3564.

47 Gertig, C., Fleitmann, L., Hemprich, C., Hense, J., Bardow, A., and Leonhard, K. (2020). Integrated in Silico Design of Catalysts and Processes based on Quantum Chemistry. In: 30th Eur. Symp. Comput. Aided Process Eng. 48. (eds. S. Pierucci, F. Manenti, G.L. Bozzano, and D. Manca), 889–894. Elsevier.

12

Pharmaceutical Industry

Challenges and Opportunities for Establishing Pharma 4.0

Wenqian Chen[a],*, Sung Joon Park[b], Thomas Nok Hin Cheng[a],
Nicholas Wai Hoi Lau[a], Liang Fa Khaw[c], Dan Lee-Lane[d],
Xizhong Chen[e], and Muhsincan Sesen[f]

[a] *Department of Chemical Engineering, Imperial College London, London, UK*
[b] *Department of Chemical Engineering, The University of Melbourne, Parkville, Victoria, Australia*
[c] *Department of Chemical & Petroleum Engineering, UCSI University, Cheras, Kuala Lumpur, Malaysia*
[d] *ABB Ltd, Oldends Lane, Stonehouse, UK*
[e] *Process and Chemical Engineering, School of Engineering and Architecture, University College Cork, Cork, Ireland*
[f] *Department of Bioengineering, Imperial College London, London, UK*
* *Corresponding author. Department of Chemical Engineering, Imperial College London, London, UK*

12.1 Pharma 4.0: Smart Manufacturing for the Pharmaceutical Industry

The progress of human civilization has been powered by the advancements in the mode of manufacturing. Industry 1.0 refers to the use of water- and steam-powered mechanical facilities for manufacturing in the 19th century. In contrast, Industry 2.0 refers to the mass production enabled by electrically powered facilities and labor division in the first half of 20th century [1]. Since the mid to late 20th century, we have been benefiting from Industry 3.0, in which manufacturing is highly automated through the use of electronics and information technology.

Powered by the rapid development in data science and cyber technologies, Industry 4.0 is a forward-thinking concept introduced by Germany in the early 2010s, in which "smart factories" are envisioned to generate customized products, that is, products that meet the needs of individual customers, based on highly flexible manufacturing capabilities. At the technical level, Industry 4.0 is envisioned to integrate cyber-physical systems into manufacturing and logistics and encompass the Internet of things (IoT) and Internet of services (IoS) in industrial processes [1]. Other countries such as the United States, the United Kingdom, China, Japan, and Singapore have followed this futuristic concept closely and introduced policies that support the development of similar intelligent manufacturing capability [2–6]. The visions of intelligent manufacturing share several desirable features, including adaptability to market changes, sustainability, and demand (and hence career opportunity) for highly skilled workforces [3]. All these features will be enabled by the integration of current automation technologies with data science and artificial intelligence based on four important principles: interconnectivity, information transparency, technical assistance, and decentralized decision-making (Figure 12.1).

Embracing the idea of Industry 4.0 in the context of pharmaceutical industry, the International Society for Pharmaceutical Engineering has introduced the concept of Pharma 4.0, which is the adoption of digitalization in operations model of a

Industry 4.0 Vision for the Supply of Energy and Materials: Enabling Technologies and Emerging Applications, First Edition. Edited by Mahdi Sharifzadeh.
© 2022 John Wiley & Sons, Inc. Published 2022 by John Wiley & Sons, Inc.

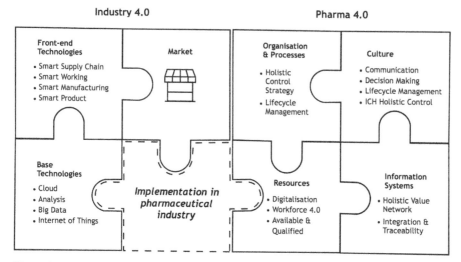

Figure 12.1   General Technology Framework for Industry 4.0. Adapted from [14].

pharmaceutical organization to enhance the exchange of information from the physical process with regulatory authorities while being capable of responding to the dynamics of the supply chain [7, 8]. In other words, Pharma 4.0 can be understood as a fusion between Industry 4.0 and quality management process based on International Council for Harmonisation of Technical Requirements for Pharmaceuticals for Human Use (ICH) Q10 guidelines [7], where ICH Q10 guidelines are published by the International Council for Harmonization of Technical Requirements for Pharmaceuticals for Human Use (ICH) on "Pharmaceutical Quality System" [9]. In Pharma 4.0, a pharmaceutical manufacturer needs to adopt a "holistic production control strategy," which is "a set of enablers and elements that provides a holistic view of production to ensure a flexible, agile, sustainable, and reliable pharmaceutical production that mitigates the risk to patients, products, processes, and the business" [10].

Within the vast scope of Pharma 4.0 [11–13], this chapter focuses only on the manufacturing aspects of Pharma 4.0. The driving forces behind the transition into Pharma 4.0 are first discussed as the background information for further discussions on key technical areas that are essential to the development of smart manufacturing in Pharma 4.0.

12.2   Regulatory Driving Force for Pharma 4.0

The pharmaceutical industry in general has a risk-averse attitude toward the adoption of new technology due to its strong emphasis on regulatory compliance. Nevertheless, being of the opinion that the transition into Pharma 4.0 will most likely follow the success of other industries, different stakeholders of the pharmaceutical industry have started to prepare for the arrival of the new age.

The most important driving force comes from the top: the US Food and Drug Administration launched the Pharmaceutical Quality for the 21st Century—A Risk-Based Approach initiative in 2002 (formerly known as the Pharmaceutical CGMP

Initiative for the 21st Century—A Risk Based Approach) to encourage early adoption of new technologies by pharmaceutical companies [15]. Within this framework, the concept of quality by design (QbD) is introduced instead of the traditional quality by testing (QbT) approach (Figure 12.2), where the qualities of the final drug products are tested against specifications approved by the FDA without understanding the causes of deviation [16]. Attempts to change the manufacturing process to address quality deviation require significant effort from both sides, as manufacturers must file supplements and obtain approval from the FDA, whereas the FDA must assess a sheer number of supplements.

To overcome these problems, QbD (Figure 12.2) emphasizes the importance of understanding the relationship between the manufacturing process, monitored through critical process parameters and the product quality through critical quality attributes. The range of critical process parameters that result in a suitable range of critical quality attributes forms the design space, within which modifications to the process can be justified and approved by the FDA. QbD necessitates an understanding of the critical process parameters' effects on the critical quality attributes of the product, which can be rigorously established with approaches such as risk assessment, design of experiment, and process analytical technology [17]. Based on this knowledge, quality consistency can be achieved with proper process design and control.

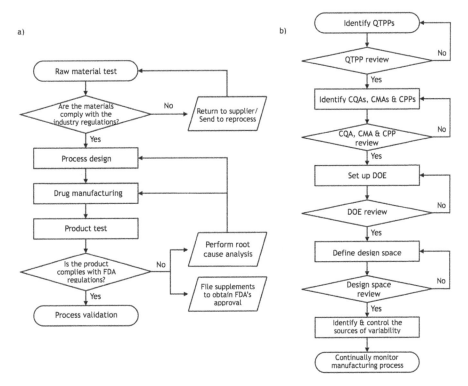

Figure 12.2 General Quality Assurance Process with (a) Quality by Test (QbT) and (b) Quality by Design (QbD) Approaches Where QTPP Is Quality Target Product Profile, CQA Is Critical Quality Attributes, CMA Is Critical Material Attributes, CPP Is Critical Process Parameters, and DoE Is Design of Experiments. Adapted from [16, 17].

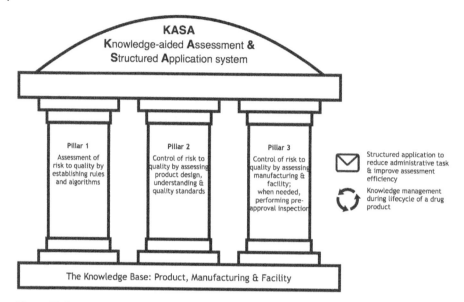

Figure 12.3 Knowledge-Aided Assessment and Structured Application (KASA) Framework Introduced by the FDA. Adapted from [18].

To improve the regulatory oversight of efficiency and consistency under the QbD framework, the FDA recently introduced the knowledge-aided assessment and structured application (KASA) in 2019, which consists of a broad knowledge base about a product, its manufacturing process and facility, and three pillars that represent different phases of development (Figure 12.3) [18]. When fully implemented, KASA will provide a set of relatively comprehensive information on each critical quality attribute of the product, including (1) inherent risk to quality, (2) control approaches, (3) a concise summary from the assessor detailing how the generalized approaches are applied in the regulatory application, and (4) links to supporting information. The FDA has supported the digitalization of pharmaceutical manufacturing through other initiatives. For instance, the 21 CFR Part 11 regulations on electronic records and signatures were introduced in 1997 to facilitate the regulatory compliance of digitalized processes [19]. With the establishment of KASA, more pharmaceutical companies are expected to embrace a holistic approach toward smart manufacturing systems over the next 5–10 years.

12.3 Smart Manufacturing in Pharmaceutical Industry

At the center of Pharma 4.0 is smart manufacturing, which is based on using the integration of physical and virtual technologies (Figure 12.1) to deliver efficiency, safety, traceability, flexibility, and sustainability [14]. Physical technologies include robots [20], modular production units [21, 22], process analytical technology (PAT) [12], and additive manufacturing technologies [23], all of which are automated. Virtual technologies include machine-to-machine communication [24], and data analytical technologies [13].

Pharma 4.0 still has a long way to go. According to a recent survey of approximately 80 pharmaceutical professionals based in Ireland, only 42% of the participants (mostly in managerial positions) were aware of Pharma 4.0 [19]. Fortunately, the same survey also showed that most of the pharmaceutical companies already have elements of smart manufacturing in place, which serve as the foundation for the development of Pharma 4.0. Such elements include PAT, production line automation, preventive maintenance, and data mining. Within the context of the pharmaceutical industry, process optimization, plant performance monitoring, and regulatory compliance are the top three motivations for the transition into Pharma 4.0 according to the survey [19]. In this section, three key technological areas (continuous manufacturing, PAT, and digitalization) are discussed in detail.

12.3.1 Continuous Manufacturing

Most of the conventional processes in the pharmaceutical industry have been operated in batch mode, leaving them with much room for improvement in terms of efficiency and their fundamental understanding [25]. Batch mode means materials move from one process to another intermittently with offline analysis often conducted at the end of each processing step. This follows the conventional QbT strategy [16]. For example, high-performance liquid chromatography (HPLC) is often conducted at the end of a reaction to quantify the product and side products within the reactor to decide if the materials should proceed to the next stage of processing or should be discarded or reprocessed (Figure 12.2).

In contrast, during continuous manufacturing, there is a continuous flow starting with materials feeding into the system, moving from one process to the next and the final product continuously leaving the system. A continuous manufacturing system can be fully continuous (all unit operations operate continuously) or semi-continuous (a combination of unit operations that operate in batch and continuous modes). Online or inline analyses are conducted constantly to monitor the critical process parameters and the critical quality attributes of products and intermediates, following the QbD strategy [16]. PAT and process design and control are particularly important aspects of continuous manufacturing, so they are discussed separately in later sections. This section will focus on continuous manufacturing as a whole, such as the motivations of developing continuous manufacturing, industrial examples, and future developments.

Although continuous manufacturing requires higher initial costs for research and system design and integration, it has several advantages over the traditional batch manufacturing [26]. First, product quality assurance is improved by avoiding batch-to-batch variation. Second, since smaller equipment is used, the cost of equipment is reduced and safety is improved, especially when processing highly potent compounds. Third, unstable intermediates formed after a reaction can be immediately purified and consumed in the next reaction to avoid degradation. Fourth, batch production occurs in multiple steps. The bulk of materials cannot be moved to the next step until the quality is confirmed. This can lead to long hold time, low utilization of equipment, and complicated process scheduling. The hold time between steps is eliminated in continuous manufacturing; therefore, the production time is reduced, and so is the likelihood of human error. Last, a continuous manufacturing system provides flexibility in terms of production scale. Scale is decoupled from system size since increasing the operation time can also increase the quantity of product.

Due to these advantages, pharmaceutical companies have developed several continuous manufacturing processes at research scale [27–36] and current good manufacturing practice (cGMP) production scales [37-40]. For example, Eli Lilly developed a continuous manufacturing system for the cGMP production of prexasertib monolactate monohydrate (Figure 12.4), which is a highly potent oncolytic drug substance [40]. The total production scale was 24 kg with a throughput of 3 kg per day. The continuous manufacturing system consisted of 3 plug flow reactors (PFRs), 1 continuous counter-current extractor, 1 continuous crystallization, and 1 semi-continuous filtration and dissolution unit (Figure 12.4). The outlet of each PFR was connected with online HPLC to quantify the reaction conversion and to determine the impurity profile. The outlets of two tubular reactors were connected to online refractive index detectors for the determination of product concentration and residence time distribution. In addition, temperature, pressure and mass flowrates were monitored by the distributed control system. Although the demonstration of continuous manufacturing at cGMP production scale was successful, the relationships between the process parameters of each unit operation and the purity of the intended product at each stage were not monitored and as such an effective control strategy was not in place. In other words, more effort was needed to turn the current QbT approach to QbD approach for the developed process.

In another example, Eli Lilly reported the continuous production of the penultimate intermediate of evacetrapib, which is an inhibitor of cholesteryl ester transfer protein for the prevention of cardiovascular events in patients with high-risk vascular disease [39]. The continuous manufacturing system consisted of a pipes-in-series PFR and a vapor–liquid separator, the liquid stream of which was connected to online HPLC

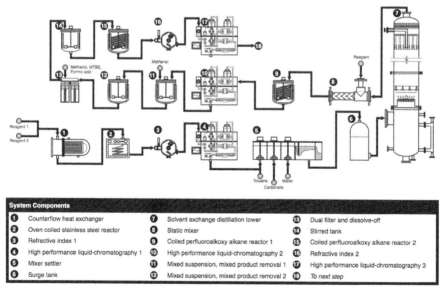

System Components		
❶ Counterflow heat exchanger	❼ Solvent exchange distillation tower	⓭ Dual filter and dissolve-off
❷ Oven coiled stainless steel reactor	❽ Static mixer	⓮ Stirred tank
❸ Refractive index 1	❾ Coiled perfluoroalkoxy alkane reactor 1	⓯ Coiled perfluoroalkoxy alkane reactor 2
❹ High performance liquid-chromatography 1	❿ High performance liquid-chromatography 2	⓰ Refractive index 2
❺ Mixer settler	⓫ Mixed suspension, mixed product removal 1	⓱ High performance liquid-chromatography 3
❻ Surge tank	⓬ Mixed suspension, mixed product removal 2	⓲ To next step

Figure 12.4 Continuous Manufacturing System for the cGMP Production of Prexasertib Monolactate Monohydrate.

(Figure 12.5). Similar to the previous example, although the demonstration of continuous manufacturing was successful, more effort was needed to establish an effective process control strategy by understanding the relationships between process parameters and product purity and thus achieve the goals of QbD.

Moving forward, continuous purification technologies such as membrane filtration [41, 42] and crystallization [43] should receive more attention, as conducting reactions in the continuous mode is relatively well understood. As shown in a previous study [44], biomolecules such as protein can be continuously crystallized in an oscillatory flow system. Using a systematic strategy of investigation, the development of continuous crystallization has three general steps (Figure 12.6): (1) vapor-diffusion screening experiments to identify suitable pH, temperature, buffer, and precipitant; (2) shaking batch crystallization to gather data on crystallization kinetics and thermodynamics; and (3) continuous crystallization experiments to gather data for process control and optimization. Using lysozyme as the model protein, the study successfully demonstrated the development of a continuous oscillatory flow crystallization system, and the frequency and amplitude were identified as critical process parameters that had significant effects on the critical quality attribute (crystal size). After screening a wide range of conditions (lysozyme concentration from 30 to 100 mg/mL and oscillatory amplitude from 5 to 30 mm and frequency from 0.1 to 1.0 Hz), the continuous

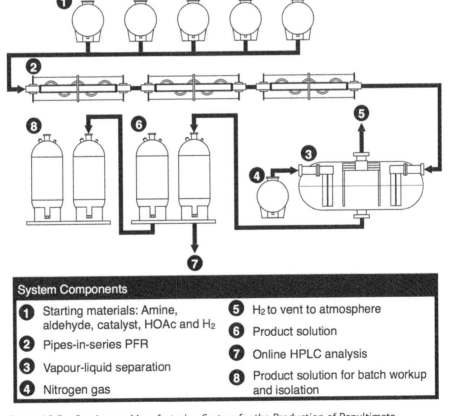

System Components	
1 Starting materials: Amine, aldehyde, catalyst, HOAc and H₂	**5** H₂ to vent to atmosphere
2 Pipes-in-series PFR	**6** Product solution
3 Vapour-liquid separation	**7** Online HPLC analysis
4 Nitrogen gas	**8** Product solution for batch workup and isolation

Figure 12.5 Continuous Manufacturing System for the Production of Penultimate Intermediate of Evacetrapib.

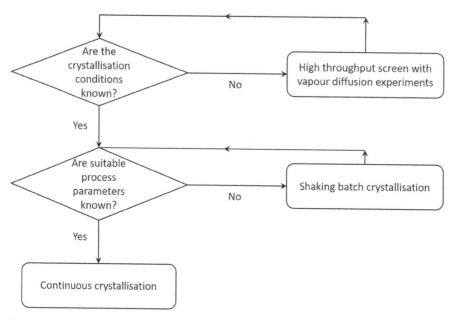

Figure 12.6 Strategy for Developing Continuous Protein Crystallization.

crystallization of lysozyme was demonstrated with a feed lysozyme concentration of 25 mg/mL, a net flow rate of 0.1 mL/min, an oscillatory amplitude of 20 mm and frequency of 0.5 Hz, and lysozyme crystals with unique size (~ 11 μm) were obtained, and the overall yield was ~ 60%.

12.3.2 Process Analytical Technologies

According to the FDA, PAT is defined as "a system for designing, analyzing and controlling manufacturing through timely measurements (i.e., during processing) of critical quality and performance attributes of raw and in-process materials and processes with the goal of ensuring final product quality" [45, 46]. Under the QbD framework, PAT plays the central role in gaining a deep understanding of the process and hence the effective design and operation of the process within the design space. As shown in Figure 12.7a, PAT in general consists of physical and virtual systems accessed by a human (i.e., operator, regulating organization, and customer). In the physical system, different types of inline (i.e., samples analyzed in the process without being removed), online (i.e., sample removed, analyzed, and then returned to the process again), at-line (i.e., samples removed and analyzed near the process without being returned to the process), or offline (i.e., samples removed and analyzed away from the process without being returned to the process) sensors are in place to measure the product and system properties (Figure 12.7b). The data are processed locally (i.e., fogging), and then the processed data are sent to the controller in the virtual system. The controller then sends signals to the control element in the physical system to effect the intended change in process parameters to ensure the system operate in the design space.

The implementation of PAT has three stages: design, analyze, and control [47]. In the design phase, the critical quality attributes of the pharmaceutical substance/product

a)

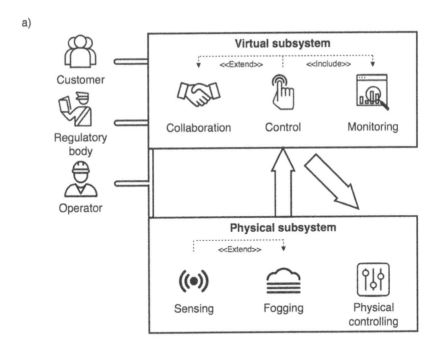

b)

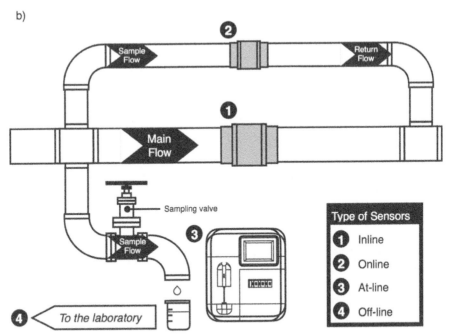

Figure 12.7 (a) Physical and Virtual Systems in PAT; (b) illustrations of Inline, Online, At-Line, and Off-Line Sensors.

and the critical process parameters that have a substantial influence on them are identified. In the analyze phase, suitable analyzers of the identified critical quality attributes and critical process parameters are selected based on their estimated ranges. Last, in the control phase, a controlling scheme is designed with suitable control elements to be incorporated into the process system.

A wide range of analytical technologies have been used in the pharmaceutical industry, including ultraviolet-visible (UV-vis) spectroscopy, chromatography (gas and liquid), Fourier transform infrared spectroscopy, and near infrared spectroscopy. Near infrared spectroscopy is one of the most popular technologies due to its versatility (e.g., analysis of raw material quality [48], powder flowability [49], dissolution rate [50], and blend uniformity [51]). When the critical quality attributes or critical process parameters cannot be measured directly, soft sensors (i.e., software sensors) can be used instead to estimate these critical quality attributes or critical process parameters using chemical or physical laws or statistical approaches based on the data of related measurable attributes and parameters [52–55].

The time lag between analysis and the implementation of controlling actions (relative to the process time of the corresponding unit operation) is one of the most important considerations in the design of PAT system. When the analysis and control implementation can be completed within a relatively short time, the PAT system should be relatively easy to design and vice versa. While some unit operations such as biosynthesis with mammalian cell culture have a long process time, some other unit operations such as preparative chromatography have a short process time, making the design of PAT difficult. In such a case, the analysis time must be reduced significantly (e.g., converting the conventional offline liquid chromatography analysis into online mode can achieve a substantial reduction in the analysis time [56]).

For Pharma 4.0, there is a continuous demand for efficient PAT, especially in inline and online modes. The development of microfluidic devices is a promising area in this context, as microfluidic devices can be coupled with different instruments for real-time analysis [57–59].

12.3.3 Emergence of Microfluidic Techniques for Continuous Manufacturing and PAT

There have been significant advances in the last decade toward making use of microfluidic solutions in continuous manufacturing and PAT [60, 61]. In microfluidics, fluids are circulated through micron-sized channels; this enables higher throughputs using lower sample volumes as well as better sensitivity with analytical detection techniques [62]. High throughput screening, the golden standard for early drug screening studies carried out by pharmaceutical companies, is a batch processing method carried out on microplates. Similarly, for protein crystallization, multiple crystallization conditions are batch screened using microplates for optimum crystal growth, especially for proteins with unknown structures [63]. High throughput screening struggles for further miniaturization because it is limited by evaporation and the accuracy of robotic dispensing [64]. Automated microfluidic platforms [65] promise higher throughputs, using smaller volumes on the order of nL to fL that significantly reduce reagent costs in a continuous manufacturing/screening scenario [66, 67]. Yet there are challenges to overcome for future integration of microfluidic technologies in Pharma

4.0; namely, development of flow chemistry transformations, handling of solids [68], interfacing with pre-existing technologies, and integration of QbD principles [69].

To briefly summarize opportunities enabled by microfluidic techniques, three main areas will be discussed: flow properties, high surface-to-volume ratio, and microfluidic PAT integration. First, due to the small size of microfluidic channels, Reynolds number (Re) values that determine the flow regime are very small (Re ≪ 1). Owing to this property, microchannel flows are always in the predictable laminar regime (i.e., no turbulence). This enables flow cytometry [70, 71], controlled mixing [72–75], formation of gradients [76, 77], and deterministic lateral displacement [78, 79]. For example, in personalized medicine, an emerging field enabled by microfluidics, laminar flow generated gradients have been used to screen three-dimensional (3D) micro-tumor models obtained from biopsies of cancer patients [77]. At the intermediate Reynolds number range (1 < Re < 100), inertial microfluidics offers very promising applications for cellular sample processing with high throughput [80, 81]. Moreover, unique flow properties of microfluidic chips have been utilized for protein separation [82], purification [83, 84], crystallization [85], isoelectric point based sorting [86], and many more applications [60].

The second advantage of microfluidics arises from the high surface-to-volume ratio in microfluidic chips due to the reduction in channel size. This property not only enhances sensitivity for detection but also gives rise to new separation techniques that are unique to or work more efficiently in this regime, including acoustofluidics [87, 88], electrophoresis [89-92], dielectrophoresis [93–95], and magnetophoresis [96]. For example, acoustofluidic mixing [97] has been used for nanoparticle synthesis [98] as well as exosome [99–101] and circulating tumor cell separation [102]. A microfluidic device was designed to independently measure electrophoretic mobility and diffusion coefficients to quantify effective charges of proteins in solution, which can be unreliable when estimated based on the chemical structure [103, 104]. Recently, there has been a renewed interest in diffusiophoresis for colloid separation as its physics rely on non-equilibrium solute–surface interactions [105, 106] that benefit from the high surface-to-volume ratio in microfluidics. It is also important to note that microfluidic chips can achieve high thermal-transfer efficiencies because of high surface-to-volume ratios; therefore, exothermic or high-temperature reactions can be performed with better control [73].

The previous examples demonstrate how microfluidics can play a key role in upgrading the gold standard batch processing methods with continuous manufacturing/screening methods for Pharma 4.0. Microfluidic detection and automation technologies are also very promising for implementing QbD protocols for the PAT framework endorsed by the FDA. This can be exemplified in three main categories: spectrophotometric techniques [82, 107], mass spectrometry [108], and electrochemical techniques [109]. Common materials used in chip microfabrication such as glass, plastics, and polydimethylsiloxane offer optical clarity allowing spectrophotometric techniques such as UV-vis [110, 111] and fluorescence [112] detection techniques to be carried out inline. Special materials or techniques can be used for other methods like Raman spectroscopy [113–116], X-ray spectroscopy [74, 117, 118], and nuclear magnetic resonance spectroscopy [119, 120]. For example, coherent anti-Stokes Raman scattering was combined with epi-fluorescence for process monitoring in a chip-based high-performance liquid chromatography system to detect polycyclic aromatic hydrocarbons with µg per mL sensitivity [121]. Coupling analytical separation and detection techniques such as chromatography [122] and mass spectrometry [108, 123] into microfluidic workflows find strong applications in protein and proteome measurements [124]. Finally, electrochemical detection is very promising because electrodes can be microfabricated on a chip using conventional integrated circuit fabrication technology. Such systems' footprint is usually much smaller and can be used for environmental monitoring [125] and point-of-care diagnostics [126].

12.3.4 Digitalization of Pharmaceutical Process

Since the emergence of computers in the 20th century, digital technologies in the form of the Internet, programmable logical controller, information technology, artificial intelligence, Internet of things devices, and digital twins have fundamentally changed a wide range of manufacturing activities [127–129]. Among these digital technologies, digital twin plays a crucial role in the development of Pharma 4.0 [127, 130, 131].

A digital twin is defined as a digital informational construct of a physical system that is created as an entity on its own and linked with the physical system [132]. Although this definition appeared in the early 2000s, the concept of a twin for a product or process can be traced back to the Apollo project of NASA in the 1960s, in which two identical space vehicles were assembled and one served as the twin to mimic the real-time movements of another vehicle [133–135]. The creation of a digital twin requires integrating multiple simulations that describe the different aspects of a process based on data from a large number of sensors [136]. Ideally, a digital twin should be a real-time representation of the physical twin, and the manipulation of the physical twin can be achieved virtually through the digital twin. However, a simplified or partial version is usually used in practice (e.g., a digital shadow of the physical system that receives data from sensors but does not have the control capability) [130].

The digital twin consists of a collection of models of the physical process and a data management system. The models can be data-driven, mechanistic, or hybrid. Mechanistic models are constructed based on the fundamental understanding of certain crucial aspects of the process and have physically understandable parameters and variables that require a relatively low level of data processing [137]. This facilitates the scientific interpretation of the prediction results. For aspects of the process that lack sufficient fundamental understanding, data-driven models can be constructed instead based on the process data [137]. However, the prediction accuracy is usually limited to the original range of test data, and thus, extrapolation of the prediction should be cautious [138]. Mechanistic and data-driven models can complement each other in terms of development and computational costs versus interpretability and generalizability; that is, the mechanistic model excels in terms of interpretability and generalizability but requires high development and computational costs and vice versa for data-driven model [139–141]. Therefore, hybrid models are often developed to strike a balance in practice [142–144].

There are commercial simulation softwares for developing these models, including gPROMS FormulatedProducts®, aspenONE, DynoChem, MATLAB, Simulink, COMSOL Multiphysics, and STAR-CCM + [127]. On the other hand, open-source platforms are also available, including OpenFOAM, Octave, LAMMPS, Dyssol, and

CAPE-OPEN. This software has built-in models for unit operations and the capacity for users to build their own models. For example, gPROMS Formulated Products has a relatively comprehensive set of models for simulating crystallization. Users can also develop new models to describe new processes, as shown in two case studies of peptide synthesis in membrane systems [41, 42]. Commercial cloud and local platforms are also available to host and develop models and support the integration of physical components and data management [127].

With the data from sensors, real-time process simulation and analysis can be conducted, including sensitivity analysis [145], design space studies [146], and optimization [139]. The analysis results can subsequently be used for process control purposes to ensure the process operates within the design space. The generated data can be visualized and processed from the remote management platform. Commercial platforms for data management and visualization, including Mindsphere, Predix, TIBCO Cloud, and TrendMiner, can ensure secure and the reliable device connectivity [127]. Data integration is at the core of data management, and its most significant problem is data heterogeneity, as sensors and software from different suppliers produce data in different formats. As a result, several languages have been proposed as standard languages, including extensible markup language, web ontology language, and structured query language [147, 148].

In pharmaceutical manufacturing, developing a digital twin requires the modeling of upstream and downstream processes and the integration of these models [149–152]. For example, synthesis in a bioreactor can be simulated in terms of system heterogeneity, intracellular biochemical pathways, and extracellular fluid dynamics to study the effect of parameters (e.g., pH, temperature, and dissolved oxygen) on the overall yield for process optimization purposes [153–156]. Similarly, in downstream processing, chromatography can be modeled mechanistically (e.g., plate model and mass balance model) to generate breakthrough and gradient elution curves [157, 158].

Figure 12.8 shows the schematic of a full flowsheet of a continuous pharmaceutical tablet manufacturing process using gPROMS FormulatedProducts® [159, 160]. All the unit operations, including API crystallization, milling, blending, and compaction processes, can be simulated within one interconnected network. This seamless and integrated model from the crystallization of API to the drug tablet compress can serve as a digital replica of the physical drug product manufacturing processes. The critical material attributes and process parameters are embedded in the flowsheet model so that their effects on the critical product quality attributes can be investigated effectively. With distributed computing and the increase in computational power of processors, virtual design experiments can be executed efficiently in silico to comprehensively survey the design space. Therefore, it provides a powerful capability for digital design and optimization of the manufacturing operations. Furthermore, a drug product performance model can also be built and connected to the end of drug production, which facilitates the direct identification and investigation of the impacts of material attributes and process parameters on the drug product's final therapeutical performance. With further advance of the model developments and computational power, the digital flowsheet model will become a part of the drug formulation and manufacturing workflow and for manufacturers to enhance process robustness and improve product quality.

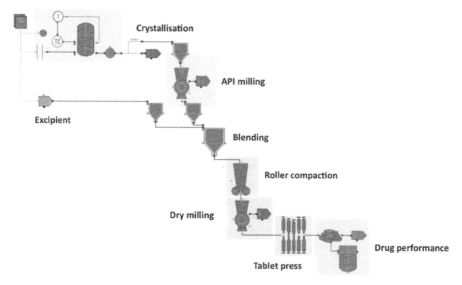

Figure 12.8 Dynamic Flowsheet Model of a Continuous Pharmaceutical Tablet Manufacturing Process Using gPROMS FormulatedProducts.® [159] / ADDoPT.

With process models validated with physical data, the design space under the QbD quality control strategy can be determined. Furthermore, the process models allow the development of effective process control strategies to advance the current continuous manufacturing into Pharma 4.0. However, future research is needed to develop fundamental understandings of both upstream and downstream processes for robust mechanistic models. Data-driven or hybrid models can play an important role in the development of a digital twin of the entire pharmaceutical manufacturing plant during this process.

In addition to process development, data science methods and tools are useful for process validation, monitoring, and continued validation [13]. The appropriate tools can help companies to manage risk associated with pharmaceutical process development and operation effectively [161]. At the practical level, one of the most critical problems for using statistical tools in the pharmaceutical manufacturing context, especially biopharmaceuticals, is the "n « p problem," which means the number of observations (i.e., "n") is much smaller than the number of variables per observation (i.e., "p") [162]. Regularization methods such as elastic net, least absolute shrinkage and selection operator, and ridge regression can be used to avoid overfitting [163, 164].

12.4 Summary and Outlook

Pharma 4.0 (i.e., accomplishing smart manufacturing in the pharmaceutical industry to deliver personalized medicine, sustainability, and more affordable therapeutics for patients) is definitely on its way, but will probably take at least another two to three decades to be fully realized with the current pace. However, the COVID-19 pandemic in 2020–22 and the race to develop vaccines within the shortest period have demonstrated a few things that are relevant to the development of Pharma 4.0.

First, the professional bodies including regulators, academic researchers, pharmaceutical companies, and health service providers must coordinate to respond to a public health crisis like a pandemic to minimize the damage. As shown during the COVID-19 pandemic, this coordination has been extremely challenging due to the vast number of organizations involved, but this situation has also clearly demonstrated the need for a smart integrated management tool that will be an important component in Pharma 4.0. Second, the leadership role of a regulator is extremely important in the response to the public health crisis. Without the effort of the regulator to speed up the vaccine trials, it would be impossible for pharmaceutical companies to develop vaccines within a year to protect the general public from the contagious virus. Similarly, for Pharma 4.0, regulators like FDA have taken an active role in the promotion of smart manufacturing in the pharmaceutical industry by introducing the Pharmaceutical Quality for the 21st Century—A Risk-Based Approach initiative almost 20 years ago and recently the KASA framework. The regulators should continue their efforts in this direction so that Pharma 4.0 can move at a steady speed with proper risk management. Third, the high flexibility and responsiveness of the pharmaceutical industry are essential in the timely management of a public health crisis as shown in 2020–22 and the pharmaceutical industry in Pharma 4.0 will have these characteristics. As one of the many side effects of globalization, pandemics will occur more frequently, creating a push for the development of Pharma 4.0. Based on these three reasons, we are hopeful that Pharma 4.0 can be a reality in the foreseeable future.

At the technical level, continuous manufacturing, PAT, and digitalization of pharmaceutical manufacturing processes will continue to advance. For example, more works will be done to address issues related to data security and cybersafety for the digitalization of pharmaceutical manufacturing processes. The key to success for Pharma 4.0 is the integration of different technologies, which requires academic researchers and industry professionals with a wide range of expertise to work closely together. Fortunately, many countries have created platforms that promote such collaborations (e.g., Pharma Innovation Programme Singapore [PIPS]), and more tangible outcomes toward the realization of Pharma 4.0 are expected to appear consistently in the coming years.

References

1 Wahlster, W., Kagermann, H., and Lukas, W.-D. (2013). Recommendations for implementing the strategic initiative INDUSTRIE 4.0 Final report of the Industrie 4.0 Working Group. National Academy of Science and Engineering (Germany). https://en.acatech.de/publication/recommendations-for-implementing-the-strategic-initiative-industrie-4-0-final-report-of-the-industrie-4-0-working-group.

2 Reif, R., Shirley, A.J., and Liveris, A. (2014). Report To The President Accelerating U.S. Advanced Manufacturing. Manufacturing USA. https://www.manufacturingusa.com/sites/prod/files/amp20_report_final.pdf.

3 Department for Trade and Industry. (2013). Future of manufacturing: a new era of opportunity and challenge for the UK - summary report. Gov. UK. https://assets.publishing.service.gov.uk/government/uploads/system/uploads/attachment_data/file/255923/13-810-future-manufacturing-summary-report.pdf.

4 Wübbeke, J., Meissner, M., Zenglein, M.J., Ives, J., and Conrad, B. (2016). *Made in China 2025: The making of a high-tech superpower and consequences for industrial countries.* MERICS Mercat. Inst. China Stud. https://merics.org/en/report/made-china-2025.

5 Fukuyama, M. (2018). *Society 5.0: Aiming for a New Human-centered Society.* Japan SPOTLIGHT. https://www.jef.or.jp/journal/pdf/220th_Special_Article_02.pdf.

6 Singapore Government. (2016). *Research Innovation Enterprise 2020 Plan.* Singapore Gov https://www.nrf.gov.sg/docs/default-source/Publications/rie2020-publication-%28final%29-%28may-2016%29.pdf.

7 International Society for Pharmaceutical Engineering. (2018). Pharma 4.0: Hype or reality? ISPE. https://ispe.org/pharmaceutical-engineering/july-august-2018/pharma-40tm-hype-or-reality.

8 Minero, T., and Augeri, A. (2019). Pharma 4.0—The new frontier for the pharma industry. ISPE. https://ispe.org/pharmaceutical-engineering/ispeak/pharma-40tm-new-frontier-pharma-industry.

9 ICH. (n.d.) Quality guidelines. https://www.ich.org/page/quality-guidelines.

10 Herwig, C., Wölbeling, C., and Zimmer, T. (2017). A holistic approach to production control from industry 4.0 to pharma 4.0. *Pharm. Eng. ISEP.* https://ispe.org/pharmaceutical-engineering/may-june-2017/holistic-approach-production-control.

11 Ding, B. (2018). Pharma Industry 4.0: Literature review and research opportunities in sustainable pharmaceutical supply chains. *Process Saf. Environ. Prot.* 119: 115–130.

12 Barenji, R.V., Akdag, Y., Yet, B., and Oner, L. (2019). Cyber-physical-based PAT (CPbPAT) framework for Pharma 4.0. *Int. J. Pharm.* 567: 118445.

13 Steinwandter, V., Borchert, D., and Herwig, C. (2019). Data science tools and applications on the way to Pharma 4.0. *Drug Discov. Today.* 24(9): 1795–1805.

14 Frank, A.G., Dalenogare, L.S., and Ayala, N.F. (2019). Industry 4.0 technologies: Implementation patterns in manufacturing companies. *Int. J. Prod. Econ.* 210: 15–26.

15 FDA. (2007). Pharmaceutical Quality for the 21st Century A Risk-Based Approach Progress Report Feb 15th, 2022. Department of Health and Human Services U.S. Food and Drug Administration. https://www.fda.gov/about-fda/center-drug-evaluation-and-research-cder/pharmaceutical-quality-21st-century-risk-based-approach-progress-report

16 Yu, L.X. (2008). Pharmaceutical quality by design: Product and process development, understanding, and control. *Pharm. Res.* 25(4): 781–791.

17 Zhang, L., and Mao, S. (2017). Application of quality by design in the current drug development. *Asian J. Pharm. Sci.* 12(1): 1–8.

18 Yu, L.X., Raw, A., Wu, L., Capacci-Daniel, C., Zhang, Y., and Rosencrance, S. (2019). FDA's new pharmaceutical quality initiative: Knowledge-aided assessment & structured applications. *Int. J. Pharm. X.* X, 1: 100010.

19 Reinhardt, I.C., Oliveira, D.J.C., and Ring, D.D.T. (2020). Current perspectives on the development of Industry 4.0 in the pharmaceutical sector. *J. Ind. Inf. Integr.* 18: 100–131.

20 Mennen, S.M., Alhambra, C., Allen, C.L., Barberis, M., Berritt, S., Brandt, T.A., Campbell, A.D., Castañón, J., Cherney, A.H., Christensen, M., et al. (2019). The evolution of high-throughput experimentation in pharmaceutical development and perspectives on the future. *Org. Process Res. Dev.* 23(6): 1213–1242.

21 Reitze, A., Jürgensmeyer, N., Lier, S., Kohnke, M., Riese, J., and Grünewald, M. (2018). Roadmap for a smart factory: A modular, intelligent concept for the production of specialty chemicals. *Angew. Chemie - Int. Ed.* 57(16): 4242–4247.

22 Wan, J., Tang, S., Li, D., Imran, M., Zhang, C., Liu, C., and Pang, Z. (2019). Reconfigurable smart factory for drug packing in healthcare industry 4.0. *IEEE Trans. Industr. Inform.* 15(1): 507–516.

23 Jamróz, W., Szafraniec, J., Kurek, M., and Jachowicz, R. (2018). 3D Printing in pharmaceutical and medical applications – recent achievements and challenges. *Pharm. Res.* 35(9): 1–22.

24 Gilchrist, A. (2016). *Industry 4.0 : The industrial Internet of things*. Springer.

25 Myerson, A.S., Krumme, M., Nasr, M., Thomas, H., and Braatz, R.D. (2015). Control systems engineering in continuous pharmaceutical manufacturing May 20–21, 2014 Continuous Manufacturing Symposium. *J. Pharm. Sci.* 104(3): 832–839.

26 Burcham, C.L., Florence, A.J., and Johnson, M.D. (2018). Continuous manufacturing in pharmaceutical process development and manufacturing. *Annu. Rev. Chem. Biomol. Eng.* 9: 253–281.

27 Hopkin, M.D., Baxendale, I.R., and Ley, S.V. (2010). A flow-based synthesis of Imatinib: The API of Gleevec. *Chem. Commun.* 46(14): 2450–2452.

28 Viviano, M., Glasnov, T.N., Reichart, B., Tekautz, G., and Kappe, C.O. (2011). A scalable two-step continuous flow synthesis of nabumetone and related 4-aryl-2-butanones. *Org. Process Res. Dev.* 15(4): 858–870.

29 Lévesque, F. and Seeberger, P.H. (2012). Continuous-flow synthesis of the anti-malaria drug artemisinin. *Angew. Chemie - Int. Ed.* 51(7): 1706–1709.

30 Murray, P.R.D., Browne, D.L., Pastre, J.C., Butters, C., Guthrie, D., and Ley, S.V. (2013). Continuous flow-processing of organometallic reagents using an advanced peristaltic pumping system and the telescoped flow synthesis of (E/Z)-tamoxifen. *Org. Process Res. Dev.* 17(9): 1192–1208.

31 Hopkin, M.D., Baxendale, I.R., and Ley, S.V. (2013). An expeditious synthesis of imatinib and analogues utilising flow chemistry methods. *Org. Biomol. Chem.* 11(11): 1822–1839.

32 Cantillo, D., Damm, M., Dallinger, D., Bauser, M., Berger, M., and Kappe, C.O. (2014). Sequential nitration/hydrogenation protocol for the synthesis of triaminophloroglucinol: Safe generation and use of an explosive intermediate under continuous-flow conditions. *Org. Process Res. Dev.* 18(11): 1360–1366.

33 Correia, C.A., Gilmore, K., McQuade, D.T., and Seeberger, P.H. (2015). A concise flow synthesis of efavirenz. *Angew. Chemie - Int. Ed.* 54(16): 4945–4948.

34 Ghislieri, D., Gilmore, K., and Seeberger, P.H. (2015). Chemical assembly systems: Layered control for divergent, continuous, multistep syntheses of active pharmaceutical ingredients. *Angew. Chemie - Int. Ed.* 54(2): 678–682.

35 Tsubogo, T., Oyamada, H., and Kobayashi, S. (2015). Multistep continuous-flow synthesis of (R)- and (S)-rolipram using heterogeneous catalysts. *Nature*. 520(7547): 329–332.

36 Lau, S.H., Galván, A., Merchant, R.R., Battilocchio, C., Souto, J.A., Berry, M.B., and Ley, S.V. (2015). Machines vs malaria: A flow-based preparation of the drug candidate OZ439. *Org. Lett.* 17(13): 3218–3221.

37 May, S.A., Johnson, M.D., Braden, T.M., Calvin, J.R., Haeberle, B.D., Jines, A.R., Miller, R.D., Plocharczyk, E.F., Rener, G.A., Richey, R.N., et al. (2012). Rapid development and scale-up of a 1 H -4-substituted imidazole intermediate enabled by chemistry in continuous plug flow reactors. *Org. Process Res. Dev.* 16(5): 982–1002.

38 Frederick, M.O., Calvin, J.R., Cope, R.F., Letourneau, M.E., Lorenz, K.T., Johnson, M.D., Maloney, T.D., Pu, Y.J., Miller, R.D., and Cziesla, L.E. (2015). Development of an NH4Cl-catalyzed ethoxy ethyl deprotection in flow for the synthesis of merestinib. *Org. Process Res. Dev.* 19(10): 1411–1417.

39 May, S.A., Johnson, M.D., Buser, J.Y., Campbell, A.N., Frank, S.A., Haeberle, B.D., Hoffman, P.C., Lambertus, G.R., McFarland, A.D., Moher, E.D., et al. (2016). Development and manufacturing GMP scale-up of a continuous ir-catalyzed homogeneous reductive amination reaction. *Org. Process Res. Dev.* 20(11): 1870–1898.

40 Cole, K.P., Groh, J.M.C., Johnson, M.D., Burcham, C.L., Campbell, B.M., Diseroad, W.D., Heller, M.R., Howell, J.R., Kallman, N.J., Koenig, T.M., et al. (2017). Kilogram-scale prexasertib monolactate monohydrate synthesis under continuous-flow CGMP conditions. *Science* 356(6343): 1144–1150.

41 Chen, W., Sharifzadeh, M., Shah, N., and Livingston, A.G. (2017). The implication of side-reactions in iterative biopolymer synthesis: The case of membrane enhanced peptide synthesis (MEPS). *Ind. Eng. Chem. Res.* 56: 6796–6804.

42 Chen, W., Sharifzadeh, M., Shah, N., and Livingston, A.G. (2018). Iterative peptide synthesis in membrane cascades: Untangling operational decisions. *Comput. Chem. Eng.* 115: 275–285.

43 Chen, W., Yang, H., and Heng, J.Y.Y. (2020). Continuous protein crystallization. In: *Handb. Contin. Cryst.*, 372–392. RSC Publishing.

44 Yang, H., Chen, W., Peczulis, P., and Heng, J.Y.Y. (2019). Development and workflow of a continuous protein crystallization process: A case of lysozyme, Crystal *Growth & Design.* 19(2): 983–991.

45 FDA. (2004). Guidance for Industry PAT. https://www.fda.gov/media/71012/download.

46 Read, E.K., Park, J.T., Shah, R.B., Riley, B.S., Brorson, K.A., and Rathore, A.S. (2010). Process analytical technology (PAT) for biopharmaceutical products: Part I. Concepts and applications. *Biotechnol. Bioeng.* 105(2): 276–284.

47 Rathore, A.S., Bhambure, R., and Ghare, V. (2010). Process analytical technology (PAT) for biopharmaceutical products. *Anal. Bioanal. Chem.* 398(1): 137–154.

48 Paris, I., Janoly-Dumenil, A., Paci, A., Mercier, L., Bourget, P., Brion, F., Chaminade, P., and Rieutord, A. (2006). Near infrared spectroscopy and process analytical technology to master the process of busulfan paediatric capsules in a university hospital. *J. Pharm. Biomed. Anal.* 41(4): 1171–1178.

49 Benedetti, C., Abatzoglou, N., Simard, J.S., McDermott, L., Léonard, G., and Cartilier, L. (2007). Cohesive, multicomponent, dense powder flow characterization by NIR. *Int. J. Pharm.* 336(2): 292–301.

50 Tabasi, S.H., Moolchandani, V., Fahmy, R., and Hoag, S.W. (2009). Sustained release dosage forms dissolution behavior prediction: a study of matrix tablets using NIR spectroscopy. *Int. J. Pharm.* 382(1–2): 1–6.

51 Moes, J.J., Ruijken, M.M., Gout, E., Frijlink, H.W., and Ugwoke, M.I. (2008). Application of process analytical technology in tablet process development using NIR spectroscopy: Blend uniformity, content uniformity and coating thickness measurements. *Int. J. Pharm.* 357(1–2): 108–118.

52 Kadlec, P., Gabrys, B., and Strandt, S. (2009). Data-driven soft sensors in the process industry. *Comput. Chem. Eng.* 33(4): 795–814.

53 Wechselberger, P., Sagmeister, P., and Herwig, C. (2013). Real-time estimation of biomass and specific growth rate in physiologically variable recombinant fed-batch processes. *Bioprocess Biosyst. Eng.* 36(9): 1205–1218.

54 Sagmeister, P., Wechselberger, P., Jazini, M., Meitz, A., Langemann, T., and Herwig, C. (2013). Soft sensor assisted dynamic bioprocess control: Efficient tools for bioprocess development. *Chem. Eng. Sci.* 96: 190–198.

55 Lu, F., Toh, P.C., Burnett, I., Li, F., Hudson, T., Amanullah, A., and Li, J. (2013). Automated dynamic fed-batch process and media optimization for high productivity cell culture process development. *Biotechnol. Bioeng.* 110(1): 191–205.

56 Rathore, A.S., Parr, L., Dermawan, S., Lawson, K., and Lu, Y. (2010). Large scale demonstration of a process analytical technology application in bioprocessing: Use of on-line high performance liquid chromatography for making real time pooling decisions for process chromatography. *Biotechnol. Prog.* 26(2): 448–457.

57 Cui, P., and Wang, S. (2019). Application of microfluidic chip technology in pharmaceutical analysis: a review. *J. Pharm. Anal.* 9(4): 238–247.

58 Lin, S.L., Lin, T.Y., and Fuh, M.R. (2014). Microfluidic chip-based liquid chromatography coupled to mass spectrometry for determination of small molecules in bioanalytical applications: An update. *Electrophoresis.* 33(4): 635–643.

59 Nge, P.N., Rogers, C.I., and Woolley, A.T. (2013). Advances in microfluidic materials, functions, integration, and applications. *Chem. Rev.* 113(4): 2550–2583.

60 Karle, M., Vashist, S.K., Zengerle, R., and Von Stetten, F. (2016). Microfluidic solutions enabling continuous processing and monitoring of biological samples: a review. *Anal. Chim. Acta.* 929: 1–22.

61 Ran, R., Sun, Q., Baby, T., Wibowo, D., Middelberg, A.P.J., and Zhao, C.X. (2017). Multiphase microfluidic synthesis of micro- and nanostructures for pharmaceutical applications. *Chem. Eng. Sci.* 169: 78–96.

62 Sesen, M., Alan, T., and Neild, A. (2017). Droplet control technologies for microfluidic high throughput screening (μHTS). *Lab on a Chip.* 17(14): 2372–2394.

63 Rupp, B. and Kantardjieff, K.A. (2008). Macromolecular crystallography. In: *Molecular Biomethods Handbook* (ed. J. Walker and R. Rapley), 821–849. Humana Press.

64 Janzen, W. (2008). High throughput screening. In: *Molecular Biomethods Handbook* (ed. J. Walker and R. Rapley), 1097–1118. Humana Press.

65 Sesen, M., and Whyte, G. (2020). Image-based single cell sorting automation in droplet microfluidics. *Sci. Rep.* 10(1): 1–14.

66 Schneider, G. (2018). Automating drug discovery. *Nat. Rev. Drug Discov.* 17(2): 97–113.

67 Neužil, P., Giselbrecht, S., Länge, K., Huang, T.J., and Manz, A. (2012). Revisiting lab-on-a-chip technology for drug discovery. *Nat. Rev. Drug Discov.* 11(8): 620–632.

68 Hartman, R.L. (2012). Managing solids in microreactors for the upstream continuous processing of fine chemicals. *Org. Process Res. Dev.* 16(5): 870–887.

69 Shallan, A.I., and Priest, C. (2019). Microfluidic process intensification for synthesis and formulation in the pharmaceutical industry. *Chem. Eng. Process. - Process Intensif.* 142: 107559.

70 Yang, R.J., Fu, L.M., and Hou, H.H. (2018). Review and perspectives on microfluidic flow cytometers. *Sens. Actuators B Chem.* 266: 26–45.

71 Stavrakis, S., Holzner, G., Choo, J., and deMello, A. (2019). High-throughput microfluidic imaging flow cytometry. *Curr. Opin. Biotechnol.* 55: 36–43.

72 Sethu, P., Anahtar, M., Moldawer, L.L., Tompkins, R.G., and Toner, M. (2004). Continuous flow microfluidic device for rapid erythrocyte lysis. *Anal. Chem.* 76(21): 6247–6253.

73 DeMello, A.J. (2006). Control and detection of chemical reactions in microfluidic systems. *Nature.* 442(7101): 394–402.

74 Hong, L., Sesen, M., Hawley, A., Neild, A., Spicer, P.T., and Boyd, B.J. (2019). Comparison of bulk and microfluidic methods to monitor the phase behaviour of nanoparticles during digestion of lipid-based drug formulations using: in situ X-ray scattering. *Soft Matter* 15(46): 9565–9578.

75 Patil, S., Pandit, A., Gaikwad, G., Dandekar, P., and Jain, R. (2020). Exploring microfluidic platform technique for continuous production of pharmaceutical microemulsions. *J. Pharm. Innov.* 1–13.

76 Li, B., Qiu, Y., Glidle, A., McIlvenna, D., Luo, Q., Cooper, J., Shi, H.C., and Yin, H. (2014). Gradient microfluidics enables rapid bacterial growth inhibition testing. *Anal. Chem.* 86(6): 3131–3137.

77 Mulholland, T., McAllister, M., Patek, S., Flint, D., Underwood, M., Sim, A., Edwards, J., and Zagnoni, M. (2018). Drug screening of biopsy-derived spheroids using a self-generated microfluidic concentration gradient. *Sci. Rep.* 8(1): 1–12.

78 McGrath, J., Jimenez, M., and Bridle, H. (2014). Deterministic lateral displacement for particle separation: A review. *Lab on a Chip.* 14(21): 4139–4158.

79 Hochstetter, A., Vernekar, R., Austin, R.H., Becker, H., Beech, J.P., Fedosov, D.A., Gompper, G., Kim, S.C., Smith, J.T., Stolovitzky, G., et al. (2020). Deterministic lateral displacement: Challenges and perspectives. *ACS Nano.* 14(9): 10784–10795.

80 Zhang, J., Yan, S., Yuan, D., Alici, G., Nguyen, N.T., Ebrahimi Warkiani, M., and Li, W. (2016). Fundamentals and applications of inertial microfluidics: A review. *Lab on a Chip.* 16(1): 10–34.

81 Di Carlo, D. (2009). Inertial microfluidics. *Lab on a Chip.* 9(21): 3038–3046.

82 Rodríguez-Ruiz, I., Babenko, V., Martínez-Rodríguez, S., and Gavira, J.A. (2018). Protein separation under a microfluidic regime. *Analyst.* 143(3): 606–619.

83 Meagher, R.J., Light, Y.K., and Singh, A.K. (2008). Rapid, continuous purification of proteins in a microfluidic device using genetically-engineered partition tags. *Lab on a Chip.* 8(4): 527–532.

84 Capuano, A., Mulloni, V., Adami, A., and Lorenzelli, L. (2018). Continuous extraction of proteins with a miniaturized electrical split-flow cell equipped with suspended splitters fabricated by dry film lamination. *Sens. Actuators B Chem.* 273: 627–634.

85 Junius, N., Jaho, S., Sallaz-Damaz, Y., Borel, F., Salmon, J.B., and Budayova-Spano, M. (2020). A microfluidic device for both on-chip dialysis protein crystallization and: In situ X-ray diffraction. *Lab on a Chip.* 20(2): 296–310.

86 Song, Y.A., Hsu, S., Stevens, A.L., and Han, J. (2006). Continuous-flow pI-based sorting of proteins and peptides in a microfluidic chip using diffusion potential. *Anal. Chem.* 78(11): 3528–3536.

87 Lenshof, A., Magnusson, C., and Laurell, T. (2012). Acoustofluidics 8: Applications of acoustophoresis in continuous flow microsystems. *Lab on a Chip.* 12(7): 1210–1223.

88 Wu, M., Ozcelik, A., Rufo, J., Wang, Z., Fang, R., and Jun Huang, T. (2019). Acoustofluidic separation of cells and particles. *Microsyst. Nanoeng.* 5(1): 1–18.

89 Novo, P., and Janasek, D. (2017). Current advances and challenges in microfluidic free-flow electrophoresis—A critical review. *Anal. Chim. Acta.* 991: 9–29.

90 Johnson, A.C., and Bowser, M.T. (2018). Micro free flow electrophoresis. *Lab on a Chip*. 18(1): 27–40.

91 Rozing, G. (2016). Recent developments of microchip capillary electrophoresis coupled with mass spectrometry. In: *Capillary Electrophoresis-Mass Spectrometry (CE-MS): Principles and Applications* (ed. G. de Jong), 67–102. Wiley-VCH.

92 Castro, E.R., and Manz, A. (2015). Present state of microchip electrophoresis: State of the art and routine applications. *J. Chromatogr. A* 1382: 66–85.

93 Xing, X., Ng, C.N., Chau, M.L., and Yobas, L. (2018). Railing cells along 3D microelectrode tracks for continuous-flow dielectrophoretic sorting. *Lab on a Chip*. 18(24): 3760–3769.

94 Chan, J.Y., Kayani, A.B.A., Md Ali, M.A., Kok, C.K., Majlis, B.Y., Hoe, S.L.L., Marzuki, M., Khoo, A.S.B., Ostrikov, K., Rahman, M.A., et al. (2018). Dielectrophoresis-based microfluidic platforms for cancer diagnostics. *Biomicrofluidics*. 12(1): 011503.

95 Cheng, I.F., Chang, H.C., Hou, D., and Chang, H.C. (2007). An integrated dielectrophoretic chip for continuous bioparticle filtering, focusing, sorting, trapping, and detecting. *Biomicrofluidics*. 1(2): 021503.

96 Pamme, N., and Manz, A. (2004). On-chip free-flow magnetophoresis: Continuous flow separation of magnetic particles and agglomerates. *Anal. Chem.* 76(24): 7250–7256.

97 Van Phan, H., Coskun, M.B., Sesen, M., Pandraud, G., Neild, A., and Alan, T. (2015). Vibrating membrane with discontinuities for rapid and efficient microfluidic mixing. *Lab on a Chip*. 15(21): 4206–4216.

98 An Le, N.H., Deng, H., Devendran, C., Akhtar, N., Ma, X., Pouton, C., Chan, H.K., Neild, A., and Alan, T. (2020). Ultrafast star-shaped acoustic micromixer for high throughput nanoparticle synthesis. *Lab on a Chip*. 20(3): 582–591.

99 Habibi, R., He, V., Ghavamian, S., De Marco, A., Lee, T.H., Aguilar, M.I., Zhu, D., Lim, R., and Neild, A. (2020). Exosome trapping and enrichment using a sound wave activated nano-sieve (SWANS). *Lab on a Chip*. 20(19): 3633–3643.

100 Wu, M., Ouyang, Y., Wang, Z., Zhang, R., Huang, P.H., Chen, C., Li, H., Li, P., Quinn, D., Dao, M., et al. (2017). Isolation of exosomes from whole blood by integrating acoustics and microfluidics. *Proc. Natl. Acad. Sci. U. S. A.* 114(40): 10584–10589.

101 Xu, L., Shoaie, N., Jahanpeyma, F., Zhao, J., Azimzadeh, M., and Al-Jamal, K.T. (2020). Optical, electrochemical and electrical (nano)biosensors for detection of exosomes: a comprehensive overview. *Biosens. Bioelectron.* 161: 112222.

102 Dao, M., Suresh, S., Huang, T.J., Li, P., Mao, Z., Peng, Z., Zhou, L., Chen, Y., Huang, P.H., Truica, C.I., et al. (2015). Acoustic separation of circulating tumor cells. *Proc. Natl. Acad. Sci. U. S. A.* 112(16): 4970–4975.

103 Charmet, J., Arosio, P., and Knowles, T.P.J. (2018). Microfluidics for protein biophysics. *J. Mol. Biol.* 430(5): 565–580.

104 Herling, T.W., Arosio, P., Müller, T., Linse, S., and Knowles, T.P.J. (2015). A microfluidic platform for quantitative measurements of effective protein charges and single ion binding in solution. *Phys. Chem. Chem. Phys.* 17(18): 12161–12167.

105 Shin, S. (2020). Diffusiophoretic separation of colloids in microfluidic flows. *Phys. Fluids* 32(10): 101302.

106 Shimokusu, T.J., Maybruck, V.G., Ault, J.T., and Shin, S. (2020). Colloid separation by CO_2-induced diffusiophoresis. *Langmuir*. 36(25): 7032–7038.

107 Rodríguez-Ruiz, I., Ackermann, T.N., Muñoz-Berbel, X., and Llobera, A. (2016). Photonic lab-on-a-chip: Integration of optical spectroscopy in microfluidic systems. *Anal. Chem.* 88 (13): 6630–6637.

108 Wang, X., Yi, L., Mukhitov, N., Schrell, A.M., Dhumpa, R., and Roper, M.G. (2015). Microfluidics-to-mass spectrometry: A review of coupling methods and applications. *J. Chromatogr. A* 1382: 98–116.

109 Sierra, T., Crevillen, A.G., and Escarpa, A. (2019). Electrochemical detection based on nanomaterials in CE and microfluidic systems. *Electrophoresis.* 40(1): 113–123.

110 Liu, Z., Ou, J., Samy, R., Glawdel, T., Huang, T., Ren, C.L., and Pawliszyn, J. (2008). Side-by-side comparison of disposable microchips with commercial capillary cartridges for application in capillary isoelectric focusing with whole column imaging detection. *Lab on a Chip.* 8(10): 1738–1741.

111 Vlčková, M., Kalman, F., and Schwarz, M.A. (2008). Pharmaceutical applications of isoelectric focusing on microchip with imaged UV detection. *J. Chromatogr. A* 1181(1–2): 145–152.

112 De Kort, B.J., De Jong, G.J., and Somsen, G.W. (2013). Native fluorescence detection of biomolecular and pharmaceutical compounds in capillary electrophoresis: Detector designs, performance and applications: A review. *Anal. Chim. Acta.* 766: 13–33.

113 Esmonde-White, K.A., Cuellar, M., Uerpmann, C., Lenain, B., and Lewis, I.R. (2017). Raman spectroscopy as a process analytical technology for pharmaceutical manufacturing and bioprocessing. *Anal. Bioanal. Chem.* 409(3): 637–649.

114 Chrimes, A.F., Khoshmanesh, K., Stoddart, P.R., Mitchell, A., and Kalantar-Zadeh, K. (2013). Microfluidics and raman microscopy: Current applications and future challenges. *Chem. Soc. Rev.* 42(13): 5880–5906.

115 Langer, J., De Aberasturi, D.J., Aizpurua, J., Alvarez-Puebla, R.A., Auguié, B., Baumberg, J.J., Bazan, G.C., Bell, S.E.J., Boisen, A., Brolo, A.G., et al. (2020). Present and future of surface-enhanced Raman scattering. *ACS Nano.* 14(1): 28–117.

116 Zhang, Q., Zhang, P., Gou, H., Mou, C., Huang, W.E., Yang, M., Xu, J., and Ma, B. (2015). Towards high-throughput microfluidic Raman-activated cell sorting. *Analyst.* 140(18): 6163–6174.

117 Ghazal, A., Lafleur, J.P., Mortensen, K., Kutter, J.P., Arleth, L., and Jensen, G.V. (2016). Recent advances in X-ray compatible microfluidics for applications in soft materials and life sciences. *Lab on a Chip.* 16(22): 4263–4295.

118 Pham, N., Radajewski, D., Round, A., Brennich, M., Pernot, P., Biscans, B., Bonneté, F., and Teychené, S. (2017). Coupling high throughput microfluidics and small-angle X-ray scattering to study protein crystallization from solution. *Anal. Chem.* 89(4): 2282–2287.

119 Hale, W., Rossetto, G., Greenhalgh, R., Finch, G., and Utz, M. (2018). High-resolution nuclear magnetic resonance spectroscopy in microfluidic droplets. *Lab on a Chip.* 18(19): 3018–3024.

120 Liu, W.W., and Zhu, Y. (2020). Development and application of analytical detection techniques for droplet-based microfluidics. *Anal. Chim. Acta.* 1113: 66–84.

121 Geissler, D., Heiland, J.J., Lotter, C., and Belder, D. (2017). Microchip HPLC separations monitored simultaneously by coherent anti-Stokes Raman scattering and fluorescence detection. *Microchim. Acta.* 184(1): 315–321.

122 De Mello, A. (2002). On-chip chromatography: The last twenty years. *Lab on a Chip.* 2(3): 48N–54N.

123 De Mello, A. (2001). Focus: Chip–MS: coupling the large with the small, *Lab on a Chip*. 1(1): 7N–12N.

124 Lazar, I.M., Gulakowski, N.S., and Lazar, A.C. (2020). Protein and proteome measurements with microfluidic chips. *Anal. Chem.* 92(1): 169–182.

125 Maguire, I., O'Kennedy, R., Ducrée, J., and Regan, F. (2018). A review of centrifugal microfluidics in environmental monitoring. *Anal. Methods.* 10(13): 1497–1515.

126 Fang, T.H., Ramalingam, N., Xian-Dui, D., Ngin, T.S., Xianting, Z., Lai Kuan, A.T., Peng Huat, E.Y., and Hai-Qing, G. (2009). Real-time PCR microfluidic devices with concurrent electrochemical detection. *Biosens. Bioelectron.* 24(7): 2131–2136.

127 Chen, Y., Yang, O., Sampat, C., Bhalode, P., Ramachandran, R., and Ierapetritou, M. (2020). Digital twins in pharmaceutical and biopharmaceutical manufacturing: a literature review. *Processes.* 8(9): 1088.

128 Oztemel, E., and Gursev, S. (2020). Literature review of Industry 4.0 and related technologies. *J. Intell. Manuf.* 31(1): 127–182.

129 Venkatasubramanian, V. (2019). The promise of artificial intelligence in chemical engineering: Is it here, finally? *AIChE J.* 65(2): 466–478.

130 Kritzinger, W., Karner, M., Traar, G., Henjes, J., and Sihn, W. (2018). Digital Twin in manufacturing: a categorical literature review and classification. *IFAC-PapersOnLine.* 51(11): 1016–1022.

131 Tao, F., Cheng, J., Qi, Q., Zhang, M., Zhang, H., and Sui, F. (2018). Digital twin-driven product design, manufacturing and service with big data. *Int. J. Adv. Manuf. Technol.* 94(9–12): 3563–3576.

132 Grieves, M., and Vickers, J. (2016). Digital twin: Mitigating unpredictable, undesirable emergent behavior in complex systems. In: *Transdiscipl. Perspect. Complex Syst. New Find. Approaches.* (ed. F. Kahlen, S. Flumerfelt, and A. Alves), 85–113. Springer.

133 Gholami Mayani, M., Svendsen, M., and Oedegaard, S.I. (2018). Drilling digital twin success stories the last 10 years. In: *Soc. Pet. Eng. - SPE Norw. One Day Semin.*

134 Schleich, B., Anwer, N., Mathieu, L., and Wartzack, S. (2017). Shaping the digital twin for design and production engineering. *CIRP Ann. - Manuf. Technol.* 66(1): 141–144.

135 Rosen, R., Von Wichert, G., Lo, G., and Bettenhausen, K.D. (2015). About the importance of autonomy and digital twins for the future of manufacturing. *IFAC-PapersOnLine.* 48(3): 567–572.

136 Glaessgen, E.H., and Stargel, D.S. (2012). The digital twin paradigm for future NASA and U.S. air force vehicles. In: *53rd AIAA/ASME/ASCE/AHS/ASC Struct. Struct. Dyn. Mater. Conf.*

137 Von Stosch, M., Oliveira, R., Peres, J., and Feyo De Azevedo, S. (2014). Hybrid semi-parametric modeling in process systems engineering: Past, present and future. *Comput. Chem. Eng.* 60: 86–101.

138 Lohninger, H. (1999). *Teach/me: Data Analysis*. Springer.

139 Eugene, E.A., Gao, X., and Dowling, A.W. (2020). Learning and optimization with Bayesian hybrid models. In: *Proc. Am. Control Conf.*

140 Bhosekar, A., and Ierapetritou, M. (2018). Advances in surrogate based modeling, feasibility analysis, and optimization: A review. *Comput. Chem. Eng.* 108: 250–267.

141 Zendehboudi, S., Rezaei, N., and Lohi, A. (2018). Applications of hybrid models in chemical, petroleum, and energy systems: A systematic review. *Appl. Energy* 228: 2539–2566.

142 Azarpour, A., Borhani, T.N.G., Wan Alwi, S.R., Manan, Z.A., and Abdul Mutalib, M.I. (2017). A generic hybrid model development for process analysis of industrial fixed-bed catalytic reactors. *Chem. Eng. Res. Des.* 117: 149–167.

143 Laursen, S.Ö., Webb, D., and Ramirez, W.F. (2007). Dynamic hybrid neural network model of an industrial fed-batch fermentation process to produce foreign protein. *Comput. Chem. Eng.* 31(3): 163–170.

144 Schäfer, P., Caspari, A., Mhamdi, A., and Mitsos, A. (2019). Economic nonlinear model predictive control using hybrid mechanistic data-driven models for optimal operation in real-time electricity markets: in-silico application to air separation processes. *J. Process Control.* 84: 171–181.

145 Wang, Z., Escotet-Espinoza, M.S., and Ierapetritou, M. (2017). Process analysis and optimization of continuous pharmaceutical manufacturing using flowsheet models. *Comput. Chem. Eng.* 107: 77–91.

146 Grossmann, I.E., Calfa, B.A., and Garcia-Herreros, P. (2014). Evolution of concepts and models for quantifying resiliency and flexibility of chemical processes. *Comput. Chem. Eng.* 70: 22–34.

147 Venkatasubramanian, V., Zhao, C., Joglekar, G., Jain, A., Hailemariam, L., Suresh, P., Akkisetty, P., Morris, K., and Reklaitis, G.V. (2006). Ontological informatics infrastructure for pharmaceutical product development and manufacturing. *Comput. Chem. Eng.* 30(10–12): 1482–1496.

148 Michels, J., Hare, K., Kulkarni, K., Zuzarte, C., Liu, Z.H., Hammerschmidt, B., and Zemke, F. (2018). The new and improved SQL:2016 standard. *SIGMOD Rec.* 47(2): 51–60.

149 Narayanan, H., Luna, M.F., Von Stosch, M., Cruz Bournazou, M.N., Polotti, G., Morbidelli, M., Butté, A., and Sokolov, M. (2020). Bioprocessing in the digital age: The role of process models. *Biotechnol. J.* 15(4): 1900172.

150 Tang, P., Xu, J., Louey, A., Tan, Z., Yongky, A., Liang, S., Li, Z.J., Weng, Y., and Liu, S. (2020). Kinetic modeling of Chinese hamster ovary cell culture: factors and principles. *Crit. Rev. Biotechnol.* 40(2): 265–281.

151 Farzan, P., Mistry, B., and Ierapetritou, M.G. (2017). Review of the important challenges and opportunities related to modeling of mammalian cell bioreactors. *AIChE J.* 63(2): 398–408.

152 Smiatek, J., Jung, A., and Bluhmki, E. (2020). Towards a digital bioprocess replica: Computational approaches in biopharmaceutical development and manufacturing. *Trends Biotechnol.* 38(10): 1141–1153.

153 Li, X., Scott, K., Kelly, W.J., and Huang, Z. (2018). Development of a computational fluid dynamics model for scaling-up Ambr bioreactors. *Biotechnol. Bioprocess Eng.* 23(6): 710–725.

154 Farzan, P., and Ierapetritou, M.G. (2018). A framework for the development of integrated and computationally feasible models of large-scale mammalian cell bioreactors. *Processes.* 6(7): 82.

155 Sokolov, M., Ritscher, J., MacKinnon, N., Souquet, J., Broly, H., Morbidelli, M., and Butté, A. (2017). Enhanced process understanding and multivariate prediction of the relationship between cell culture process and monoclonal antibody quality. *Biotechnol. Prog.* 33(5): 1368–1380.

156 Kotidis, P., Jedrzejewski, P., Sou, S.N., Sellick, C., Polizzi, K., Del Val, I.J., and Kontoravdi, C. (2019). Model-based optimization of antibody galactosylation in CHO cell culture. *Biotechnol. Bioeng.* 116(7): 1612–1626.

157 Behere, K., and Yoon, S. (2020). Chromatography bioseparation technologies and in-silico modelings for continuous production of biotherapeutics. *J. Chromatogr. A* 1627: 461376.

158 Kumar, V., and Lenhoff, A.M. (2020). Mechanistic modeling of preparative column chromatography for biotherapeutics. *Annu. Rev. Chem. Biomol. Eng.* 11: 235–255.

159 Benito, M.M., Doshi, P., Davies, C., and Braido, D. (2019). Solid drug product and process design using multi-scale interconnected flowsheet modelling and global system analysis. ADDoPT Digit. Des. Showc. Event. https://www.addopt.org/downloads/presentations/10_marta_benito.pdf

160 Rogers, A.J., Hashemi, A., and Ierapetritou, M.G. (2013). Modeling of particulate processes for the continuous manufacture of solid-based pharmaceutical dosage forms. *Processes.* 1(2): 67–127.

161 Kannt, A., and Wieland, T. (2016). Managing risks in drug discovery: Reproducibility of published findings, Naunyn. Schmiedebergs. *Arch. Pharmacol.* 389(4): 353–360.

162 Zahel, T., Marschall, L., Abad, S., Vasilieva, E., Maurer, D., Mueller, E.M., Murphy, P., Natschläger, T., Brocard, C., Reinisch, D., et al. (2017). Workflow for criticality assessment applied in biopharmaceutical process validation stage 1. *Bioengineering.* 4(4): 85.

163 Rasmussen, M.A., and Bro, R. (2012). A tutorial on the Lasso approach to sparse modeling. *Chemom. Intell. Lab. Syst.* 119: 21–31.

164 Ferreira, A., Menezes, J.C., and Tobyn, M. (2018). *Multivariate Analysis in the Pharmaceutical Industry.* Academic Press.

13

Additive Manufacturing

A Game-Changing Paradigm in Manufacturing and Supply Chains

Mojtaba Molla-Hosseini[a], Marjan Hosseini[b], Gholamreza Vossoughi[a], and Mahdi Sharifzadeh[c],*

[a] Mechanical Engineering Department, Sharif University of Technology, Tehran, Iran
[b] Civil Engineering Department, Tehran University, Tehran, Iran
[c] Sharif Energy, Water and Environment Institute (SEWEI), Sharif University of Technology, Tehran, Iran
* Corresponding author. Sharif Energy, Water and Environment Institute (SEWEI), Sharif University of Technology, Tehran, Iran

13.1 The Additive Manufacturing Concept

Additive manufacturing (AM), also known as three-dimensional (3D) printing, is defined as a process in which materials are joined to make a 3D structure. This process is usually performed layer upon layer in three dimensions. This technology is further extended to the so-called four-dimensional (4D) printing in which time is also added as the fourth dimension. Nonetheless, the basic concept of both 3D and 4D printing is the same. The key feature of 4D printing technologies is the ability to utilize programmable smart materials, as their shape changes over time under an external stimulus, such as water and heat [1, 2]. The smart materials will be discussed. A photo of a 3D printer is shown in Figure 13.1 [3].

13.2 Category

ASTM[1] 529000:2015standard [4] classifies additive manufacturing processes in seven categories. According to Figure 13.2, AM processes are classified as (1) binder jetting (BJ), (2) direct energy deposition (DED), (3) material extrusion (ME), (4) material jetting (MJ), (5) powder bed fusion, (6) sheet lamination, and (7) vat photopolymerization, as discussed in the following.

13.2.1 Binder Jetting

Binder jetting is one of the AM processes. It consists of two essential materials: a powder base and a liquid binder. The key steps in BJ 3D printing are as follows:

1. A roller metal spreads powder over the build platform.
2. Liquid binder is added to adhere the particles.
3. The build platform is slightly transmitted to lowers height for the next layer.
4. Another layer of powder is added to the platform, and the binder is added again in an iterative pattern to complete the building cycle [4, 5].

---

1 American Society for Testing and Materials.

Industry 4.0 Vision for the Supply of Energy and Materials: Enabling Technologies and Emerging Applications, First Edition. Edited by Mahdi Sharifzadeh.
© 2022 John Wiley & Sons, Inc. Published 2022 by John Wiley & Sons, Inc.

Figure 13.1   A Photo of a 3D Printer [3], figure 1b (p.03) / with permission of JOHN WILEY & SONS, INC.

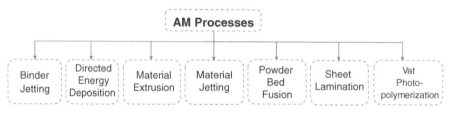

Figure 13.2   AM Processes Classification According to ISO/ASTM 529000:2015 Based on [4].

At the end of the printing process, the unused powder is removed, and then the built platform is sintered in a kiln to remove polymeric qualities. During this procedure, called the post-sintering process, the printed parts are substantially shrunk [4, 5].

The advantages of binder jetting include (1) design freedom, (2) applicability for large printing volumes, (3) relatively low costs, and (4) high print speed, which poses them as one of the most popular methods in additive manufacturing.

Commonly used materials applied in the BJ technique include ceramic, metal, polymer, sand, and glass with many commercial products such as soda-lime glass, bonded tungsten, zircon, and ceramic beads. Some more specific examples of these materials and applications are discussed in the following.

Stainless steel with high tensile strength as well as heat and corrosion resistance are suitable for building parts exposed to highly abrasive environments such as mining equipment, parts for down-hole drilling, and pump components.

Another printing material is *ceramic beads* with high permeability and desirable thermal expansion. They are compatible with all binders in 3D printing. These features make them suitable for casting steel alloys and printing cores exposed to high thermal stress conditions.

Inconel alloy, which is highly dense with mechanical properties, is applied in steam generators in pressurized nuclear water reactors, pressure vessels in the aerospace industry, turbine blades, and seals.

Iron with excellent wear-resistance and good mechanical properties is the most widely used material for various industrial applications. It is applicable in many areas such as automotive components, decorative hardware, machine tools, and so on. A schematic illustration of the BJ process is shown in Figure 13.3 [6].

13.2.2 Direct Energy Deposition

In DED, energy is sent into a small region so that it provides heat to melt materials being deposited. DED is classified into three groups in terms of the energy source used for melting materials: laser-based DED, electron beam-based DED, and plasma or electric arc-based DED. Manipulating both the energy source and the material feed nozzle can be performed by using a gantry system or robotic arm [7]. The resolution of the printed part is directly affected by the quantity of metal powder deposited in the DED process. As opposed to the conventional way, in DED there is a laser with high power-density focused on a continuous stream of metal powder. The resolution of the DED process is directly connected to its energy source. For example, laser energy is more than that of electron beams. Additionally, the fabrication speed of a DED process is affected by the deposition rate. Due to the nature of 3D printing processes, a DED process is capable of repairing damaged parts such as turbine blades by adding materials to them. Cost-effective materials used as a metal powder in DED are steels, aluminum, titanium, copper, cobalt, tin, and nickel. Figure 13.4 shows a schematic illustration of the laser-based DED process [8].

13.2.3 Material Extrusion

ME is another process in which a nozzle pushes out heated material with constant pressure and speed. When the material coming out of the nozzle is solidified on the substrate, the 3D object is printed. In this process, the material coming out from the nozzle needs to bind with previous material, while it keeps its structure [4, 9]. The main materials applied in this process are discussed in the following.

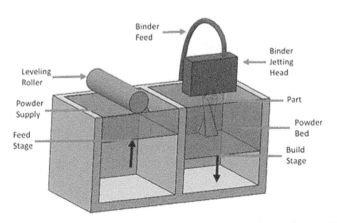

Figure 13.3 A Schematic Illustration of the Binder Jetting Process [6].

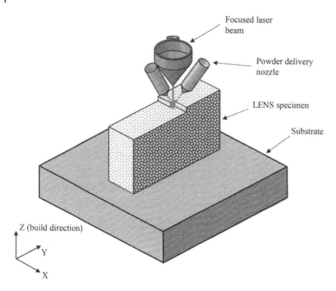

Figure 13.4 A Schematic Illustration of the Laser-Based DED Process [8].

1- **Acrylonitrile butadiene styrene**: Featuring desirable properties such as high toughness and strength, acrylonitrile butadiene styrene is applicable in automotive, medical, and aerospace devices.

2- **Acrylonitrile styrene acrylate**: Mechanical strength and ultraviolet (UV) stability are the main features of acrylonitrile styrene acrylate. This material is applied to produce parts for outdoor use under the sun and electrical housings, as well as automotive prototypes.

3- **Nylon 12**: Nylon 12 has high fatigue and chemical resistance and high impact strength. Some applications of this product are in custom production tooling, antenna covers, and aerospace industries in which high fatigue endurance and impact-protective components are in demand.

4- **Polycarbonate**: Having flexural and high tensile strength are two important features of polycarbonates. Polycarbonate is applicable in tooling and fixtures and functional prototypes. It is also used in blow-molding masters in the aerospace industry.

5- **Polyphenylsulfone**: The main features of polyphenylsulfone are high chemical and heat resistance and excellent mechanical strength. This material is capable of withstanding various sterilization methods including ethylene oxide, radiation, and autoclaving. Such advantages have made it applicable in sterilizable medical devices and automotive prototypes.

6- **Polyetherimide**: Chemical and thermal stability, excellent mechanical properties, and being biocompatible are three important features of polyetherimide. Due to its high strength-to-weight ratio, it is suitable for rapid prototyping and advanced tooling applications such as in the aerospace, medical, food, and automotive industries.

7- **Polylactic acid**: The important features of polylactic acid are good tensile strength and surface quality. It is applicable in models and prototypes for both home and office when aesthetic detail is needed.

8- **Thermoplastic polyurethane**: The important properties of thermoplastic polyurethane are high impact strength and hardness as well as excellent tear and wear resistance. Thermoplastic polyurethane is a very versatile material with both rubber and plastics properties. It has also corrosion resistance to many industrial oils and chemicals. These properties make it applicable in a variety of industrial applications like automotive, film and sheet, food processing equipment, and flexible tubing [10].

Figure 13.5 shows a schematic representation of the material extrusion process [11].

13.2.4 Material Jetting

MJ is another 3D printing process that is defined through four steps as follows:

1. The liquid-formed resin is heated to 30–60 Celsius degrees to achieve optimal viscosity for printing.
2. Then the printhead moves over the build platform and hundreds of small photopolymer's droplets are deposited (jetted) to the desired locations.
3. An UV light source, which is attached to the printhead, cures and solidify the deposited material to create the first layer of the part.
4. After completion of the first layer, the build platform moves downward a distance called one layer height and the process repeats in a sequence to complete a building cycle.

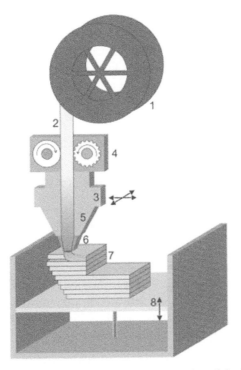

Figure 13.5 A Schematic Representation of the Material Extrusion Process [11].

The main materials used in this process and their applications are as follows:

1- **VeroWhitePlus**: A rigid and durable, it has high dimensional accuracy. These features are desirable for industrial applications such as medical devices, building workpieces that have complicated features, and electronic housing.

2- **Digital acrylonitrile butadiene styrene**: The outstanding feature of digital acrylonitrile butadiene styrene is high heat deflection temperature, making it applicable in building injection molds, manufacturing tools, durable presentation models, electronics enclosures, functional prototypes, and engine parts.

3- **Fullcure RGD 720**: Fullcure RGD 720 is highly transparent with its smooth surface finish. It is deployed in some applications such as visualization of liquid flow, eye-wear, and artistic modeling, as well as fit testing of see-through parts such as glass and color dying.

4- **Rigur RGD 450**: Rigur RGD 450 is a bright white material that has properties similar to polypropylene. It is suitable for packaging, consumer goods, electronics, and living hinges. In automotive industries, this material is deployed in flexible closures and reusable containers.

5- **Rubber-like materials** (e.g., the polyjet family of rubber-like materials includes tangogray, tangoblack, tangoplus, and tangoblackplus [12]): The important feature of rubber-like materials is various shore elastomers. This material is used in simulated gaskets, electronic button covers, o-rings, and keypads.

6- **High-temperature materials** (e.g., face-centered-cubic René 142, hexagonal close-packed Ti-6Al-4 V [13]): High-temperature materials have heat resistance and stability. They are ideal for testing applications such as hot-air flow in pipes and faucets or hot-water flow testing.

7- **Biocompatible materials** (e.g., polymer, metal [14]): Biocompatible materials are rigid and transparent with high dimensional stability. They are suitable for applications in which short-term (<24 hours) and long-term (>30 days) skin contact is required.

8- **Dental material**: Dental material needs to be durable and highly accurate with good strength. Examples of materials with such features are VeroGlaze, VeroDentPlus, and VeroDent that are functional in dental applications such as veneer try-ins, in-mouth placement, and diagnostic wax-ups.

In Figure 13.6, a schematic illustration of the material jetting process is shown [15].

13.2.5 Powder Bed Fusion

In the powder bed fusion process, a thermal source (i.e., laser) is used to make a partial or full fusion between powder particles. Sintering and melting are two kinds of binding mechanisms for this process.

Sintering is a partial melting process [16]. In other words, in solid-state sintering, particles are fused only at the surface. As a result, they inherit the porosity of the original materials. On the other hand, in liquid-state melting, all particles are fully melted, fusing together that results in a dense substance. Depending on which mechanism is applied, the produced material properties, as well as fabrication speed, vary.

Powder bed fusion is mainly based on a thermal heating process in which there is a highly energized laser or electron beam as the energy source. Three Examples of the powder bed fusion process are selective laser sintering, selective laser melting [17, 18],

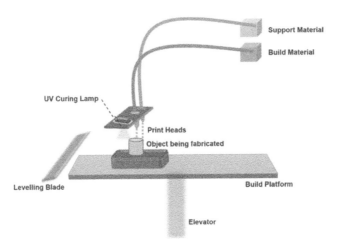

Figure 13.6 A Schematic Illustration of the Material Jetting Process [15].

and electron beam melting [19, 20]. Some materials used during this process and their applications are as follows:

1- **Titanium**: Titanium is a biocompatible, corrosion-resistant material with high strength, low density, and low thermal expansion. It is suitable for a wide range of applications such as medical technology, marine and chemical industry, automotive, aerospace, and orthopedic implants and prosthesis.

2- **Stainless steel:** The important properties of stainless steel are the ability to harden, having high ductility, and high resistance to wear, tear, and corrosion. In applications such as mechanical engineering, medical technology, toolmaking, and the automotive industry, stainless steel could be functional.

3- **Aluminum:** Desirable properties such as good processability, good alloying properties, low material density, and electrical conductivity have made aluminum suitable for a wide range of industrial applications including the automotive industry, aerospace engineering, and prototype construction especially thin-wall components with complex geometries.

4- **Cobalt-chrome:** Cobalt-chrome is a biocompatible material with very high hardness, high strength, corrosion resistance, and high ductility. It is applicable in medical and dental technologies. It is functional in fields that operate at high temperatures, such as jet engines.

5- **Nickel-based alloys:** Nickel-based alloy is a material with the ability to harden and several desirable properties such as good mechanical strength, high weldability, and corrosion resistance. It is used in toolmaking, aerospace engineering, and fields with high-temperature.

Figure 13.7 shows a schematic illustration of the powder bed fusion process [21].

13.2.6 Sheet Lamination

There are two main sheet lamination processes: ultrasonic additive manufacturing (UAM) and laminated object manufacturing (LOM). In the UAM process, sheets of metal bonded together by ultrasonic welding are applied. This process requires additional

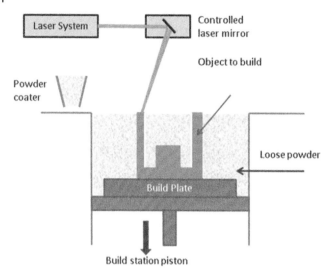

Figure 13.7 A Schematic Illustration of the Powder Bed Fusion Process [21].

computer numerical control (CNC) machining and removal of the unbound metal, mainly during the welding process. The UAM is a hybrid sheet lamination process that consists of ultrasonic metal welding and CNC machining [22]. LOM is a similar layer-by-layer approach, although it uses paper as material and adhesive for bonding sheets. In this process, there are diverse bonding mechanisms between the sheets including adhesive, clamping, and thermal bonding. Laminated objects by LOM are often useful for visual and aesthetic models and are not suitable for structural uses [4].

Metals sheets, plastic sheets, and paper are the main materials used in the sheet lamination process. Figure 13.8 shows a basic schematic illustration of the LOM process [23].

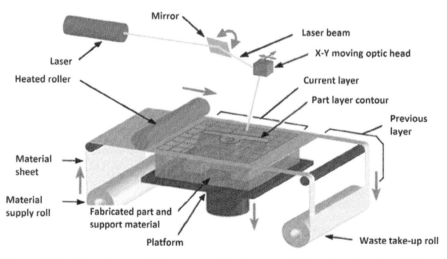

Figure 13.8 A Basic Schematic Illustration of the Laminated Object Manufacturing (LOM) Process [23].

13.2.7 Vat Photopolymerization

Vat photopolymerization generally includes stereolithography and its related processes. In stereolithography, photocurable resins exposed to laser beam, undergo a chemical reaction to become a solid part. This chemical reaction is called photopolymerization in which small monomers are linked into chain-like polymers. It is a combination of many chemical compounds such as additives, photo-initiators, and reactive monomers/oligomers. These components are called photopolymers, most of them are curable in the UV range. The main issue is to conduct crosslinking polymers sufficiently so that polymerized molecules do not dissolve into the liquid monomers. On the other hand, it must have sufficient strength to withstand various forces. Specific components in vat photopolymerization are photo-initiators, solvent, reactive diluents, monomers, and additives (e.g., stabilizers, flexibilizers). Figure 13.9 shows a schematic illustration of the stereolithography printing process [6].

Special materials used in this process and their applications are as follows.

1- **DC 100:** High accuracy and low shrinkage are two important features of DC 100. They are ideal materials for the direct casting of models with accurate shapes and smooth surfaces such as jewelry patterns.

2- **DC 500:** DC 500 burns easily like wax materials. It is applicable in thin wire-shaped patterns of jewelry that are impossible to be created using rubber molding methods.

3- **DL 350:** This material has properties similar to polypropylene with high flexibility and strength. It is applicable in producing functional parts for a wide range of applications and industrial design.

4- **DL 360:** Having clear transparency and good strength are two important features of DL 360.

5- **AB 001:** The properties of AB 001 are similar to acrylonitrile butadiene styrene mentioned in the material extrusion process. Its application is in producing functional parts that are strong and smooth.

6- **GM 08:** The properties of GM 08 are similar to rubber with high transparency and flexibility. It is applicable in producing ready-to-use functional parts that are durable, flexible, and smooth.

7- **DM 210:** This material has a high surface quality that is comparable to ceramics. It is functional in both thin and thick rubber models of jewelry with liquid silicone. Its properties make it easily removable from the rubber.

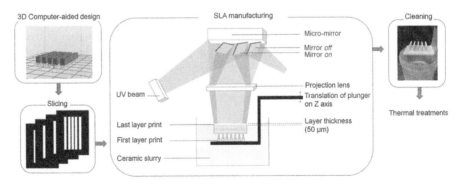

Figure 13.9 A Schematic Illustration of the Stereolithography Printing Process [6].

8- **DM 220:** DM 220 is a nano-filled ceramic with a smooth surface. Its application is in rubber master models of jewelry with liquid silicone where a higher resistance to temperature is needed.

9- **PlasTM range:** PlasTM range is a chemically resistant material with high resolution that is applicable for producing mechanical assemblies, enclosures, jigs, and fixtures, and generally for models with great durability and surface finish.

10- **SuperCAST:** This material is a range of resins that are direct-cast. It is suitable for jewelry and dental casting in which great precision is demanded.

11- **SuperWAX:** SuperWAX is a material that burns out easily, similar to waxes. It is the first 3D printing photopolymer. SuperWAX melts like wax and liquefies at 50 degrees Celsius.

12- **Rigid polyurethane:** This material has desirable properties such as being tough, stiff, and abrasion-resistant. It has also good mechanical strength, toughness, and hardness. Its applications are in automotive electronic devices and industrial components.

13- **Flexible polyurethane:** Flexible polyurethane is a flexible and impact-resistant material with moderate stiffness. It has an acceptable balance between stiffness and fatigue and abrasion resistance and flexibility for impact. This property makes flexible polyurethane, which is applicable in components requiring toughness to withstand repetitive stresses.

14- **Elastomeric Polyurethane:** It is a highly elastic as well as resilient material. High elasticity makes it resilient under stress and enables applications in which high impact, elasticity, and wear resistance are required;

15- **Cyanate ester:** Cyanate ester is a stiff material with high strength, hardness, thermal stability, and temperature resistance. This material has a wide range of applications in industrial components and electronics;

The energy formulations of all the aforementioned processes are discussed in [4].

13.3   The Pros and Cons of Seven AM Processes

In the following the pros and cons of each previously mentioned AM process are presented:

13.3.1   Binder Jetting

For the binder jetting method, a wide range of materials has the potential to be used. Furthermore, it takes low specific energy during imaging. On the other hand, the materials produced by this method are mostly rough with a grainy appearance and poor strength. Another disadvantage is that in binder jetting, post-processing is needed to remove moisture to improve the material's strength.

13.3.2   Direct Energy Deposition

In this process, material deposition is performed at a high rate. It has also high efficiency for repairing and adding more features to the structure. The materials used in this process (mainly metal) are suitable for producing large components. In this process, a thin layer of wear-resistant metals is deposited on the components, which enhances the performance of the product. On the contrary, this process is not capable of producing parts with high complexity. Besides that, the parts produced by this process have poor dimensional accuracy, resolution, and surface finish. There is also a small number of materials that could be used in this process.

13.3.3 Material Extrusion

The printing machines used in this process are entry-level, making it a low-cost process. Furthermore, various raw materials are available to be used in this process. This process is also easy to handle and customizable. However, it takes a long time to produce parts by this process. Besides, the produced parts have a low level of precision. Sharp external corners could not also be built by this process.

13.3.4 Material Jetting

In this process, the amount of wasted materials is negligible which is an important advantage. Besides, parts produced with this method have high resolution and accuracy. Using material jetting, it is possible to produce parts consisting of multiple materials and colors as well.

In the post-processing stage, there is the probability of damaging small and thin features. It should also be mentioned that the produced structures could not be recycled.

13.3.5 Powder Bed Fusion

The material used in this process (i.e., polymer and metal) could be recycled. A wide variety of materials are used in the powder bed fusion method. The produced parts have high complexity, and those of metals have good accuracy and resolution. A limitation of this method is that the build rate is relatively low and only small to medium parts can be produced. In addition, the printing machines used in this process are expensive, and the produced parts from polymer have a rough surface finish.

13.3.6 Sheet Lamination

In the sheet lamination process, the fabrication rate is high. In this process, produced parts could contain multi-materials and multi-colors. Furthermore, low internal stress is forced on the parts during processing. However, the material waste is high in this method. In addition, due to the thermal cutting process, noxious fumes are produced. The lamination has also the possibility of warping due to the heat of the laser.

13.3.7 Vat Photopolymerization

In the vat photopolymerization process, the fabrication rate is high. Moreover, specific energy related to imaging is low. Many materials are used in this process. The produced parts have a high resolution and accuracy as well as a good surface finish. On the other hand, post-processing (i.e., cutting) is required to enhance the strength of parts produced.

13.4   New Materials Used in 3D Printing

Thanks to recent developments in 3D printing techniques, several newer materials have been developed for AD. These types of materials are categorized as follows [3].

13.4.1   Digital Materials

Digital materials are advanced composite materials including two or three photopolymers in specific microstructures. Such materials are useful for producing functional prototypes with different characteristics such as colors, superficial hardness, and textures. Each of the characteristics could be tuned. The latest Stratasys J750 is an example of this material that could be a combination of over 360,000 colors and use up to six materials simultaneously in a single build process without exchanging canisters.

Additionally, digital materials could act similar to industrial parts such as elastomers and standard plastics. Several photorealistic details could be created with digital materials that are applicable in industries such as manufacturing tooling, medical models, communication models, and functional prototyping. The design of digital materials for the 3D printing process has been studied in [24]. Digital materials are capable of producing multi-material objects with microscale accuracy [24].

13.4.2   Smart Materials for 4D Printing

Smart materials are capable of transforming their geometric shape when an external stimulus is applied to them ([25–27]). They are used in 4D printing, which was discussed earlier. For instance, Ge et al. used the concept of 4D printing in order to design and fabricate the active origami, where a flat plate with active hinges could be folded into a 3D component [28]. Smart materials have been used in the fabrication of complex self-evolving structures. They are capable of transforming into a predetermined shape [29].

Shape memory polymer which could return from a deformed shape to its original shape is a polymeric smart material. 4D printing of multi-material architectures which employ shape memory polymer has been introduced in [30]. This new approach applies the so-called projection micro-stereolithography method.

Biomimetic 4D printing is one aspect of 4D printing. In recent years, Gladman et al. presented a new approach for biomimetic 4D printing [31]. In this approach, the alignment of cellulose fibrils controls composite hydrogel architectures alongside prescribed four-dimensional paths. It is necessary to find out that the efficacy of the biomimetic printing concept could only be discovered based on the local control of the orientation of cellulose fibrils in the hydrogel composite.

13.4.3   Ceramics Materials

Metals and polymers could be fused using heat applied at the melting temperature. In contrast, ceramics and concrete cannot be fused together using this approach. As a result, providing heat in the melting point of ceramics materials is a critical challenge in 3D printing. Several methods of additive manufacturing of ceramics materials have been reviewed in [32]. New advanced methods are capable of producing ceramics components in which there are no cracks or large pores. These methods are based on optimizing the parameters of the additive manufacturing (AM) process, in which the mechanical properties of the materials could be tuned to be similar to those of ceramics parts fabricated traditionally. Powder-based technology in AM methods is one of the most economic ways for producing ceramics parts because of the low cost of raw material, the possibility of parallel processing of multiple parts, manufacturing scalability issues, and opportunities of 3D printing of ceramics materials [33].

One of the limitations of 3D printing of these materials is the availability of starting materials. This issue has been addressed in [34], in which specific pre-ceramics monomers that mix with UV photo-initiator have been used in the process.

13.4.4 Electronic Materials

In recent years, significant advancements in electronic materials that are used in 3D printing techniques have emerged [34, 35]. In new AM technologies, there is the possibility of producing functional electronic devices such as inductors, antennas, and capacitors. One of the advantages of these technologies is that no post-processing is required [36].

Inkjet and aerosol jet printing are two common methods, in which there is no direct contact with the nozzles and 3D-printed electronic devices. As a result, they are categorized in non-contact 3D printing methods. Fabrication of thin-film transistors using self-synthesized silver ink has been addressed in [34]. Methods of printing resistors with higher resistance value and repeatability have been demonstrated in [37].

Several materials are proved to have the ability to be used in 3D printing electronic device components with active properties. Examples of these materials are quantum dot nanoparticles, metal leads, conducting polymers, and elastomeric matrix [38, 39]. For instance, quantum dot nanoparticles could be used in the fabrication of diodes exhibiting tunable color emission properties. In [39], it is concluded that AM is a versatile method of fabricating electronic devices that a wide range of materials could be used for this purpose. Fabrication of emitters with low cost using silver nanoparticle ink was addressed in [40]. These emitters are applicable in patch antennas used in wireless communication industries. This method could be a functional as well as a low-cost solution for prototyping emitters through 3D printing.

3D printing technique has also a large number of applications in robotics. As shown in the Chapter 3 (Robotics a Key Driver of Industry 4.0), there are soft robots against conventional ones, which are a combination of rigid materials and electronic components. Fabrication and assembly of these parts require a multi-step process. In contrast, soft robots can be 3D printed using multi-material embedded 3D printing techniques. No electronic components exist in soft robots that are 3D printed. Instead of electronic devices, soft robots could be controlled through other methods such as microfluidic logic, which regulates all movements of the soft robot [41].

13.4.5 Biomaterials

In recent years, in biocompatible materials, 3D bioprinting of functional living tissues has become possible, which opens opportunities for new applications such as regenerative medicine used in organs transplantation [42–45]. There are a limited number of materials capable of being used in 3D bioprinting such as modified diblock copolymers, chitosan, and sodium alginate [46–49]. However, material selection is extremely critical in this field [42].

From recent developments in this field is an integrated tissue-organ printer, which fabricates human-scale tissue structures with the desired shape and mechanical stability [50]. The important feature of this printer is producing multi-material freeform shape and multiple types of biomaterials and cells such as calvarial bone, skeletal muscle, and cartilage.

Recently, 3D bioprinting has contributed to animal studies in laboratories as an alternative to the conventional method. One example is organs-on-chips (known as microphysiological systems) mimicking the function of native tissues.

13.5 Current Challenges and Future Trends

13.5.1 Volume and Cost of the Production

Current metal AM technologies are most suitable for applications in which a customized product demands high complexity geometries. On the other hand, there is no need to produce these parts on large scale [51, 52]. In fact, for manufacturing parts in large numbers, the production cost of metal AM is higher than that with conventional manufacturing methods because of the material cost and the cycling time required for components with similar geometry [53]. Many companies have introduced higher production volume in AD techniques, which remains a research frontier with a high potential for commercial utilization [54].

13.5.2 The Quality of the Product

Variations in mechanical properties, which lead to the lack of precision, are another serious challenge in this field [55]. It is observed that metal AM products may have heterogeneity and anisotropy in their microscopic structure [56]. Material properties and behavior may change during AD manufacturing process due to cyclic thermal loading conditions [57]. As a result, inspections are needed after AD manufacturing to identify defects such as undesirable grain characteristics, porosity, and voids. Post-processing operations such as treatments using heat may help to reach the required mechanical properties [56].

Another problem is the deviation of components from actual size and geometry [58]. The dimensional accuracy of the printed part may also be affected by residual stresses [59, 60]. Quality control methods and standards should be developed to come up with suitable solutions for these problems [54].

13.5.3 Post-Processing

Post-processing such as stress relief, machining operations, and support structure removal may be required. This post-processing might vary depending on the metal AM technology. These processes may help to reduce defects such as lack of fusion and porosity. On the other hand, they increase the cost of the process [61]. Among these operations, a stress-relieving operation is of more importance because it is required in many AM technologies and also residual stress can generate part distortion defects such as cracks and wraps that reduce the functionality of end-use metal parts [62, 63]. In stress relieving operations, usually, residual stresses are developed by the rapid thermal cycles.

Hot isostatic pressing is a commonly used way to release stress. It improves mechanical properties as well as decreases existing porosity after being built [64]. It was shown that hot isostatic pressing reduces the tensile strength due to the microstructure changes. On the other hand, it improves the ductility of the produced parts. As a result, a heat treatment operation after the hot isostatic pressing is recommended to reach more effectiveness of post-processing [65, 66]. In addition, dimensional accuracy may deteriorate due to post-processing, and it is suggested to heat parts before finishing operations [67]. To guarantee the printed parts' dimensional accuracy, some techniques such as chemical etching, machining, and vibrating may be required [67]. This action causes extra cost and time in AM processes. Minimizing the time and cost remains a challenge that has gained significant attention. Developing methods including the development of technologies such as design software and using hybrid AM systems is probably effective.

13.5.4 Repairs and Maintenance

A serious challenge in AM field is the challenges associated with disassembling components after the process. Disassembling is an important step in the repair and maintenance field [68]. It was shown that disassembling and recycling multi-material assemblies are difficult tasks [69]. The total costs of consolidation with AM were analyzed, and it was concluded that it is not economical to print consolidated and complex components by AM techniques than conventional manufacturing [70]. As a result, it is necessary to develop maintenance methodologies, optimization techniques, and design strategies to reduce the costs.

13.5.5 Low Variety of Materials

Despite the emergence of new materials, the number of ingredients used for AM technologies commercially is limited. Some of these materials and structures that are not compatible with current AM techniques are magnetic alloys [71], functionally graded materials [72], high-entropy alloys [73], bulk metallic glasses [74], nano-architected metals [74], and novel metal composite structures [75, 76]. Developments in AM systems will lead to the adoption of AM techniques to a broader range of materials [77].

13.5.6 3D Printing Technology AM Costs

In general, AM processing is an expensive technique for which the costs should be evaluated before starting manufacturing. The price of metal 3D printers varies from USD115,000 to USD1.9 million [54]. Prior to large-scale investment, AM production costs must be estimated carefully, and evaluate whether the conventional manufacturing techniques such as forging are more economical than AM. The cost of AM processes consists of gas lines, industrial compressors, electrical works, heat treatment furnaces, post-processing machines, and so on. The development of a cost analysis model for estimating the total expenses of a metal 3D printed part compared with conventional techniques is another future contribution in this field.

13.6   Conclusion

In additive manufacturing technique, three-dimensional parts are produced. In this process, materials are added gradually layer by layer to the parts. Metals, polymers, and so forth are applied as raw materials in AM techniques. Every production of AM has specific properties and performance that depend on the required application. Desirable properties could be precisely designed, through proper selection of the processing type and material compositions. In this chapter, seven types of AM processes including binder jetting, direct energy deposition, material extrusion, material jetting, powder bed fusion, sheet lamination, and vat photopolymerization were briefly explained. In each process, special materials used in the process were introduced, and their applications in diverse industries were reviewed. Finally, several remaining challenges in this field for future contributions were introduced.

References

1 Tibbits, S. (2014). 4D printing: multi-material shape change. *Archit. Des.* 84 (1): 116–121.

2 Ge, Q., Qi, H.J., and Dunn, M.L. (2013). Active materials by four-dimension printing. *Appl. Phys. Lett.* 103 (13): 131901.

3 Lin, L., Fang, Y., Liao, Y., Chen, G., Gao, C., and Zhu, P. (2019). 3D printing and digital processing techniques in dentistry: a review of literature. *Adv. Eng. Mater.* 21 (6): 1801013.

4 Lee, J.-Y., An, J., and Chua, C.K. (2017). Fundamentals and applications of 3D printing for novel materials. *Appl. Mater. Today* 7: 120–133.

5 Jackson, B. (2017). GE teases details of prototype H1 binder jet 3D printer – 3D Printing Industry. https://3dprintingindustry.com/news/ ge-teases-details-prototype-h1-binder-jet-3d-printer-126072.

6 Methani, M.M., Revilla-León, M., and Zandinejad, A. (2020). The potential of additive manufacturing technologies and their processing parameters for the fabrication of all-ceramic crowns: a review. *J. Esthet. Restor. Dent.* 32 (2): 182–192.

7 Engineeringproductdesign. (2017–2021). What is direct energy deposition, its types and pros and cons. https://engineeringproductdesign.com/knowledge-base/ direct-energy-deposition.

8 Razavi, S.M.J., and Berto, F. (2019). Directed energy deposition versus wrought Ti-6Al-4V: a comparison of microstructure, fatigue behavior, and notch sensitivity. *Adv. Eng. Mater.* 21 (8): 1900220.

9 DREAMS. (2021). Material Extrusion. http://seb199.me.vt.edu/dreams/ material-extrusion.

10 Omnexus. (2021).Complete Guide on Thermoplastic Polyurethanes (TPU). https:// omnexus.specialchem.com/selection-guide/thermoplastic-polyurethanes-tpu.

11 Spoerk, M., Holzer, C., and Gonzalez-Gutierrez, J. (2020). Material extrusion-based additive manufacturing of polypropylene: A review on how to improve dimensional inaccuracy and warpage. *J. Appl. Polym. Sci.* 137 (12): 48545.

12 RapidModel. (2020). Rubber-like, Polyjet Material. https://www.rapidmodel.com.my/portfolio-item/rubber-like.

13 Murr, L.E., and Li, S. (2016). Electron-beam additive manufacturing of high-temperature metals. *MRS Bull.* 41 (10): 752–757.

14 Biocompatible 3D Printing Materials Market (MarketsandMarkets™). (2018). Biocompatible 3D printing materials market by type (polymer, metal), application (implants & prosthesis, prototyping & surgical guides, tissue engineering, hearing aid), form (powder, liquid), and region - global forecast to 2023. https://www.marketsandmarkets.com/Market-Reports/biocompatible-3d-printing-material-market-236523134.html.

15 Sireesha, M., Lee, J., Kiran, A.S.K., Babu, V.J., Kee, B.B.T., and Ramakrishna, S. (2018). A review on additive manufacturing and its way into the oil and gas industry. *RSC Adv.* 8 (40): 22460–22468.

16 Kruth, J., Mercelis, P., Van Vaerenbergh, J., Froyen, L., and Rombouts, M. (2005). Binding mechanisms in selective laser sintering and selective laser melting. *Rapid Prototyp. J .* 11 (1): 26–36.

17 Yap, C.Y., Chua, C.K., and Dong, Z.L. (2016). An effective analytical model of selective laser melting. *Virtual Phys. Prototyp.* 11 (1): 21–26.

18 Yap, C.Y., et al. (2015). Review of selective laser melting: Materials and applications. *Appl. Phys. Rev.* 2 (4): 41101.

19 Tan, X., et al. (2016). Revealing martensitic transformation and α/β interface evolution in electron beam melting three-dimensional-printed Ti-6Al-4V. *Sci. Rep.* 6 (1): 1–10.

20 Kok, Y.H., Tan, X.P., Loh, N.H., Tor, S.B., and Chua, C.K. (2016). Geometry dependence of microstructure and microhardness for selective electron beam-melted Ti–6Al–4V parts. *Virtual Phys. Prototyp.* 11 (3): 183–191.

21 Rieder, H., Dillhöfer, A., Spies, M., Bamberg, J., and Hess, T. (2014). Online monitoring of additive manufacturing processes using ultrasound. In: *11*[th] *European Conference on Non-Destructive Testing (ECNDT)*, October 6–10, Prague: Czech Republic.

22 Gibson, I., Rosen, D., and Stucker, B. (2010). *Additive Manufacturing Technologies: Rapid Prototyping to Direct Digital Manufacturing*. Springer, New York, 459.

23 Dermeik, B., and Travitzky, N. (2020). Laminated object manufacturing of ceramic-based materials. *Adv. Eng. Mater.* 22 (9): 2000256.

24 Hiller, J., and Lipson, H. (2009). Design and analysis of digital materials for physical 3D voxel printing. *Rapid Prototyp. J.* 15 (2): 137–149.

25 An, J., Chua, C.K., and Mironov, V. (2016). A perspective on 4D bioprinting. *Int. J. Bioprinting* 2 (1): 3–5.

26 Khoo, Z.X., et al. (2015). 3D printing of smart materials: a review on recent progresses in 4D printing. *Virtual Phys. Prototyp.* 10 (3): 103–122.

27 Leist, S.K., and Zhou, J. (2016). Current status of 4D printing technology and the potential of light-reactive smart materials as 4D printable materials. *Virtual Phys. Prototyp.* 11 (4): 249–262.

28 Ge, Q., Dunn, C.K., Qi, H.J., and Dunn, M.L. (2014). Active origami by 4D printing. *Smart Mater. Struct.* 23 (9): 94007.

29 Raviv, D., et al. (2014). Active printed materials for complex self-evolving deformations. *Sci. Rep.* 4 (1): 1–8.

30 Ge, Q., Sakhaei, A.H., Lee, H., Dunn, C.K., Fang, N.X., and Dunn, M.L. (2016). Multimaterial 4D printing with tailorable shape memory polymers. *Sci. Rep.* 6 (1): 1–11.

31 Gladman, A.S., Matsumoto, E.A., Nuzzo, R.G., Mahadevan, L., and Lewis, J.A. (2016). Biomimetic 4D printing. *Nat. Mater.* 15 (4): 413–418.

32 Deckers, J., Vleugels, J., and Kruth, J.-P. (2014). Additive manufacturing of ceramics: a review. *J. Ceram. Sci. Technol.* 5 (4): 245–260.

33 Zocca, A., Colombo, P., Gomes, C.M., and Günster, J. (2015). Additive manufacturing of ceramics: issues, potentials, and opportunities. *J. Am. Ceram. Soc.* 98 (7): 1983–2001.

34 Eckel, Z.C., Zhou, C., Martin, J.H., Jacobsen, A.J., Carter, W.B., and Schaedler, T.A. (2016). Additive manufacturing of polymer-derived ceramics. *Science (80-.).* 351 (6268): 58–62.

35 Saengchairat, N., Tran, T., and Chua, C.-K. (2017). A review: Additive manufacturing for active electronic components. *Virtual Phys. Prototyp.* 12 (1): 31–46.

36 Tan, H.W., Tran, T., and Chua, C.K. (2016). A review of printed passive electronic components through fully additive manufacturing methods. *Virtual Phys. Prototyp.* 11 (4): 271–288.

37 Jung, S., Sou, A., Gili, E., and Sirringhaus, H. (2013). Inkjet-printed resistors with a wide resistance range for printed read-only memory applications. *Org. Electron.* 14 (3): 699–702.

38 Kong, Y.L., et al. (2014). 3D printed quantum dot light-emitting diodes. *Nano Lett.* 14 (12): 7017–7023.

39 Lewis, J.A., and Ahn, B.Y. (2015). Three-dimensional printed electronics. *Nature* 518 (7537): 42–43.

40 Goh, G.L., Ma, J., Chua, K.L.F., Shweta, A., Yeong, W.Y., and Zhang, Y.P. (2016). Inkjet-printed patch antenna emitter for wireless communication application. *Virtual Phys. Prototyp.* 11 (4): 289–294.

41 Wehner, M., et al. (2016). An integrated design and fabrication strategy for entirely soft, autonomous robots. *Nature* 536 (7617): 451–455.

42 Murphy, S.V., and Atala, A. (2014). 3D bioprinting of tissues and organs. *Nat. Biotechnol.* 32 (8): 773–785.

43 Lee, J.M., Sing, S.L., Tan, E.Y.S., and Yeong, W.Y. (2016). Bioprinting in cardiovascular tissue engineering: a review. *Int. J. Bioprinting* 2 (2).

44 An, J., and Chua, C.K. (2016). An engineering perspective on 3D printed personalized scaffolds for tracheal suspension technique. *J. Thorac. Dis.* 8 (12): E1723.

45 Suntornnond, R., An, J., and Chua, C.K. (2017). Bioprinting of thermoresponsive hydrogels for next generation tissue engineering: a review. *Macromol. Mater. Eng.* 302 (1): 1600266.

46 Ng, W.L., Yeong, W.Y., and Naing, M.W. (2016). Polyelectrolyte gelatin-chitosan hydrogel optimized for 3D bioprinting in skin tissue engineering. *Int. J. Bioprinting* 2 (1): 53–62.

47 Derby, B. (2012). Printing and prototyping of tissues and scaffolds. *Science 80:* 338 (6109): 921–926.

48 Li, H., Liu, S., and Lin, L. (2016). Rheological study on 3D printability of alginate hydrogel and effect of graphene oxide. *Int. J. Bioprinting* 2 (2): 54–66.

49 Suntornnond, R., Tan, E.Y.S., An, J., and Chua, C.K. (2016). A mathematical model on the resolution of extrusion bioprinting for the development of new bioinks. *Materials (Basel).* 9 (9): 756.

50 Kang, H.-W., Lee, S.J., Ko, I.K., Kengla, C., Yoo, J.J., and Atala, A. (2016). A 3D bioprinting system to produce human-scale tissue constructs with structural integrity. *Nat. Biotechnol.* 34 (3): 312–319.

51 Yamazaki, T. (2016). Development of a hybrid multi-tasking machine tool: Integration of additive manufacturing technology with CNC machining. *Procedia Cirp.* 42: 81–86.

52 Gupta, N., Weber, C., and Newsome, S. (2012). Additive manufacturing: status and opportunities. *Sci. Technol. Policy Inst. Washingt. DC:1–29.*

53 Thompson, A., Maskery, I., and Leach, R.K. (2016). X-ray computed tomography for additive manufacturing: a review. *Meas. Sci. Technol.* 27 (7): 72001.

54 Vafadar, A., Guzzomi, F., Rassau, A., and Hayward, K. (2021). Advances in metal additive manufacturing: a review of common processes, industrial applications, and current challenges. *Appl. Sci.* 11 (3): 1213.

55 Anandan, H., and Kumaraguru, S. (2019). Distortion in metal additive manufactured parts. In: *3D Printing and Additive Manufacturing Technologies* (ed. L.J. Kumar, P.M. Pandey, and D.I. Wimpenny), 281–295. Springer.

56 Kok, Y., et al. (2018). Anisotropy and heterogeneity of microstructure and mechanical properties in metal additive manufacturing: a critical review. *Mater. Des.* 139: 565–586.

57 Seifi, M., et al. (2017). Progress towards metal additive manufacturing standardization to support qualification and certification. *Jom* 69 (3): 439–455.

58 Xiao, J., Anwer, N., Durupt, A., Le Duigou, J., and Eynard, B. (2018). Information exchange standards for design, tolerancing and additive manufacturing: a research review. *Int. J. Interact. Des. Manuf.* 12 (2): 495–504.

59 Oropallo, W., and Piegl, L.A. (2016). Ten challenges in 3D printing. *Eng. Comput.* 32 (1): 135–148.

60 Ding, D., Pan, Z., Cuiuri, D., and Li, H. (2015). Wire-feed additive manufacturing of metal components: technologies, developments and future interests. *Int. J. Adv. Manuf. Technol.* 81 (1): 465–481.

61 Lewandowski, J.J., and Seifi, M. (2016). Metal additive manufacturing: A review of mechanical properties. *Annu. Rev. Mater. Res.* 46: 151–186.

62 Prashanth, K.G., et al. (2014). Microstructure and mechanical properties of Al–12Si produced by selective laser melting: Effect of heat treatment. *Mater. Sci. Eng. A* 590: 153–160.

63 Bartlett, J.L., and Li, X. (2019). An overview of residual stresses in metal powder bed fusion. *Addit. Manuf.* 27: 131–149.

64 Herzog, D., Seyda, V., Wycisk, E., and Emmelmann, C. (2016). Additive manufacturing of metals. *Acta Mater.* 117: 371–392.

65 Cai, C., et al. (2016). Effect of hot isostatic pressing procedure on performance of Ti6Al4V: Surface qualities, microstructure and mechanical properties. *J. Alloys Compd.* 686: 55–63.

66 Tammas-Williams, S., Withers, P.J., Todd, I., and Prangnell, P.B. (2016). The effectiveness of hot isostatic pressing for closing porosity in titanium parts manufactured by selective electron beam melting. *Metall. Mater. Trans. A* 47 (5): 1939–1946.

67 Oyelola, O., Crawforth, P., M'Saoubi, R., and Clare, A.T. (2016). Machining of additively manufactured parts: implications for surface integrity. *Procedia Cirp.* 45: 119–122.

68 Campbell, R.I., Jee, H., and Kim, Y.S. (2013). Adding product value through additive manufacturing. In: *ICED 13: 19th International Conference on Engineering Design. Proceedings Volume DS 75-4. Design for Harmonies: Volume 4: Product, Service and Systems Design* (ed. U. Lindeman et al.), Seoul, Korea, August 19–22.pp. 25–268.

69 Thompson, M.K., et al. (2016). Design for additive manufacturing: Trends, opportunities, considerations, and constraints. *CIRP Ann.* 65 (2): 737–760.

70 Knofius, N., Van Der Heijden, M.C., and Zijm, W.H.M. (2019). Consolidating spare parts for asset maintenance with additive manufacturing. *Int. J. Prod. Econ* 208: 269–280.

71 Yang, X., et al. (2019). Effect of remelting on microstructure and magnetic properties of Fe-Co-based alloys produced by laser additive manufacturing. *J. Phys. Chem. Solids* 130: 210–216.

72 Wu, D., Gao, W., Hui, D., Gao, K., and Li, K. (2018). Stochastic static analysis of Euler-Bernoulli type functionally graded structures. *Compos. Part B Eng.* 134: 69–80.

73 Chew, Y., et al. (2019). Microstructure and enhanced strength of laser aided additive manufactured CoCrFeNiMn high entropy alloy. *Mater. Sci. Eng. A* 744: 137–144.

74 Bordeenithikasem, P., Stolpe, M., Elsen, A., and Hofmann, D.C. (2018). Glass forming ability, flexural strength, and wear properties of additively manufactured Zr-based bulk metallic glasses produced through laser powder bed fusion. *Addit. Manuf.* 21: 312–317.

75 Ryder, M.A., Lados, D.A., Iannacchione, G.S., and Peterson, A.M. (2018). Fabrication and properties of novel polymer-metal composites using fused deposition modeling. *Compos. Sci. Technol.* 158: 43–50.

76 Martin, J.H., Yahata, B.D., Clough, E.C., Mayer, J.A., Hundley, J.M., and Schaedler, T.A. (2018). Additive manufacturing of metal matrix composites via nanofunctionalization. *MRS Commun.* 8 (2): 297.

77 Ngo, T.D., Kashani, A., Imbalzano, G., Nguyen, K.T.Q., and Hui, D. (2018). Additive manufacturing (3D printing): A review of materials, methods, applications and challenges. *Compos. Part B Eng.* 143: 172–196.

Glossary

Three-dimensional (3D) printing Also known as additive manufacturing (AM). A production technique that produces three-dimensional parts. In this process, materials are added gradually, layer by layer to the part.

3rd Generation Project Partnership (3GPP) A global joint organisation composed of mobile and telecom standard development companies that collaborate and create cellular communications standards. It provides technical reports and specifications for cellular network technologies and mobile systems development.

Four-dimensional (4D) printing An advanced production technique in which time is also added as the fourth dimension. The key feature of (four-dimensional (4D) printing technologies is the ability to utilize programmable smart materials, as their shape changes over time under an external stimulus.

5G New Radio (NR) A set of new radio access technology standards that uses OFDMA, MIMO, licensed/unlicensed spectrum, and beamforming to assist 5G wireless communication growth. The air interface specification of 5G NR and its requirements for connectivity are developed by 3GPP.

Accessibility The capability to access the information as needed, at any time and place using any device with an Internet connection.

Actuator A device deployed to move and control a mechanism or system, for example, by opening a valve. Its system consists of control signals and a source of energy.

Adaptive neuro-fuzzy inference system (ANFIS) A kind of artificial neural network based on Takagi–Sugeno fuzzy inference system. The technique was developed in early 1990s. ANFIS integrates both neural networks and fuzzy logic principles, it has the potential to capture the benefits of both in a single framework.

Additive manufacturing See three-dimensional (3D) printing.

Advanced metering infrastructure A system on the demand side that is responsible for measuring, processing, analyzing, communicating, and storing energy consumption data. It supports two-way power flow monitoring, bidirectional data flow between the consumer and the utility, and also enables energy-aware programs.

Application Programming Interface (API) A type of software interface that connects computers or pieces of software to each other.

Industry 4.0 Vision for the Supply of Energy and Materials: Enabling Technologies and Emerging Applications, First Edition. Edited by Mahdi Sharifzadeh.
© 2022 John Wiley & Sons, Inc. Published 2022 by John Wiley & Sons, Inc.

Artificial intelligence (AI) A branch of computer science that enables the utilization of machine capabilities to perform human tasks like perceiving, reasoning, learning, and problem-solving.

Artificial neural network (ANN) Computing systems inspired by the biological neural networks that constitute animal brains. Based on a collection of connected units or nodes called artificial neurons, which loosely model the neurons in a biological brain. Each connection, like the synapses in a biological brain, can transmit a signal to other neurons. An artificial neuron that receives a signal then processes it and can signal neurons connected to it. The "signal" at a connection is a real number, and the output of each neuron is computed by some nonlinear function of the sum of its inputs.

Asset Administration Shell (AAS) This provides a standardized digital presentation of the asset. The digital twin in the context of Industry 4.0 is implemented using an AAS.

American Society for Testing and Materials (ASTM) An international standards organization that develops and publishes voluntary consensus technical standards for a wide range of materials, products, systems, and services.

Augmented reality (AR) A technology that replaces a real-world environment with a fully digital alternative and integrates perceptual information and digital objects such as visual/audio information over real-world objects. Offers an interactive experience of the physical world to users that is achieved by the usage of the device camera and its sensors.

Battery electric vehicle (BEV) A vehicle that is fully powered by electric power using a rechargeable battery, which is charged by plugging into an external electricity source.

Battery swapping A developing concept for EV charging that replaced the depleted battery with a fully charged one without waiting for charging.

Big data High-volume data with high frequency, variety, and variability.

Binder jetting One of the additive manufacturing processes in which a a liquid binder is utilized to adhere the particles.

Block header Used for block identification and includes block hash, timestamp, Merkle root, and Nonce block.

Block A set of recorded data including block header and transaction records.

Blockchain Blockchain is a set of chained information blocks for storing transaction records that are shared between all peers distributed all over a network. It consists of a system for recording information that chains the data blocks together to form a chronological single-source-of-truth for the data that is difficult or almost impossible to change and hack.

Blockchain technology A combination of a digital data structure and a decentralized consensus mechanism to maintain the consistency and precision of the data stored in a ledger where the ledger is distributed through a peer-to-peer system. A chain of blocks can be created, maintained, and stored by multiple entities in the network, and each entity is able to verify that the chain order and data are not tampered with.

Cloud computing Services including servers, storage, databases, networking, software, analytics, and intelligence over the Internet instead of their computers.

Cloud customer The user of cloud services.

Cloud provider The company that provides cloud services.

Community cloud: A type of cloud computing implementation model to share some services to a limited number of organizations.

Computer vision A scientific field related to how computers can gain high-level features from digital images or videos. In engineering fields, it is used to understand and automate tasks that are performed by the human visual system.

Computer-aided molecular design (CAMD) A promising technique that can accelerate the discovery of new molecules, such as solvents, refrigerants, and pharmaceutical products. CAMD problems are defined as given a set of building blocks and a specified set of target properties, determine the molecule or molecular structure that matches these properties.

Conductive charging The direct connection of an electric vehicle to the electrical outlet at home or a charging station through a standard cable.

Connected and autonomous vehicle (CAV) Different types of vehicles that use artificial intelligence for operation using various human inputs and can communicate with the driver, other cars on the road (V2V), roadside infrastructures (V2I), the pedestrians (V2P), and the cloud (V2C).

Consensus A distributed mechanism in blockchains for reaching an agreement among all users, regarding new transactions and corresponding blocks.

Continuous manufacturing A mode of manufacturing where raw materials enter and final products leave the manufacturing system uninterruptedly.

Cryptography The technique for secure communication in the presence of third parties or the public to protect users' privacy and assets by converting information from a readable state to unintelligible nonsense.

Cybersecurity Also known as information technology security. Refers to a series of actions for protecting electronic devices, communicating networks, and data from cyber-attacks.

Cyberattack A n attempt to damage, disrupt, disable, destroy, or take control of information, networks, infrastructures, or personal computer devices via cyberspace.

Cyber-physical system (CPS) A system that integrates networks of computational and physical components, and interconnects them to the Internet and each other. Emerging systems of intelligence that range from simple devices to complex systems-of-systems where their operations are coordinated and monitored by a computing and communicating core.

Decentralized architecture Does not need central coordination, and peers can directly communicate for automatic control.

Degrees of freedom The number of independent motions in a robot, which could be angular or translational.

Demand response Refers to all intentional actions of end users to modify electricity consumption patterns by changing the timing, level of instantaneous use, or total electricity consumption.

Digital materials Advanced composite materials including two or three photopolymers in specific microstructures. Such materials are useful for producing functional prototypes with different characteristics each of them could be tuned. Could act similar to industrial parts such as elastomers and standard plastics.

Digital twin A virtual representation that serves as the real-time digital counterpart of a physical process.

Digitalization The conversion of physical information into digital information for the purpose of process design, control, and optimization, for example in the context of pharmaceutical manufacturing.

Direct energy deposition (DED) One of the additive manufacturing processes in which energy is sent into a small region so that it provides heat to melt materials being deposited. Classified into three groups in terms of the energy source used for melting materials: laser-based DED, electron beam-based DED, and plasma or electric arc-based DED.

Distributed energy resources Distributed energy resources are electric generation units located near the end-users. They are usually in the range of 3kW to 50 MW.

Distributed ledger A distributed ledger is a common database that is distributed and stored over different nodes of a network of computers

Edge computing A specific architecture for cloud computing in which computations are conducted close to the data sources.

Elasticity The scalability of the cloud to automatically increase or decrease its services in order to meet the user demands.

Electric vehicle (EV) Vehicle that is partially or fully powered by electric power for moving.

Energy hub An energy hub is the core of multi-energy systems. It is a place where different forms of energy can be converted, conditioned, or stored.

Energy router The core of an Internet of energy (IoE) system. Provides two-way flows of information and energy between the IoE components to facilitate the integration of renewable energy sources and storage devices into the conventional power grid and manage energy optimally.

FedRAMP Federal Risk and Authorization Management Program.

Fog computing A specific architecture for cloud computing that utilizes IoT devices as a data acquisition system and processes IoT data in real time.

Food and Drug Administration (FDA) The U.S. federal government agency responsible for inspecting, testing, approving, and setting safety standards for food, drugs, chemicals, cosmetics, and household and medical devices.

Genetic algorithm (GA) A metaheuristic inspired by the process of natural selection that belongs to the larger class of evolutionary algorithms (EA). Commonly used to generate high-quality solutions to optimization and search problems by relying on biologically inspired operators such as mutation, crossover, and selection.

Genetic programming (GP) Inspired by biological evolution and its fundamental mechanisms, these systems implement an algorithm that uses random mutation, crossover, a fitness function, and multiple generations of evolution to resolve a user-defined task. Can be used to discover a functional relationship between features in data (symbolic regression), to group data into categories (classification), and to assist in the design of electrical circuits, antennae, and quantum algorithms.

Greenhouse gas Absorbs and emits infrared radiation.

Grid computing The practice of using a widely distributed network of computers for a common specific purpose.

Group contribution methods (GCM) Uses the principle that some simple aspects of the structures of chemical components are always the same in many different molecules. The smallest common constituents are atoms and bonds. A group-contribution method is used to predict properties of pure components and mixtures by using group or atom properties. This reduces the number of needed data dramatically. Instead of needing to know the properties of thousands or millions of compounds, only data for a few dozens or hundreds of groups must be known.

Hash cryptography A mathematical one-way function to transform block data into a specific length called hash output to enhance security in blockchain. In the hash cryptographic, it is almost impossible to regenerate the original input data from the hash output.

Heavy-duty cars Defined as any passenger car with more than eight seats or freight vehicles that can carry more than 3.5 tons.

High throughput screening (HTS) A screening method where a large number of testing is performed and the data is collected for analysis.

Human–machine interaction and interface The applications in which there are interactions with humans. Usually involve the risk of injury.

Hybrid cloud A combination of private and public clouds as an extensive network.

Hybrid electric vehicle (HEV) A type of EV that is powered partially by the electric power of a battery and burns fossil fuel as the main power source.

Inductive charging Defined as a charging method that uses an electromagnetic field for transferring power to the electric vehicle without any physical contact.

Industrial IoT (IIoT) An aspect of the IoT in industrial sectors and applications that focuses on the connection of production control with instruments, production, and industrial processes to enable and enhance smart industrial operations. Aims at real-time data analysis to support advanced predictive analytics and preventive maintenance.

Industry 4.0 Also called the Fourth Industrial Revolution. The current trend of automation of manufacturing and industrial practices using modern smart technologies.

Infrastructure as a Service (IaaS) A type of cloud-based service that provides basic services (server, processor, storage, and other resources) as needed.

Intelligent transportation system (ITS) A transportation system equipped with data processing and communication technologies to improve the traffic management, safety, efficiency, and sustainability of people and goods mobility.

Interactive autonomous robots Used to connect with humans through verbal or nonverbal communication.

Internal combustion engine (ICE) vehicle Converts the chemical energy of burning fossil fuels into motion energy to run the vehicle.

International Electrotechnical Commission (IEC) An international standard body that sets and publishes international standards for all electrical, electronic, and related technologies. Provides a more consistent core standard for a vast range of electronic and technical products.

Internet Engineering Task Force (IETF) A premier world organization made up of network operators, vendors, and researchers that develop, promote, and define standard protocols and specifications governing the mechanism behind the Internet.

Internet of things (IoT) The network of physical objects (i.e., things) embedded with sensors, software, and other technologies for the purpose of connecting and exchanging data with other devices and systems over the Internet.

Kinematics A subfield in physics related to classical mechanics. Describes the motion bodies or points without considering the external forces that the system is exposed to.

Knowledge-aided assessment and structured application (KASA) Itroduced by the U.S. Food and Drug Administration (FDA) to capture and manage knowledge during the life cycle of a drug product; to establish rules and algorithms for risk assessment, control, and communication; to perform computer-aided analyses of applications to compare regulatory standards and quality risks across applications and facilities; and to provide a structured assessment that minimizes text-based narratives and summarization of provided information.

Laminated object manufacturing (LOM) One of the main sheet lamination processes that is a layer-by-layer approach similar to ultrasonic additive manufacturing (UMV), although it uses paper as material and adhesive for bonding sheets. In this process, there are diverse bonding mechanisms between the sheets, including adhesive, clamping, and thermal bonding.

Light-duty cars Any vehicle for transporting passengers and goods with a gross weight of less than 10,000 pounds.

Long-term evolution (LTE) A standard wireless communication for mobile systems that relies on the GSM/EDGE and UMTS/HSPA technologies. Proper for low latency applications such as IoT systems and offers higher network capacity, speed, and quality of services for mobile systems. Has an important role in the standard development of 5G new radio (NR).

Machine type communication (MTC) A modality of data communication between two entities (things) without the involvement of a human. Allows machines to communicate and share information with each other. Typically between an MTC device and an MTC server.

Material extrusion (ME) One of the additive manufacturing (AM) processes in which a nozzle pushes out heated material with constant pressure and speed. When the material that comes out of the nozzle is solidified on the substrate, the three-dimensional (3D) object is printed.

Material jetting (MJ) One of the additive manufacturing (AM) processes in which an ultraviolet (UV) light source is used to cure and solidify the deposited material.

Media access control (MAC) address A unique hardware address that identifies every active node of a network and is used as a network address to communicate within the segment of a network. In this context, MAC protocol is used to gain access to the physical layer of a LAN (i.e., network medium).

Merkle tree A data organization tool that includes the Merkle root in each block data structure. The leaves are the block hash values, and the non-leaf nodes are the cryptographic hash values composed of of their child's hash. The combination of hashes is stored within the Merkle root.

Microgrid A distribution level energy system that provides energy for local loads. Consists of various small power generating sources (micro sources) that make it very efficient and highly flexible. Key components are distributed generators, loads, and storage systems. Can work in two modes of connected and island.

Millimeter wave (mmWave) The band of extremely high frequency (EHF) that lies within the spectrum range of 30–300 GHz and enables a massive data rate in mobile and wireless networks. Considered for various applications in 5G that require a faster and enhanced capacity for communication.

Multiple input, multiple output (MIMO) A technology for wireless communication that takes advantage of multipath effects in a wireless environment and allows sending and receiving more data signals on the same channel and at the same time. Uses multiple antennas at sources and their destinations to send and receive data through a number of different paths. Results in fewer potential errors in data transmissions, enhanced data rates of the communication link, and optimized data speeds.

Multi-robot systems (MRS) Have several agents and benefit from collaborations between agents.

Multivariate linear regression (MLR) Also known simply as multiple regression. A statistical technique that uses several explanatory variables to predict the outcome of a response variable. Goal is to model the linear relationship between the explanatory (independent) variables and response (dependent) variables. In essence, the extension of ordinary least-squares (OLS) regression because it involves more than one explanatory variable.

Multi-vector energy system I ntegrates different forms of energy including electricity, gas, heat, and hydrogen to improve energy utilization efficiency and system resilience.

Multi-vector Internet of energy A multi-energy system based on a smart grid that is unified with the Internet of things (IoT) to enable access to distributed renewable energies in a large area and improve energy usage efficiency.

NAMUR Open Architecture (NOA) Aims to make production data easily and securely usable for plant and asset monitoring as well as optimization.

National Institute of Standards and Technology (NIST) A nonregulatory government agency that promotes technology, metrics, and standards for private sector organizations in the United States to drive innovation and industrial competitiveness, produces standards and guidelines, and assists in protecting their information and information systems through cost-effective programs. Guidance documents and recommendations provided through its special publications.

Nonce A random number added to hash that can be used only once and ensures the required difficulty level.

Open platform communications (OPC) A series of standards and specifications for industrial telecommunication. The original standard was named Object Linking and Embedding (OLE) for process control). OPC standards define an open and royalty-free set of standards for accessing and sharing real-time data from control systems and sensors.

Open process automation framework (OPAF) A standard developed by the Open Group, following an initiative led by Exxon Mobil with the aim of building an "open, interoperable, secure-by-design" process automation architecture.

Partial least-squares (PLS) regression A statistical method that has some similarities to principal components regression (PCR). In PLS instead of finding hyperplanes of maximum variance between the response and independent variables (that is done in PCR), it finds a linear regression model by projecting the predicted variables and the observable variables to a new space. Because both the X and Y data are projected to new spaces, the PLS family of methods are known as bilinear factor models.

Particle swarm optimization (PSO) A population-based optimization technique inspired by the motion of bird flocks and schooling fish. Shares many similarities with evolutionary computation techniques. The system is initialized with a population of random solutions, and the search for the optimal solution is performed by updating generations. Unlike GA, PSO has no evolution operators, such as crossover and mutation.

Pay-as-you-use In the context of cloud computing, refers to consumers pay providers based on their usage.

Peer-to-peer (P2P) network Interconnected nodes ("peers") share resources without the use of a centralized administrative system.

Pharma 4.0 The adoption of Industry 4.0 in pharmaceutical manufacturing.

Platform as a service (PaaS) A type of cloud computing service in which the service provider offers a platform service to customers.

Plug-and-play A set of operating system standards that enable automatic detection and configuration of a device as soon as it is connected.

Plug-in hybrid electric vehicle (PHEV) A type of EV capturing motion power from batteries that can be recharged through plugging into an electric power source. Also contains a small conventional internal combustion engine for sustainable supply of electricity to the battery and extends the driving range.

Post-processing Includes stress relief, machining operations, and support structure removal that may be required after any three-dimensional (3D) printing technique. Might vary depending on the metal additive manufacturing (AM) technology. These processes may help to reduce defects such as lack of fusion and porosity.

Powder bed fusion One of the additive manufacturing (AM) processes in which there is a highly energized laser or electron beam as the energy source. Sintering and melting are two kinds of binding mechanisms for this process.

Power line communication (PLC)–based energy router Uses modulation and multiplexing techniques such as time division multiple access to transfer both energy flows and information flows through the same transmission line. They reduce the cost and device volume.

Principal component regression (PCR) A regression analysis technique that is based on principal component analysis (PCA). More specifically, it is used for estimating the unknown regression coefficients in a standard linear regression model. In PCR, instead of regressing the dependent variable on the explanatory variables directly, the principal components of the explanatory variables are used as regressors.

Private cloud In contrast to public clouds, computing services are offered to specific organizations.

Process analytical technology (PAT) A technology that monitors the process parameters and/or the quality attributes of the products and/or intermediates in a manufacturing process.

Public cloud A platform that provides PaaS, IaaS, and SaaS to multiple customers over a public network.

Quality by design (QbD): A manufacturing approach where an understanding of the relationships among the critical process parameters and critical quality attributes is established to monitor and explain the causes of deviations.

Quality by testing (QbT) A manufacturing approach where the qualities of the final drug products are tested against specifications without the need of understanding the causes of deviation.

Quantitative structure–property relationships (QSPR) A powerful analytical method for breaking down a molecule into a series of numerical values describing its relevant chemical and physical properties. Involve regression or classification models used in the chemical and biological sciences and engineering. Like other regression models, these relate a set of "predictor" variables X, which are known as descriptors to the potency of the response variable, Y. Descriptors consist of physicochemical properties or theoretical molecular descriptors of chemicals.

Quantum mechanics (QM) A fundamental theory in physics that provides a description of the physical properties of nature at the scale of atoms and subatomic particles. It is the foundation of all quantum physics including quantum chemistry, quantum field theory, quantum technology, and quantum information science. Allows the calculation of the properties and behaviour of physical systems. Typically applied to microscopic systems: molecules, atoms, and subatomic particles.

Radio access network (RAN) The technology that uses radio links to provide device connectivity to different parts of networks and delivers specific services. A major aspect of the mobile networks. Used from 1G to 5G.

RAMI architecture Defines a service-oriented architecture where application components provide services through a communication protocol over a network. The basic principles are independent of vendors, products, and technologies.

Rehabilitation robotics A field of research related to understanding and improving rehabilitation through robotic devices.

Repairs and maintenance A combination of functional repairing or replacing of necessary devices, servicing, equipment, machinery, building infrastructure, and supporting utilities in industrial places.

Sensor Part of a robotic system used to detect events and changes in the environment and send the sensed data to a processor.

Serial and parallel industrial robots In the case in which there is only one loop, the structure of industrial robot is serial. In contrast, when more than one loop exists, the structure is called parallel.

Serverless services A cloud computing model in which the servers are in a cloud service and dynamically manage resource allocation.

Service level agreement (SLA) The level of service which a customer expects from a service provider.

Servo or non-servo Robots apply servo or non-servo control strategies. In non-servo robots, there are no closed control loops, and only open-loop devices exist and mechanical orders are predetermined. In contrast, in servo robots, a control loop exists and the actuation is performed with regard to online responses.

Sheet lamination One of the additive manufacturing (AM) processes. One of two main additive manufacturing (AM) processes: ultrasonic (UAM) and laminated object (LOM).

SigFox A type of global network operator that connects low-power objects such as utility smart meters, home appliances, and smartwatches. SigFox connects devices that need to be in continuous operation and devices that transmit small amounts of data in time intervals.

Sintering A partial melting process. In solid-state sintering, particles are fused only at the surface. As a result, they inherit the porosity of the original materials. On the other hand, in liquid-state melting, all particles are fully melted, fusing together that results in a dense substance.

Smart contract A flexible programmable contract that automate agreement procedures between contracts corresponds

Smart grid An electrical grid that uses digital communications to detect and responds to local changes in usage.

Smart manufacturing Enabled by data science and artificial intelligence, with advanced features such as adaptability to constant market changes, sustainability, and demand for highly skilled workforces.

Smart materials Smart materials are special materials with the capability of transforming their geometric shape when an external stimulus is applied to them. Used in four-dimesional (4D) printing process.

Soft robots Inspired by biological systems and consist of soft materials. In contrast to conventional robots made with hard materials such as steel, titanium, or aluminum, they are made from hyper-elastic materials such as polymer, silicone, rubber, and other flexible materials.

Soft sensor Also called virtual sensor.A common name for software where several measurements are combined to represent a missing measurement.

Software as a service (SaaS) Most visible (to end-users) type of services of cloud computing that offers a software distribution model.

Software-defined networking (SDN) An architecture to monitor and manage a network optimally. Separates the data plane from the control plane, and increases the flexibility of the network for global management. Adds several features and capabilities to the network such as programmability, protocol independence, traffic identification at any flow, packet-level determination, and controlling the quality of service.

Solid-state transformer An AC-to-AC converter with a smaller size and more efficiency than the conventional converters. Operates at high frequency and actively regulates the voltage and current.

Statistical associating fluid theory (SAFT) A theoretical group contribution model based on perturbation theory. Illustrates a statistical theory for the thermodynamic characterization of a fluid by incorporating the effects of association and the various interactions taking place between its molecules. Can present perfect results to model the behavior of the associating (self-association) fluids and non-associating chains.

Stereolithography A process in which photocurable resins exposed to a laser beam undergo a chemical reaction to become a solid part. This reaction is called photopolymerization, in which small monomers are linked into chain-like polymers. It is a combination of many chemical compounds such as additives, photo-initiators, and reactive monomers/oligomers.

Supply chain management (SCM)   The management of the flow of goods and services between businesses and locations. Includes the movement and storage of raw materials, work-in-process inventory, and finished goods as well as end-to-end order fulfillment from point of origin to point of consumption. Interconnected, interrelated, or interlinked networks, channels, and node businesses combine in the provision of products and services required by end customers in supply chain management.

Support vector machines (SVMs)   Also called support-vector networks. Supervised machine learning models with associated learning algorithms that analyze data for classification and regression analysis. Developed at AT&T Bell Laboratories by Vladimir Vapnik with colleagues. One of the most robust prediction methods, being based on statistical learning frameworks.

Team heterogeneity   A team of mobile robots handling complex tasks a lot more efficiently than a single robot owing to the diverse capabilities of its members.

The Industrial, Scientific and Medical (ISM) bands   Defined by the International Telecommunication Union (ITU) radio regulations as a range of radio frequency spectrum reserved internationally for industrial, scientific, and medical purposes other than telecommunications.

Timestamp   Provides the data associated with the time of each transaction, to distinguish between different records in the blockchain and to provide historical data for future reference.

Ultrasonic additive manufacturing (UMV)   One of the main sheet lamination processes in which sheets of metal bonded together by ultrasonic welding are applied. This process requires additional computer numerical control (CNC) machining and removal of the unbound metal, mainly during the welding process.

Ultraviolet   A type of electromagnetic radiation with a wavelength from 10 nm to 400 nm, which is shorter than visible light but longer than X-rays.

Unmanned aerial vehicle (UAV)   Also known as a drone. An aircraft with no human pilot, crew, or passengers onboard. Flight may be performed by a human operator or autopilot.

Vat photopolymerization   One of the additive manufacturing (AM) processes. Generally includes stereolithography and its related processes.

Vehicle-to-everything (V2X)   Technologies that employ the energy in the batteries of EVs for other purposes than transportation.

Virtual reality (VR)   A technology or method that uses computer technology to create an artificial three-dimensional simulated environment or experience that is almost identical to reality. The interactive three-dimensional (3D) space can be modified and allows users to feel virtual places with their senses and interact within the synthetic environment as they are really there. Although VR is often associated with AR, the two technologies have some principal differences.

Water monitoring   The process of observation and analysis of water quality in permanent points, for example, across distribution networks, as well as data acquisition and processing. It also involves analysis of the chemical condition of water and sediments to determine contamination and other key constituents.

Wireless local area anetwork (WLAN) A LAN that relies on radio transmissions rather than wired ethernet connections. Connects devices of the network within a limited geographic area to the public Internet.

Wireless personal area network (WPAN) Designed for short distances wireless communication between a group of devices close to a person or object. Assists in conveying information among devices of a person's workspace or an intimate group of participant devices.

Wireless sensor network A wireless network that is self-configured and infrastructure-less and is applied to monitor and recode physical conditions of the environment and forward the collected data to a central location where the data can be monitored, processed, and analyzed.

Wireless sensor networks (WSN) Interconnected group of sensors distributed in the space that communicate wirelessly to monitor and record the physical conditions of the surrounding environment.

Zebra battery: A rechargeable battery type that consists of sodium-nickel chloride cells. Stands for Zeolite Battery Research Africa project, which started in 1985 in South Africa.

Index